Synthese Library

Studies in Epistemology, Logic, Methodology, and Philosophy of Science

Volume 506

The aim of *Synthese Library* is to provide a forum for the best current work in the methodology and philosophy of science and in epistemology, all broadly understood. A wide variety of different approaches have traditionally been represented in the Library, and every effort is made to maintain this variety, not for its own sake, but because we believe that there are many fruitful and illuminating approaches to the philosophy of science and related disciplines.

Special attention is paid to methodological studies which illustrate the interplay of empirical and philosophical viewpoints and to contributions to the formal (logical, set-theoretical, mathematical, information-theoretical, decision-theoretical, etc.) methodology of empirical sciences. Likewise, the applications of logical methods to epistemology as well as philosophically and methodologically relevant studies in logic are strongly encouraged. The emphasis on logic will be tempered by interest in the psychological, historical, and sociological aspects of science. In addition to monographs *Synthese Library* publishes thematically unified anthologies and edited volumes with a well-defined topical focus inside the aim and scope of the book series. The contributions in the volumes are expected to be focused and structurally organized in accordance with the central theme(s), and should be tied together by an extensive editorial introduction or set of introductions if the volume is divided into parts. An extensive bibliography and index are mandatory.

Selene Arfini
Editor

Scientific Cognition, Semiotics, and Computational Agents: Essays in Honor of Lorenzo Magnani - Volume 2

Springer

Editor
Selene Arfini
Department of Humanities – Philosophy Section
University of Pavia
Pavia, Italy

ISSN 0166-6991 ISSN 2542-8292 (electronic)
Synthese Library
ISBN 978-3-031-96687-3 ISBN 978-3-031-96688-0 (eBook)
https://doi.org/10.1007/978-3-031-96688-0

This Springer imprint is published by the registered company Springer Nature Switzerland AG.
The registered company address is: Gewerbestrasse 11, 6330 Cham, Switzerland

Contents

Chapter 1
From Intelligence to Creativity: The Quest for Automated Creative Cognition

Atocha Aliseda

Abstract We argue that the guiding idea of the artificial intelligence research programme, which we call the quest for automated intelligence, is not the appropriate one for giving an account of the relationship between humans and the computational devices they design and use. We should instead change the very question of intelligence by one of creativity, and provide a model for automated creative cognition, as a quest for it. Such a model is found in the work of Lorenzo Magnani. We begin by looking into characterizations of intelligent tasks, such as problem-solving and learning, both already present in the work of Alan Turing. Concerning creativity, two are the key concepts: abduction and heuristics; the former proposed by Charles S. Peirce and the latter promoted by many, and most notably, by Herbert Simon. The new guiding idea is a quest for automated creative cognition, one in which abduction takes a main role, and can be represented in in the Eco-cognitive model of abduction devised by Magnani, where there a continuous interplay with various internal and external supports.

Keywords Intelligence · Creativity · Abduction · Automated intelligence · Automated creative cognition

1.1 Introduction

The field which came to be known as artificial Intelligence—AI for short—emerged in the fifties of the twentieth century with the appearance of computers. The mathematical model of these artifacts was created by Alan M. Turing. Nowadays this research program is referred to as Good Old Fashion Artificial Intelligence (GOFAI)

A. Aliseda (✉)
Philosophy Department, Universidad Nacional Autónoma de México, Ciudad de México, Mexico
e-mail: atocha@filosoficas.unam.mx

S. Arfini (ed.), *Scientific Cognition, Semiotics, and Computational Agents: Essays in Honor of Lorenzo Magnani - Volume 2*, Synthese Library 506,
https://doi.org/10.1007/978-3-031-96688-0_1

and belongs to the core of the Cognitive Sciences, a set of disciplines that converge in the study of the mind: philosophy, computer science, cognitive psychology, linguistics and, more recently, neuroscience. This enterprise is indeed a research program—in the sense of Lakatos (1978)—in which the guiding idea is the quest for automated intelligence. That is, the construction of computer programs which simulate intelligent behavior.

In this paper, we argue that given recent developments of this research programme, the search for intelligence should be changed into the quest for creativity. And there are several reasons for this move. Besides the fact that we have not been able to describe what human intelligence is in the first place, our claim is that computers power should be tested with respect to the potential for generating new elements, being this information, ideas or even new discoveries. Ina nutshell, what is needed is a model for automated creative cognition.

In order to motivate our claim, we look first into what is it to be computer intelligent, such as the ability to solve problems and of learning. As for creativity, we turn into philosophy and rely on two key concepts: abduction and heuristics. The former was identified and subsequently characterized by Charles S. Peirce; the latter has its roots in Greek geometry and much more recently was promoted, most notably, by Herbert Simon. The quest for automated creative cognition, is one in which abduction takes a main role, such as found in the Eco-cognitive model of Lorenzo Magnani (1992, 2001, 2017 to list some of the many references around). Under his view, there is a continuous interplay with various internal and external cognitive supports.

Moreover, conceptions about what kind of artifacts computers are, in virtue of being carriers of simulated intelligence—or creativity—via computer programs, will depend largely on which theory of mind we hold when modeling these phenomena. On the one hand, in the GOFAI paradigm the human mind is conceived as a computer, either as a Turing machine represented as universal automata or as a connectionist machine with a neural network configuration. From the perspective of embodied cognition, on the other hand, the human mind extends through these computer programs, it is through interaction with humans these artifacts come to life. Thus, one may either conceive artifacts exhibiting intelligent behavior as artificial agents, or it is humans who disembody themselves by extending their mind by merging with external representations.

In Sect. 1.2 the original research programme as the quest for an automated intelligence will be presented, from the ideas of Alan Turing to those of machine learning nowadays. A key distinction will be presented, that of weak vs. strong AI. While there is no doubt that the former has shown significant advances, such as in the field of image processing for early breast cancer detection, the latter seems closer to science fiction than to an academic discipline.

In Sect. 1.3 what we conceive as creative reason will be presented rooted in the ideas of Charles S. Peirce, highlighting the concepts of abduction and of heuristics. In Sect. 1.4, the Eco cognitive model of Magnani will be briefly described, highlighting the continuous interplay with various internal and external supports. As we shall see, it shows significant differences with GOFAI, and it is in complete accord with the embodied cognition paradigm. In the final Sect. 1.5, a discussion and conclusions are offered.

We end this introduction with a disclaimer. The analysis we offer in this paper is essentially theoretical. It is an inquiry into the quest for automated creative cognition, one which does not need to align to any psychological theory on creative cognition (such as those reviewed in Lebuda & Benedek, 2023) nor convey any practical implementation such as found in applications of generative AI already circulating in the market. It remains to be seen how is that computer scientists may give form to the ideas shown here, but this challenge is beyond the scope of this paper.

1.2 The Quest for Automated Intelligence

The first article on AI was published by Turing himself in a philosophy journal in under the title Computing machinery and intelligence (Turing, 1950). Turing is considered the father of AI; more particularly, it is the seed of what we now call machine learning—the big buzzword of the day—that machinery that underlies the so-called generative artificial intelligence some many of which is represented in Large Language Models (LLMs) such as Chat GPT. However, in what sense do these programs exhibit "intelligence"? Although they do not understand a word of what appears at the prompt (a sign that the program is in the mode of receiving input), they provide similar answers to those a human would produce.

The question with which Turing opens his pioneering text is the following one: Can a machine think? a question that in turn implies this other one: can a machine be intelligent? Turing makes it even more precise: Can a machine behave as if it were intelligent? Note that Turing's question is about the possibility of simulating intelligent behavior and not about whether a machine is intelligent. This is a key point in the foundation of AI, which is often overlooked. (For a broader description of the origin of this question, see Aliseda, 2013 and Leavitt, 2007).

1.2.1 Weak Artificial Intelligence Versus Strong Artificial Intelligence

One of the distinctions at the heart of GOFAI as a research program, and that should be highlighted nowadays, is one that sharply separates a weak from a strong AI. The former conceives intelligent behavior as in problem solving tasks in areas like planning and games. Well—known examples of this kind are programs that play chess. As is well known, Deep Blue won several chess games against the best chess player in the world, Garry Kasparov. Later on, AlphaGo beat Fan Hui in GO, a Japanese game much more sophisticated than chess. These world champions electricity—dependent made of silicon and running under the order of computer code behave as if they were intelligent, and many of them are smarter than their human opponents. Are they really smart? The problem is not so much the answer to this question but rather what presupposes, namely that we humans fully understand what being smart and intelligent is, something which is not the case even nowadays. As one of the reviewers pointed out: “We can solve this dilemma by answering they are “machine-smart” in the same way as airplanes fly. They do not fly like birds. They need not. Airplanes and computers are our human technologies that serve our human purposes. Some machines wash the dishes, not as we do but as we made them to do the task. The problem with “intelligent” machinery unlike dishwasher is that we actually do not know what thinking is, what intelligence and cognition are. We are learning and as we learn those concepts are changing”. Indeed, one of the pioneers of AI, John McCarthy, was already aware of our human ignorance in this respect:

> He [John McCarthy] believed that we would have to know much more about how human intelligence works before being able to duplicate it in machines, writing that “[unfortunately we] understand human mental processes only slightly better than a fish understands swimming (Nilsson, 2012: p. 15).

One negative answer of whether machines are smart, is given by Terry Winograd, founder of the field of human-computer interaction research, who described artificial intelligence in a very fine way: “artificial intelligence is that magnificent computer program that at the same time it is beating the best player in the world at chess, it inhabits a room that is on fire.” (personal communication). In other words, programs in weak AI are aimed at solving specific tasks and often outperform expert humans, but they are not intelligent in the broad sense of what we would like to conceive as human intelligence.

Let’s now try another characterization of Intelligence, one pointing out to the ability to adequately react to unforeseen situations. In his respect, programs within computer bodies cannot be isolated from the world, and therefore a device for interaction must be provided. We are here in the territory of computer cognition, but before we elaborate on this conception, we shall say a few words about strong or robust artificial intelligence, a much more ambitious goal than that of weak AI. Promoters of strong AI—perhaps inspired by science fiction—would like to build machines with consciousness, reminding us of HAL 900, that computer from Space Odyssey 2001 based on a story by Arthur C. Clarke and brought to the screen by Stanley Kubrick. HAL is aware it will be disconnected, a consequence of a programming error that led to an announcement the ship would fail. Not only does HAL rebelled against the crew, but she seemed to reveal emotions when is gradually disconnected: “I know that all has not been well with me” and finally says: “My mind is fading, I can feel it”. The capacity for self-awareness is a cognitive process of a higher order. Whoever possesses it, not only knows but also knows that she knows. Thus, strong, or robust AI enters the territory of mind studies and thus encounters the hard problem of consciousness. As is well known in the philosophy of mind, the philosopher David Chalmers posed the challenge of understanding qualia or phenomenal consciousness, what our subjective experiences are made of. He introduced the distinction between hard and soft problems. The latter can be reduced to functional explanations, whether behavioral or mechanistic, such as the abilities and tasks to discriminate and integrate information of various kinds. The way phenomenal consciousness manifests itself cannot be explained in a functional way and is therefore classified as a hard problem. Twenty-five years ago, the neuroscientist Christof Koch challenged David Chalmers that within this time frame it would be possible to explain how consciousness emerges. Today, neuroscience has yet to decipher this phenomenon and only a few months ago, Koch had to accept that he had lost the bet (Lenharo, 2023).

These two projects—weak and strong—AI are complementary and there is consensus on the great success and progress of the former, but not so of the latter. In any case, when talking about AI, it is always important to clarify which of these two projects is the one, we are analyzing.

1.2.2 The Mind as a Computer

During the 1950s, a time of first contact between philosophy and computer science, theories of mental representation emerged out of the "mind as a computer" metaphor introduced by the philosopher Jerry Fodor. This idea served as a bridge between philosophy of mind and artificial intelligence. The former provided the conceptual basis and the latter the tools to represent and manipulate knowledge.

One of the fundamental theses of the cognitive sciences that underlies the metaphor is the following: thought is the result of mental representations and computational processes that operate on these representations. This thesis assumes that the mind possesses representations analogous to data structures and computational procedures analogous to algorithms. That is, human thinking processes can be modeled by computational programs. This is of course an elaborated view of thinking as a problem-solving task. However, there cannot be a single computational model of the mind since there is no single computational model. The functionalist approach, on the one hand, assumes that the mind is a Turing machine, one that operates on structures in the form of transition rules between states (the automata model devised by Turing). On the other hand, the connectionist approach proposes a representation mimicking neurons and their connections as inspirations of data structures—neurons firing—and neuron activations as inspirations of algorithms (Rummelhart, 1989).

The metaphor of the mind as a computer has been extremely useful in the development of cognitive theories for problem solving. There are at least three types of problem solving that have interested the cognitive sciences both at conceptual level as well as at the application level: problems of planning, decision making and explanation. Let's describe the first one of these.

Consider a robot with an artificial body. In addition to having a clear representation of the world around it, it must be able to perform planning actions. For example, the robot is instructed to go from the room where it is currently located to the other end of the building. It then makes use of a mental map of the place and forms a plan of action to move towards its goal. Moreover, in carrying out the plan, the robot must be able to make appropriate decisions in case it has more than one option for approaching its target. Finally, to provide an explanation of both its plans and their execution, the robot constructs explanations of its action plans. If something goes wrong and the robot cannot continue its way, it must be at least able to construct the explanation as to why it has encountered an obstacle, leading to a modification of an action plan. In short, a robot must be able to learn new information, and ideally be able to produce new knowledge. Within the problem-solving territory as applied to robotics, planning is performed in the head—so to speak—mind and brain are merged in the head.

1.2.3 The Extended Mind

When mind processes are conceived outside the head and are rather situated in the interaction with the world, the mind extends beyond the body, and we are thus in the territory of embodied cognition. This is a paradigm developed in the second half of last century involving several disciplines: psychology, neuroscience, ethology, philosophy, linguistics, robotics, and artificial intelligence. Embodied cognition is now labelled as 4E cognition, reflecting a finer-grain analysis of four cognition types: situated, extended, distributed and enactive (see Newen et al., 2018, for an in-depth analysis of each one and for an exposition of this research programme at large).

In this paper, it is enough we refer to the extended mind approach. Let me elaborate on some illustrative examples in this direction. Consider calculators, the antecedents of computers, in virtue of being assistants in human cognitive tasks. These artifacts extend our calculation potential concerning speed. Now let's consider another artifact, the telescope. It also enhances a human ability, that of observation. However, what we can observe via a telescope, could not be seen with the naked eye. Indeed, it was thanks to the telescope that Galileo discovered that Jupiter has four moons. Thus, while a calculator accelerates our calculation speed, a telescope extends our vision potential. Computers can assist us in both ways. On the one hand, they may speed up cognitive processes, and on the other hand, they can assist us in tasks we could not perform without them, such as in machine learning programs. In both cases "minds are 'extended' and artificial in themselves" as Magnani (2017, p. 71) asserts.

In order to appreciate machines with learning abilities, let's start by describing the simplest type of computer program in which no learning takes place, one that implements a deterministic algorithm, that is a series of instructions that manipulate input data in view of a result. A simple but illustrative example is a program implemented in a washing machine: open water tap, run the water basin inside the drum, run drain, open water tap, run drain, return to point 1 and repeat the cycle once more if the program for very dirty clothes has been chosen … and finally end with the spinning process, finishes and gives notice of it by turning on a green light.

According to what was already stated at the beginning of this chapter, when intelligence is characterized as the adequate ability to react to unforeseen situations, and there is some learning of new information as a result, it is clear that the label intelligent to programs like the ones implemented in washing machines is misleading (though we know this is so for marketing reasons), for these artefacts behave according to a set of pre-determined programs to be selected at a human will.

The notion of computer program becomes interesting when machine learning applications are on focus, but what does it mean for a program to learn? "Machine learning is the technique that improves system performance by learning from experience via computational methods. In computer systems, experience exists in the form of data, and the main task of machine learning is to develop learning algorithms that build models from data." Zhou (2021, p. 2).

There are several learning methods as well as ways to classify them. A typology is offered by Michalski et al. (1983): rote learning and direct implanting of new knowledge, learning from instruction, and learning from observation and discovery. Given the focus of this paper, the last one will be developed further.

Learning by discovery is found in research fields such as computational creativity or philosophy of computational science, the latter coined by Paul Thagard, is a way of doing philosophy of science with computational tools. Herbert Simon and collaborators (Langley et al., 1987; Langley, 2000), Thagard himself (1988) and others, designed and implemented computational programs for scientific discovery simulation. Thus, the philosophy of science imports computational tools and accounts with them for classical notions such as explanation, confirmation, refutation, and discovery, and in general, for the modeling of the dynamics of scientific theories. Although many of these discoveries are not genuinely new but are rather simulations of discoveries recorded in the history of science, there are other representative cases that have managed to generate totally new knowledge. Such is the case of new laws on second structure proteins in chemistry, taxonomic laws in astrophysics, as well as qualitative laws in cancer research in biomedicine (Gillies, 1996; Langley, 2000).

1.3 On Creativity and Abduction

1.3.1 Creative Reason

Before getting into the logical workings of abduction and claim its place as the logic of synthetic reasoning by revealing some of its creativity powers, we must address the possibility of creative reason. At first glance, it seems that what is creative cannot be logical and whatever logical cannot be a model of a creative process.

Creative Reason has a place in the process of invention when we conceive of an idea for the first time or discover a new scientific theory. It is, however, necessary to establish a position with regard to the following question: What do we understand by creativity? The lack of even minimal agreement on the meaning of creativity as a concept is notorious. For the purpose of this work we shall adopt, although tentatively, a characterization proposed in the field of cognitive psychology (see Sternberg & Lubart, 2003, p. 3) which defines a creative product as one that is both novel (original, unexpected) and useful (appropriate, adaptable). Moreover, when referring to Creative Reason, the assumption is that there is a certain method involved. There are, of course, other reasons identified in philosophy and related disciplines: Kant's Pure Reason, the Historical Reason of Ortega y Gasset and, of course, the Poetical Reason of María Zambrano. Questions regarding its origin and justification are related to inquiries into synthetic reasoning, which go back to Immanuel Kant and were reformulated by Charles S. Peirce, as the following passage testifies:

"According to Kant, the central question of philosophy is 'How are synthetical judgments a priori possible?' But antecedently to this comes the question how synthetical judgments in general, and still more generally, how synthetical reasoning is possible at all. When the answer to the general problem has been obtained, the particular one will be comparatively simple. This is the lock upon the door of philosophy". (CP 5.348, cited in Hookway, 1992: 18).

Peirce's analysis of ampliative reasoning or, as he preferred to call it, synthetic reasoning has two fundamental aims: on the one hand, to justify its possibility; on the other, to develop a method to put it into practice. Pierce found the answer to the first goal in his metaphysics and the second is a matter addressed by contemporary formal research, referred to as abductive logic. The Kantian question of the mere possibility of synthetic reasoning is more recently captured by the question of the place of logic in Creative Reason, and put forward by Simon as the following question Is there a logic of scientific discovery? (Simon, 1977).

This question finds its origin in Ancient Greece and traverses much of the history of philosophy, it was revived in cognitive psychology by Simon himself and later adopted in contemporary philosophy of science. Later, we shall go into this question more in depth, drawing on the work of Peirce, who conferred a logical formulation upon abduction, and which I shall review in detail in relation to the question of whether this is effectively a logic of synthetic reasoning and in what sense its products are novel.

As I have argued elsewhere (Aliseda, 2021) logic and creativity are not mutually exclusive. This is true when we take this thesis in a somewhat restrictive way. As regards creativity, we shall be concerned only with the generation of new scientific hypotheses and, as far as Logic is concerned, the limit is the formal interpretation to Peirce's formulation of abduction. This modest analysis still enables us to appreciate both the scope and the limitations of this thesis.

1.3.2 Abduction: The Logic of Synthetic Reasoning

As well known, the first philosopher to propose a logical formulation for abductive reasoning was Charles Sanders Peirce (1839–1914). His influence on those who represented the "received view" was, however, minimal, or even non-existent. His importance was recognized by certain philosophers who reacted against the prevailing current, precisely because of their concern to investigate matters of relevance to the context of discovery (a prominent case being that of Hanson, 1971). Nevertheless, Peirce's notion of abduction is difficult to analyze, since his conception of it evolved throughout his philosophical trajectory, and several different versions appear as his thinking evolved. At the same time, his conception is interlaced with epistemological and semiotic aspects of his philosophy that are equally complex.

For Peirce, abductive reasoning is fundamental in every human act of inquiry. Abduction plays a role in perception: "The abductive suggestion comes to us as a flash" (CP 5.181). It also plays a fundamental role in the generation of hypotheses and new ideas in scientific activity: "Abduction is the process of forming an explanatory hypothesis. It is the only logical operation which introduces any new ideas".

As proposed by Anderson (1986), abduction seems to be both "an act of intuition and one of inference", which suggests a double aspect: both intuitive and rational. The presence of these two aspects, apparently antagonistic, has invariably caused confusion in scholars of Peirce. In general, only one of these is considered for the analysis of abduction (Kapitan, 1990; Sharp, 1970; Thagard, 1977). Some other critics have interpreted this duality as Peirce's dilemma and concluded that the philosopher lacked a coherent vision of the nature of abduction (Frankfurt, 1958). There also exists another approach seeking to give sense to both aspects, the intuitive and the rational (Ayim, 1974; Paavola, 2005); a view endorsed here by the claim that logic and creativity are not mutually exclusive. The abductive formulation is as follows (CP 5.189):

The surprising fact, C, is observed.
But if A were true, C would be a matter of course.
Hence, there is reason to suspect that A is true.

This formulation has been the starting point for many studies in both philosophy and computing. In general, its formal interpretation is as follows:

$$\begin{array}{l} C \\ A \rightarrow C \\ A \end{array}$$

Implicit is the extra-logical specification that the status of the conclusion is tentative; otherwise, one would incur in the fallacy of affirming the consequent. Likewise, it is assumed that the second premise is part of the background theory in question.

Aside from its explanatory capacity, Peirce considers a further two criteria for a successful hypothesis, namely, testability and economy. An abductive hypothesis can be regarded as explanatory if it accounts for the facts in accordance with the formulation; its status remains, however, tentative until it is put to the test, thus achieving empirical corroboration, and yielding place to the third criterion: that of economy. The latter answers to the practical problem of managing an endless series of hypotheses that satisfy the former criteria. Some approaches attempt to capture Peirce's economy criterion as a subsequent process for choosing "the best explanation", since there may be many formulae that fit this logical formulation and that are all subject to empirical testing, without being ideal hypotheses. For the interpretation of abduction as inference to the best explanation, see Harman (1965) and Douven (2011). For a discussion on whether Peirce's notion of abduction is one of inference to the best explanation, see Campos (2011).

One question that has arisen by Pierce scholars has to do with the idea of novelty, in this formulation. What counts as something completely new? As I have shown in more detail elsewhere (Aliseda, 2021) according to Frankfurt (1958), the conclusion does not offer anything genuinely new, since it is already present in the premises. However, if the creative act consists in identifying something already known A → C (a natural law, for example) and, on this basis, suggest A is the case (as when looking for the cause of C), then we affirm that the creativity lies precisely in establishing new connections between the elements that present themselves to us, in this case as a surprising fact C and by the associated law (A → C) in order to propose A.

Another more general question concerning this formulation has to do with analyzing the scope of the (formalization of the) original formulation offered by Peirce. In the first place, it could well be argued that it fails to capture in itself either the fact of C being novel, or any of the other criteria proposed by Peirce (see Aliseda, 2006 for an in-depth discussion). Furthermore, note that the formalization of the second premise need not be classical material implication; it could be a logical implication of some other kind, or even a (computational) process with input A and output C. Indeed, nowadays there as many logical as there are computational systems that implement the abductive formulation in somewhat sophisticated ways. For example, in adaptive logics, when an abductive conclusion is generated, it is added as a line of the proof with a tentative status. If in later lines new information is added that refutes this hypothesis, a strategy enters into operation that marks the line in which it appears (and perhaps other lines that depend on that) and blocks it until it chances to be "resuscitated" as a product of later inferences. Other abductive approaches exist whose aim is to construct algorithms in order to generate abductive hypotheses. For an introduction and panorama of the logic of abduction in which three different logical approaches are reviewed (as logical inference, as computational process and as epistemic change), see Aliseda (2017).

As regards the question whether these formal systems produce genuinely new elements, we must acknowledge that the abductive formulation, even in its most sophisticated forms, reveals an intrinsic characteristic of Creative Reason in its formal version: new ideas are always a product of combinations of already existing ideas. This vision is manifested clearly as a "syntactic restriction of language", in the sense that in abductive logic, new elements can only be generated in the framework of the language in question. At first sight this reflects a limitation; at least it underlines the fact that in formal models of creativity there is no place for the generation of new concepts, a phenomenon usually associated with Kuhn's Revolutionary Science. In so far as they are creative hypotheses, abductions modeled in formal frameworks refer to facts or relations between them; but when a hypothesis is based on a new theory that involves the creation of a new concept, such models have neither the language nor the mechanisms to generate totally new products. Even so, the generation of new facts and of relations between empirical facts that explain a phenomenon are authentic creative products, inasmuch as they arise from new connections and combinations between existing ideas. No few scientific discoveries have been the product of new connections. Discoveries of this sort represent the type of advance in science described by Kuhn as Normal Science (Thagard, 1992).

Regarding the intuitive aspect of creativity, for Peirce scholars it is well-known that abduction is rooted in an animal instinct, but at the same time, it is somewhat puzzling that it is also manifested as a human competence to conjecture true hypotheses:

> In regard to instinctive considerations, I have already pointed out that it is a primary *hypothesis* underlying all abduction that the human mind is akin to the truth in the sense that in a finite number of guesses it will light upon the correct hypothesis (CP 7.220, my emphasis).

Thus, the cognitive faculty that confers abduction on us is manifested as a rational instinct, as characterized by Ayim (1974), more precisely as the guessing instinct, the term used by Peirce himself. This hypothesis is similar to Chomsky's regarding the acquisition of language, according to which human ability for spoken language can be imputed to an innate competence.

In any case, creativity combines as many intuitive as logical aspects. According to Peirce, the intuitive part is necessary for setting in train that innate capacity for guessing; but intuition is not the only aspect that counts for explaining success in the generation of abductive hypotheses. It is also necessary, although not by itself sufficient, that abductive hypotheses conform to a certain logical structure such as that provided by Peirce's formulation.

Therefore, as witnessed in Peirce's abductive formulation, logic has a privileged place in Creative Reason, and more generally, logic and creativity are not mutually exclusive, as I have put forward. The following quote by Herbert Simon captures very well the conception of creativity defended so far:

> New representations, like new problems, do not spring from the brow of Zeus, but emerge by gradual —and very slow— stages [...]. Most scientific activity goes on within the framework of established paradigms. Even in revolutionary science, which creates those paradigms, the problems and representations are rooted in the past; they are not created out of whole cloth. (Simon, 1977: 301–302)

1.3.3 Heuristics in Cognitive Psychology

The subject of scientific discovery has found a privileged position beyond the philosophy of science in cognitive science. I shall briefly mention the position of a pioneer, that of Herbert Simon. In his essay "Does Scientific Discovery Have a Logic?" Simon sets out to refute Popper's general argument, re-construed for his purposes as follows: "If 'there is no such thing as a logical method of having new ideas', then there is no such thing as a logical method of having small new ideas" (Simon, 1973, p. 327).

The ambitious project of a logic of scientific discovery, one that reveals in all its complexity the process I am trying to capture, becomes, in Simon's hands, a much less ostentatious process. Before detailing his proposal, he tells us (in reference to the examples he cites): "Their modesty as instances of discovery will be compensated by their transparency in revealing underlying process." (Ibid.)

This approach, as brilliant as it is humble, enables Simon both to establish the types of problems that are to be analyzed and to specify the methods to be used. For Simon and his followers, "scientific discovery is a problem-solving activity". If this slogan affirms something that seems obvious nowadays when referring to scientific practice, it is in fact an idea that the methodology of science imported from cognitive psychology. When Simon characterizes scientific discovery as a "problem-solving activity", this translates into computational programs with tools of their own for modeling scientific discovery. Such programs exemplify the possibility of automatic simulation of scientific discoveries, oriented towards modeling fundamental mechanisms that point towards conditions for creative inferences sufficient for leading to the generation of new knowledge. In the context of this conception, a distinction is provided between two types of problems —well-structured vs. ill-structured ones—, and the aim of finding a logic of discovery concentrates on well-structured problems. The latter type is one for which a clearly defined criterion of testing exists and for which there is, at least, a problem space in which the initial state and the final state can be represented, and where all the remaining intermediate states can be reached with the appropriate transitions between them. An ill-structured problem lacks at least one of the above conditions. Although there is no effective method by means of which to achieve scientific discovery as a problem-solving activity, it can be characterized by way of strategies. The key concept in all this is heuristics. Heuristic methods of discovery are characterized by selective searches with fallible results. That is to say, although they do not guarantee success in arriving at a solution, the search in the problem space is not blind, but selective according to a predefined strategy. The authors also distinguish between weak and strong methods of discovery. The former comprise the type of problem solving used in novel fields; they are characterized by their generality, since they do not demand deep knowledge of the particular field. In contrast, the strong methods are used for cases in which our knowledge of the field is extensive and are especially designed for a specific structure. The weak methods include heuristic methods of generation and testing, and analysis of ends and means, in order to build up explanations and solutions for given problems. For details of the approach of Simon and his followers, see Langley et al. (1987).

Popper and Simon's postulations have both convergences and differences. Both subscribe to the vision of science as a problem-solving activity, though differ on just what they consider to be the Logic of scientific discovery. While for Popper ideas are generated by the method of Conjectures and Refutations, which can be interpreted as a method of 'blind search', Simon and his team develop a theory for upholding the position that ideas are generated by means of a method of 'selective search'. The latter approach enables the way in which new ideas can be generated to be captured as such, and concerns heuristics. (For a deeper analysis comparing these approaches, see Aliseda 2006).

The vision of scientific discovery as an instance of Simon's problem-solving activity as well as the conception of a computational philosophy of science (briefly described in previous section), conceive the inquiry into the processes of scientific discovery as part of the methodology of science. The interaction between the different disciplines that make up this study promotes the importation of computational tools into the philosophy of science, with the aim of modeling the dynamics of scientific knowledge, including its generation and development.

1.4 The Quest for Automated Creative Cognition

In this section, we shall attempt to present some of Magnani's ideas in the context of our discussion. First, Magnani adopts Peirce's inquiry system where abduction, induction, and deduction constitute each one a phase in an inquiry cycle, rather than existing in isolation as an inference of its own. This is witnessed in his model STm for medical diagnosis, therapy planning and monitoring, as Barés-Gómez and Fontaine (2022) point out:

> [S]elective abduction is the making of a preliminary guess that introduces a set of plausible diagnostic hypotheses, followed by deduction to explore their consequences, and by induction to test them with available patient data (Magnani, 1992, p. 25).

> According to the STm, medical reasoning begins with the abductive phase, in which hypotheses are selected on the basis of the patient's data. Following this, the deduction-induction phase deals with the process of evaluation. Deduction is used for predicting expected consequences and evolution (e.g., prognosis) (Barés-Gómez & Fontaine, 2022, p. 11).

Second, as for abduction, Magnani is aware of its double aspect, as a rational inference as well as creativity engine. To cover such a broad scope, he introduces some distinctions, a key one being between theoretical and manipulative abduction. The former is identified with processes of hypotheses generation, some of which may produce a single fact (sentential) and some others a whole new theory (model-based). Manipulative abduction represents the extra-theoretical behavior, it aims to capture thought in action, as when building external models for problem solving (diagrams, maps, equations, programs). He holds a distributed cognition approach that extends to logic, as when quoting Jaako Hintikka: "The most basic deductive logic is nothing but experimental model construction" (1997, pp. 70, quoted in Magnani, 2017, p. 155). Another key reference here is that of dynamic epistemic logic, a logical framework to model information processing amongst agents (van Benthem, 2011). In regard to scientific discovery, Magnani's writes:

> If we see scientific discovery like a kind of opportunistic ability of integrating information from many kinds of simultaneous constraints to produce explanatory hypotheses that account for them all, then manipulative abduction will play the role of eliciting possible hidded constraints by building external suitable experimental structures (2001, pp. 59).

Theoretical and manipulative abduction represent the internal and the external, respectively, and are both part "of the same epistemic ecology." (Magnani, 2001, pp. 59). Magnani endorses an Eco-cognitive perspective which is further developed in a later book (Magnani, 2017).

Finally, as third point, there is Magnani's notion of cognition, key in his thinking. To start with, his conception contrasts with computationalism, the view of mind-as-computer: "Contrary to the old view in cognitive science, not all cognitive processes happen in an internal model of the external environment." (Magnani, 2017, p. 142). His view goes along with an embodied cognition approach, as the one proposed originally by Clark and Chalmers (2010). However, he prefers to refer to refer to it as mind disembodiment, in the following sense:

> I have illustrated that the disembodiment of the mind refers to the fact the mind "disembodies" itself by projecting representations on the external environment, that is, it refers to the cognitive interplay between internal and external representations, mimetic and, possibly creative (in this last case, they are not necessarily mimetic), where the continuous interaction between online and off-line intelligence can properly be addressed (2017, p. 142).

Note that under this perspective, the object of cognition is neither a human nor an artifact, but it emerges from the interplay of internal and external representations. Humans extend themselves through devices such as calculators, telescopes, and computers, and with these they can create a mimetic representation, but it is also possible to extend the mind by creating something new. As Magnani rightly points out as the entry of "hybrid abducers", in his lexicon of abductive cognition in science:

> The various abductive activities human agents are engaged in drastically rely on external supports. For instance, human agents create mimetic external representations, which mirror concepts and problems that are already represented in the brain and need to be enhanced, solved, further complicated, etc. so they can sometimes creatively give rise to new concepts and meanings. In the process of distributing the mind into the material world around them, human agents develop new abductive skills, which would not be exhibited without the continuous interplay with the various external supports. In doing so, human agents can be described as hybrid abducers. Organic agents are spontaneous inducers and abducers but human beings also construct logical and computational (ideal) systems both able to mimic human inductions and abductions and to create new and more "rational" ways of inducing and abducing. These systems are in turn used by human agents: they consequently have to be seen as hybrid reasoners (Magnani, 2017, p. 211).

The extended mind idea is not just imported into Magnani's approach; it is further enriched by incorporating abduction into the picture, the creative engine of his cognitive model. As for agency, we can also appreciate that an artefact by itself is nothing. It is us humans who extend ourselves through artificial devices, becoming hybrid abducers. This approach allows to mimic our organic ability to perform induction and abduction by implementing it into a computer system opening up the possibility to create new things by extending ourselves through these computational devices.

1.5 Discussion and Conclusions

Let's begin by recapitulating of our points so far. In Sect. 1.2, the quest for automated intelligence was presented. Alan Turing's seminal paper "computing machinery and intelligence" (Turing, 1950) marks the birth of machine learning and therefore the foundation of artificial intelligence as the research programme which came to be known as GOFAI (good old fashion AI). The distinction between weak versus robust artificial intelligence was then described, followed by two main conceptions of the mind: the mind as a computer and the extended mind, the former being the one the GOFAI paradigm presupposes. The quest for automated intelligence was then illustrated as a problem-solving task and as a learning challenge. In Sect. 1.3 the notion of creative reason was presented via the work of Charles S. Peirce, who proposed abduction as the logic for synthetic reasoning. The concept of heuristics in cognitive psychology was then presented as one capturing some of Peirce's ideas made real in the computational realm. Section 1.4, the quest for an automated creative cognition was put forward. Lorenzo Magnani's distinction between theoretical and manipulative abduction was presented as a pillar of his Eco cognitive mind model, in which these two abductive types interact depicting us, humans, as hybrid abducers.

In this paper we argued for replacing the quest for automated intelligence by the quest for automated creative cognition, one in which abduction takes a main role, such as in the Eco-cognitive model proposed by Lorenzo Magnani. His model provides a cognitive ontology in accord with the view of the extended mind, conceived here as a continuous interplay with various internal and external supports, generating mimetic representations as well as creative products. His conception contrasts with that in computationalism, in which the mind is mind is a computer, and is instead in accord with the view that the mind extends beyond the body.

Nowadays the question is not whether humans can build expert systems belonging to weak or strong artificial intelligence, but instead, whether we can design interactive systems able to learn and generate new knowledge in collaboration with us. Some of these artifacts are very simple, such as calculators. Some others are sophisticated, such as programs guiding a robot in a physical environment. But in all of them internal as well as external representations interact.

Computational artefacts make us humans hybrid abducers rather than just designers of digital objects performing independently from us. We end this piece with Lorenzo Magnani's own words regarding what model-based abduction accounts for, the core concept in the quest for automated creative cognition, the research programme Lorenzo has set out for us:

> That is, a considerable part of the hypothesis creation and selection is model-based: it is occuring in the middle of a relationship between brains and model-based aspects of external objects and tools that have received cognitive and/or epistemological delegations (Magnani, 2017, p. 213).

Acknowledgements I am very grateful for have been invited to the session in tribute to my dear colleague and friend Lorenzo Magnani at the Model-Based Reasoning in Science and Technology conference, that took place in Rome in 2023. This is where the initial ideas of this paper were presented. Thank you to Selene Arfini for the invitation to participate in this volume and to two anonymous referees, whose questions and criticisms allowed me to clarify my ideas and, hopefully to improve this text. Any remaining errors and of course mine.

References

Aliseda, A. (2006). *Abductive reasoning: Logical investigations into discovery and explanation. Synthese Library, volume 330*. Springer.

Aliseda, A. (2013). ¿Inteligencia mecánica? La pregunta de Alan Turing. Ciencia. *Revista de la Academia Mexicana de Ciencias*. Octubre-diciembre 2013. *64*(4), 10–17. ISSN 1405-6550.

Aliseda, A. (2017). The logic of abduction: An introduction. In L. Magnani & T. Bertolotti (Eds.), Handbook of model-based science (section The Logic of hypothetical reasoning, abduction, and models). Springer.

Aliseda, A. (2021). The place of logic in creative reason. In J. R. Shook & S. Paavola (Eds.), *Abduction in cognition and action: Logical reasoning, scientific inquiry, and social practice* (pp. 149–160). Springer. Studies in Applied Philosophy, Epistemology and Rational Ethics. ISBN 978-3-030-61772-1.

Anderson, D. (1986). The evolution of Peirce's concept of abduction. *Transactions of the Charles S. Peirce Society, 22*(2), 145–164.

Ayim, M. (1974). Retroduction: The rational instinct. *Transactions of the Charles S. Peirce Society, 10*(1), 34–43.

Barés-Gómez, C., & Fontaine, M. (2022). Medical reasoning and the GW model of abduction. In L. Magnani (Ed.), *Handbook of abductive cognition*. Springer. https://doi.org/10.1007/978-3-030-68436-5_82-1

Campos, D. (2011). On the distinction between Peirce's abduction and Lipton's inference to the best explanation. *Synthese, 180*, 419–442.

Clark, A., & Chalmers, D. J. (2010). The extended mind. In R. Menary (Ed.), *Life and mind: Philosophical issues in biology and psychology*. MIT Press.

Douven, I. (2011). Abduction. In *The Stanford encyclopedia of philosophy*. Available via http://plato.stanford.edu/archives/spr2011/entries/abduction/. Accessed 5 Sept 2018.

Frankfurt, H. (1958). Peirce's notion of abduction. *The Journal of Philosophy, 55*(14), 593–597.

Gillies, D. (1996). *Artificial intelligence and scientific method*. OUP.

Hanson, N. R. (1971). *Observation and explanation: A guide to philosophy of science*. Allen and Unwin.

Harman, G. H. (1965). The inference to the best explanation. *The Philosophical Review, 4*(1), 88–95.

Hintikka, J. (1997). The place of C.S. Peirce in the history of logical theory. In Lingua Universalis vs. Calculus Ratiocinator. *An ultimate presupposition of twentieth-century philosophy* (pp. 140–161). Springer, Heidelberg/Berlin. Jaako Hintikka Selected Papers, Volume 2. *Transactions of the Charles S. Peirce Society*, *34*, 503–533.

Hookway, C. (1992). *Peirce*. Routledge.

Kapitan, T. (1990). In What way is abductive inference creative? *Transactions of the Charles S. Peirce Society, 26*(4), 499–512.

Lakatos, I. (1978). In J. Worrall & G. Currie (Eds.), *The methodology of scientific research programmes (philosophical papers I)*. Cambridge University Press.

Langley, P. (2000). The computational support of scientific discovery. *International Journal of Human-Computer Studies, 53*, 393–410.

Langley, P., Simon, H., Bradshaw, G., & Zytkow, J. (1987). *Scientific discovery. Computational explorations of the creative processes*. MIT Press/Bradford Books.
Leavitt, D. (2007). *The man who knew too much. Alan Turing and the invention of the computer*. Phoenix.
Lebuda, I., & Benedek, M. (2023). A systematic framework of creative metacognition. *Physics of Life Reviews, 46*, 161–181.
Lenharo, M. (2023). Decades-long bet on consciousness ends — and it's philosopher 1, neuroscientist 0. *News, Nature, 619*, 14–15. https://doi.org/10.1038/d41586-023-02120-8
Magnani, L. (1992). Abductive reasoning: Philosophical and educational perspectives in medicine. In D. Evans et al. (Eds.), *Advanced models of cognition for medical training and practice* (pp. 21–41). Springer.
Magnani, L. (2001). *Abduction, reason, and science: Processes of discovery and explanation*. Kluwer Academic Publishers.
Magnani, L. (2017). *The abductive structure of scientific creativity, Vol. 37 SAPERE*. Springer.
Michalski, R. S., Carbonell, J. G., & Mitchell, T. M. (Eds.). (1983). *Machine learning: An artificial intelligence approach*. Tioga.
Newen, A., De Bruin, L., & Gallagher, S. (2018). *The Oxford handbook of 4E cognition*. Oxford University Press.
Nilsson, N. J. (2012). *John McCarthy (1927–2011). A biographical Memoir*. National Academy of Sciences. United States of America.
Paavola, S. (2005). *Peircean abduction: Instinct or inference? Semiotica 153–154, 363–374*. Mouton de Gruyter.
Peirce, C. S. (1931–1958). In C. Hartshorne, P. Weiss, & A. W. Burks (Eds.), *Collected Papers (CP)* (Vol. 1–8). Harvard University Press. [References to Peirce's papers will be designated by CP, followed by volume, period, paragraph number].
Rummelhart, D. (1989). The architecture of mind: A connectionist approach. In M. I. Posner (Ed.), *Foundations of cognitive science*. MIT Press.
Sharp, R. (1970). Induction, abduction, and the evolution of science. *Transactions of the Charles S. Peirce Society, 6*(1), 17–33.
Simon, H. A. (1973). *Does scientific discovery have a logic?* In Simon, H. A. (1977), pp. 326–337 (originally published in Philosophy of Science 40, 471–480 (1973)).
Simon, H. A. (1977). *Models of discovery and other topics in the methods of science. Boston studies in the philosophy and history of science 54*. Reidel.
Sternberg, R. J., & Lubart, T. I. (2003). The concept of creativity: Prospects and paradigms. In R. J. Sternberg (Ed.), *Handbook of creativity*. Cambridge University Press. 3rd reprint.
Thagard, P. (1977). The unity of Peirce's theory of hypothesis. *Transactions of the Charles S. Peirce Society, 13*(2), 112–123.
Thagard, P. (1988). *Computational philosophy of science*. MIT Press.
Thagard, P. (1992). *Conceptual revolutions*. Princeton University Press.
Turing, A. M. (1950). Computing machinery and intelligence. *Mind, 59*(236), 433–460.
van Benthem, J. (2011). *Logical dynamics of information and interaction*. Cambridge University Press.
Zhou, Z. H. (2021). *Machine learning*. Springer Nature.

Chapter 2
A Scale of Quantity. Kant and Peirce on Mathematical Reasoning

Francesco Bellucci

Abstract The paper examines Peirce's interpretation of Kant's philosophy of mathematics, with special reference to Kant's notion of "construction", and its transformation into a theory of diagrammatic reasoning. For Peirce, all deductive or mathematical reasoning proceeds by the construction and manipulation of diagrams. Peirce does not accept Kant's idea that only concepts of quanta and quantitas can be constructed. Logical concepts and relations, although they are not quantitative, can be diagrammatized, namely "constructed" in Kant's sense. What is quantitative in logical construction is not the object that is constructed, but the means of constructing it. This shift from the Kantian idea that the object is quantitative to the idea that the means to construct it is quantitative is operated by Peirce's notion of a "scale of quantity", that is, the idea that in mathematical reasoning an intermediate, quantitative construction is introduced and removed afterwards. The paper considers two examples of "construction" of syllogistic logic, Euler diagrams and Boolean algebra. Both are "scales of quantity", that is, quantitative objects having quantitative relations that serve to the representation of non-quantitative objects having non-quantitative relations.

Keywords Construction · Diagram · Kant · Logic · Peirce · Quantity

The last section of Lorenzo Magnani's *Philosophy and Geometry* (2001) is devoted to logical diagrams. Magnani discusses the then recent computational program "Hyperproof" (Barwise & Etchemendy, 1993; Barwise & Etchemendy, 1994), which was called a system of "heterogeneous" logic because it employed both first-order languages and diagrams to represent deductive reasoning. Magnani's instructive comment on the very idea of doing logic with diagrams is that "[t]he geometrical cases described by Kant, heavy with transcendental meanings, schematism,

F. Bellucci (✉)
Department of the Arts, University of Bologna, Bologna, Italy
e-mail: francesco.bellucci4@unibo.it

S. Arfini (ed.), *Scientific Cognition, Semiotics, and Computational Agents: Essays in Honor of Lorenzo Magnani - Volume 2*, Synthese Library 506,
https://doi.org/10.1007/978-3-031-96688-0_2

imagination, synthetic a priori judgment, and so on, are simply deduction, in the formal and modern sense of the term, if plunged into an appropriate logical system endowed with appropriate syntax and semantics" (2001, pp. 208–209). In other words, modern theories of diagrammatic reasoning and diagrammatic logics are an application to logic of Kant's theory of mathematical construction: "geometrical constructions become rigorous deductive inferences" (*ibid.*).

Now, as it is known, Charles S. Peirce—whom Magnani especially praises for his conception of abduction—not only developed a system of logical graphs for first-order logic (the so-called "existential graphs"[1]), but also had a *theory* of diagrammatic reasoning. This theory derived from Kant's theory of mathematical construction and constitutes an extension of that theory outside the realm of quantity.

Kant holds that all mathematical knowledge is knowledge by the construction of concepts, and constructing a concept means exhibings that concept a priori in intuition (A713/B741).[2] A consequence of this, Kant explains, is that only concepts of *quanta* and *quantitas* can be constructed (A720/B748; A718/B745). In Peirce, all of this becomes the famous docrine of diagrammatic reasoning: all deductive (i.e. mathematical) reasoning proceeds by the construction and manipulation of diagrams; diagrammatic construction and manipulation explain why in mathematical reasoning it is possible to discover truths of the object of reasoning that were not contained in the instructions to construct the diagram (CP 2.279). The Kantian source of this doctrine is direct and Peirce explicitly recognizes it in several places.

However, as soon as the direct derivation of Peirce's theory of diagrammtic reasoning from Kant's theory of construction is recognized, a problem emerges that has rarely been discussed in the literature but which has nonetheless to be addressed in order to fully appreciate Magnani's idea that "geometrical constructions become rigorous deductive inferences". The problem is the following: Does Peirce accept Kant's idea that only concepts of *quanta* and *quantitas* can be constructed? If not, in what sense can a non-quantitative concept be diagrammatized and made object of deductive reasoning, according to Peirce? In particular, in what sense can logical concepts and relations be expressed in a diagrammatic notation? What it is, precisely speaking, that is "constructed a priori in intuition" in Euler diagrams or Boole's algebra of logic? How can constructions, which are about *quanta*, about magnitudes, become rigorous deductive inferences, which do not necessarily concern quantity?

---

[1] See Roberts (1973), Pietarinen (2011, 2015), Bellucci and Pietarinen (2016a, 2016b, 2021).

[2] In what follows, I use the standard abbreviation for the first (A) and second (B) edition of the *Critique of Pure Reason* (Kant, 1781), which I quote in the Guyer-Woods translation. The following abbreviations for Peirce's works will also be used: R, followed by manuscript number and page number, stands for Peirce's manuscripts (Charles S. Peirce Papers, 1787–1951, MS Am 1632, Houghton Library, Harvard University) as catalogued in Robin (1967) and Robin (1971). (Citations of the form "R # CSP #" refer to Peirce's own pagination); CP, followed by volume and paragraph number, stands for Peirce (1932–1958); NEM, followed by volume and page number, stands for Peirce (1976); LoF, followed by volume and page number, stands for Peirce (2019–2024).

The paper is divided as follows. Section 2.1 gives a brief but, I hope, sufficiently accurate portrayal of Kant's theory of mathematical construction, especially as it emerges in the *Disziplin*. Section 2.2 examines some works by Peirce written in the 1890s in which he criticizes the traditional definition of mathematics as the science of quantity and explicitly adopts Kant's definition of mathematical knowledge as based on the method and not on the object: the method of mathematical (i.e. deductive) reasoning is (in Peirce's opinion) correctly described by Kant by means of the notion of "construction", which Peirce generalizes to that of "diagram". However, while mathematical reasoning is not necessarily concerned with quantity, yet quantity plays a special role in mathematical reasoning; Peirce says that the construction or diagram in it is a "scale of quantity". Section 2.3, then, considers the examples that Peirce offers of "scales of quantity" in mathematical reasoning, while Sect. 2.4 explains in what sense the geometrical and algebraical representations of syllogistic logic may be considered as "scales of quantity". The explanation in Sect. 2.4 is also the answer to the problem pointed out above, namely, in what sense logical concepts and relations, which are not quantitative, can be constructed in the Kantian sense. Finally, Sect. 2.5 considers the Kantian distinction between ostensive and symbolic construction, which Peirce accepted, and argues that Peirce's "symbolic" interpretation of symbolic construction (which anticipates some important modern commentators) may not be the correct interpretation of Kant, and that this does not undermine his notion of a scale of quantity.

2.1 Kant on Mathematical Construction

In the *Disziplin der reinen Vernunft im domgatischen Gebrauch* at the end of the first *Kritik*, Kant presents his view of mathematical cognition. Mathematical cognition is for him cognition based on the construction of concepts:

> *Philosophical* cognition is *rational cognition* from *concepts*, mathematical cognition that from the *construction* of concepts. But to *construct* a concept means to exhibit *a priori* the intuition corresponding to it. For the construction of a concept, therefore, a *non-empirical* intuition is required, which consequently, as intuition, is an *individual* object, but that must nevertheless, as the construction of a concept (of a general representation), express in the representation universal validity for all possible intuitions that belong under the same concept. (A713/B741).

An empirical concept cannot be constructed, i.e. exhibited a priori in an intuition, because empirical concepts correspond to empirical intuitions, which are possible only a posteriori. Pure or non-empirical concepts may be related to intuition in two ways: either the concept contains a pure intuition, in which case it can be constructed; or the concept only contains the rule for a synthesis of possible empirical intuitions, in which case the concept cannot be constructed and can only be applied *discursively* to empirical intuitions (A719–720/B747–748). A concept that contains a pure intuition is a pure sensible (i.e. mathematical) concept, while a concept that only contains the rule for a synthesis of possible empirical intuitions is a pure concept of the understanding. No singular intuition could ever display the pure concept of substance or of cause. Pure concepts of the understanding only contain the rule of a possible synthesis of empirical intuitions, but those intuitions themselves are given a posteriori, not a priori (A722/B750).

The fact the mathematical cognition considers the concept in an individual intuition does not affect the characters of necessity and universality of the propositions deduced by means of it. Kant says that although the intuition through which the concept is exhibited is individual, it nevertheless must "express in the representation universal validity for all possible intuitions that belong under the same concept" (A713/B741). The individual intuition is "determined under certain general conditions of construction" (A714/B742), i.e. only those features of it are pertinent which belong to the concept and to the general conditions of its constructions. The concept of triangle, which is the definition of a plane figure enclosed by three straight line segments, is exhibited in the individual image of a triangle. The individual image has certain features (*metrical* features, e.g. sides and angles of a certain magnitude) which do not belong to the concept and which play no role in the proof of a theorem about triangles in general. As the exhibition of that general concept, the intuition is undetermined as to conditions of constructions other than the "general conditions of construction" of the triangle in general.

Mathematics is knowledge through the construction of concepts; constructivity is the "form" of mathematical knowledge. As Kant explains in the *Disziplin*, those who regard mathematics as the science of quantity have taken the effect for the cause. The *object* of mathematical cognition is the *consequence*, not the cause, of the *form* of mathematical cognition: mathematics deals with quantity *because* it is constructive, and not vice versa.[3] The reason is that "only the concept of magnitudes can be constructed, i.e. exhibited *a priori* in intuition, while qualities cannot be exhibited in anything but empirical intuition" (A714/B742). Why can only concepts of magnitudes be constructed? The reason is that in order for a concept to be constructed its corresponding intuition must be capable of being given a priori. But a priori only the form of possible intuition can be given, i.e. space and time; hence, the only concepts that can be constructed are the concepts of these, or *quanta*. Kant says:

> Now of all intuition none is given *a priori* except the mere form of appearances, space and time, and a concept of these, as *quanta*, can be exhibited a priori in pure intuition, i.e. constructed, together with either its quality (its shape) or else merely its quantity (the mere synthesis of the homogeneous manifold) through number (A720/B748).

In mathematics "we can determine our concepts *a priori* in intuition, for we create the objects themselves in space and time through homogeneous synthesis, considering them merely as *quanta*" (A723/B751). A mathematical concept "contains" (A719/B748) or "already applies to" (A724/B752) an a priori intuition. This can

[3] Cf. "Mathematics is not distinguished from philosophy by means of the object, but rather through the *modum cognoscendi*. This however determines the distinction of the objects" (Kant, 2005, p. 248).

mean nothing else that such a concept *is the concept of* an a priori intuition. But a priori intuition is space and time; thus, a mathematical concept is the concept of space and time considered as *quanta*.

Mathematical concepts can be constructed because they are concepts of a priori intuitions, i.e. are concepts of *quanta*. Here a distinction between undetermined (unconceptualized) space and determined (conceptualized) space has to be made. Kant says in the Aesthetic that, as the condition of all intuition, space is unique, and the whole of space is given before, and presupposed by, the determination of its parts: "one can only represent a single space, and if one speaks of many spaces, one understands by that only parts of one and the same unique space. And these parts cannot as it were precede the single all-encompassing space as its components (from which its composition would be possible), but rather are only thought in it" (A24–25/B39). In the Axioms of Intuition, by contrast, Kant maintains that all intuitions are extensive magnitudes, and that an extensive magnitude is one for which "the representation of the parts makes possible the representation of the whole (and therefore necessarily precedes the latter)" (A162/B203). Since space and time are intuitions, and thus extensive magnitudes, there seems to be a contradiction between the claim in the Aesthetic that the whole of space is presupposed by the parts of space, and the claim in the Axioms that the representation of space presupposes the representation of its parts. But the contradiction disappears as soon as we recognize that the Axioms are concerned not with space and time *in general*, but with *determined* space and times.[4] As Melnick puts it, "the Axioms are concerned with the construction of figures (the construction of determinate spaces) and not with the construction of the space itself within which figures are constructed" (1973, p. 18). Space itself, as a condition of intuition, is not *constructed*; rather, it is the *condition* of the construction of determined spaces (spatial figures): if space as pure intuition were itself something which we construct, *where* such construction should take place? On the contrary, all construction of determined spaces presupposes that the whole of space, as an intuition, is already given before its parts are determined. Thus, the apparent contradiction disappears as soon as we consider that while that which is presupposed by the determination of its parts is the whole of space given in intuition, that which presupposes the representation of its parts is the representation of determined spaces (spatial figures).

In the *Prolegomena*, Kant says that it is the understanding that determines the space as intuition into determined spaces: "that which determines space into the figure of a circle, a cone, or a sphere is the understanding, insofar as it contains the basis for the unity of the construction of these figures" (Kant, 2002, p. 114). In the Axioms, Kant says that a determined space (a spatial figure) is obtained "through the composition of that which is homogeneous and the consciousness of the synthetic unity of this manifold (of the homogeneous)" (B202).

[4] This manner of solving the apparent contradiction has been proposed by several commentators; see Melnik (1973) and Sutherland (2021).

Determination involves synthesis. In the B-version of the Axioms of Intuition, Kant says:

> [appearances] cannot be apprehended [...], i.e., taken up into empirical consciousness, except through the synthesis of the manifold through which the representation of a determinate space or time are generated, i.e., through the composition of that which is homogeneous and the consciousness of the synthetic unity of this manifold (of the homogeneous). Now the consciousness of the homogeneous manifold in intuition in general, insofar as through it the representation of an object first becomes possible, is the concept of a magnitude (*Quanti*). (B202–203)

Determined spaces and times are magnitudes or *quanta*, and are generated by the action of the understanding on intuition. The category of quantity or magnitude is the "category of the synthesis of the homogeneous in an intuition in general" (B 162). Thus, determined spaces and times are generated through the composition of a homogeneous manifold and through the consciousness of the synthetic unity of this manifold. The composition of a homogeneous manifold is brought to unity by the understanding, i.e. through the concept of a magnitude. In an important footnote added to the B-edition, Kant calls the synthesis performed by the understanding to represent determined spaces and times the "synthesis of composition": "[The synthesis of composition] is the synthesis of a manifold of what does not necessarily belong to each other, as e.g., the two triangles into which a square is divided by the diagonal do not of themselves necessarily belong to each other, and of such a sort is the synthesis of the homogeneous in everything that can be considered mathematically (B201n). This is echoed in a fragment written c. 1878–1880s:

> Mathematics has the peculiarity that it applies only to those objects that can be represented *coram intuitu*, and therefore always has empirical confirmation. It is possible as synthetic *a priori* cognition because there are two *a priori intuitus*, space and time, in which an *a priori* synthesis of composition is possible. These two objects are *quanta* and indeed *originaria*, and the mere synthesis of them is quantity. All concepts of *quantis* can be constructed in them, i.e., given *a priori* in intuition, likewise all concepts of quantity, i.e., number, which requires time as well as space. (Kant, 2005, p. 248)

Mathematical concepts are concepts of the composition of a homogeneous manifold of intuition (i.e. are "syntheses of composition"). Since the manifold of intuition is given a priori, such concepts can be constructed, because to construct a concept means to exhibit a priori the corresponding intuition. More precisely, to construct a mathematical concept is to construct the concept of a determined space or time, and the concept of a determined space or time is the synthesis of composition of a homogeneous manifold in intuition.

In the *Disziplin* Kant distinguishes two kinds of construction:

> But mathematics does not merely construct magnitudes (*quanta*) as in geometry, but also mere magnitude (*quantitatem*), as in algebra, where it entirely abstracts from the constitution of the object that is to be thought in accordance with such a concept of magnitude. In this case it chooses a certain notation for all construction of magnitudes in general (numbers), as well as addition, subtraction, extraction of roots, etc., [...] it then exhibits all the procedures through which magnitude is generated and altered in accordance with certain rules in intuition [...] and thereby achieves by a symbolic construction (*symbolische Konstruktion*) equally well what geometry does by an ostensive or geometrical construction

(of the objects themselves), which discursive cognition could never achieve by means of mere concepts. (A718/B745)

How a geometrical *quantum* is ostensively constructed is clear from Kant's example of the triangle: the concept of triangle is constructed by determining space into a two-dimensional figure enclosed by three straight lines. The individual figure (image) exhibits the concepts without affecting the generality of that concept, because the schema for its construction only comprises "general conditions of construction" (A714/B742), i.e. those conditions that hold for any particular falling under the concept of triangle.

In the case of arithmetical and algebraical concepts, it is not so clear how the construction is performed. Kant is silent about arithmetic in the passage from the *Disziplin*. Thompson thinks that arithmetical concepts (i.e. particular numbers) are constructed through a symbolic construction: "A numeral is thus a symbolic construction of a number, and in arithmetic by further symbolic constructions as from numerals (by calculation) we discover further numerical properties of numbers, just as in geometry by further ostensive constructions from figures we discover further geometrical properties of figures" (1992, p. 98).[5] This would require regarding the numeral "5" as the image of the number five just as the individual diagram of the triangle is the image of the concept of triangle. But this is not what Kant says about images of numbers in the Schematism: "if I place five points in a row, …, this is an image of the number five" (A140/B179). Just as with geometric constructions, the image is overdetermined with respect to the concept, and thus not fully adequate to it, because the concept is the concept of a collection of five units, while the image is the image of a collection of five *points*. That which allows the concrete image to serve as the image of the concept of the number five is the scheme, i.e. the rule for the construction of the image, which in the same passage Kant characterizes as "a method whereby a multiplicity, […] may be represented in an image in conformity with a certain concept" (*ibid.*). Indeed, it is only by operating with *images* of numbers that arithmetical propositions are shown to be synthetic a priori propositions. This is what Kant says in the Introduction:

One must go beyond these concepts, seeking assistance in the intuition that corresponds to one of the two, one's five fingers, say, or (as in Segner's arithmetic) five points, and one after another add the units of the five given in the intuition to the concept of seven. For I take first the number 7, and, as I take the fingers of my hand as an intuition for assistance with the concept of 5, to that image of mine I now add the units that I have previously taken together in order to constitute the number 5 one after another to the number 7, and thus see the number 12 arise (B15–16)

As Sgarbi (2010) has persuasively shown, the reference to Segner in this context is instructive. In his *Elementa arithmeticae et geometriae* (1739) Segner had exemplified arithmetical sums by means of dots. In Kant's terminology, a collection of five points is the image of the number five, while the numeral "5" is its symbol. In

[5] Friedman (1990, p. 223) also argues in favor of including both arithmetic and algebra under the heading of symbolic construction.

arithmetical numerals numbers are not constructed, but only symbolized. The kind of construction that we find in elementary arithmetic, as far as numerals are employed, is therefore symbolic, not ostensive. Young (1992) seems to make a similar point when he suggests that a sequence of tokens of the numerals "1" through "12" constitutes a collection of twelve perceptible objects just like a collection of twelve points is. The difference between the two cases, however, is that each such sequence can be indexed or symbolized by a single one of its member, namely the last member: "Because each numeral can serve as the index of the cardinality of the sequence beginning with "1" and ending with itself, moreover, we have the capacity to state arithmetical truths in ways that are economical and readily understood. In writing "7 + 5 = 12", for example, we are asserting that any collection of the same cardinality as that which results from joining perceptible tokens of "1" through "7" with perceptible tokens of "1" through "5" is of the same cardinality as a collection of tokens of "1" through "12" (1992, p. 163). The arithmetical formula "7 + 5 = 12" would count as the symbolic construction of the number twelve, while the actual composition of the numerals "1" through "7" with the numerals "1" through "5", which is equivalent to the composition of an image of seven points with an image of five points, would count as its ostensive construction.[6]

As to algebra, Kant clearly says in the *Disziplin* that the construction is in this case "symbolic". Given that to construct a concept is to exhibit that concept in intuition, how are we to understand algebraical symbolic constructions? There is a tendency among commentators to consider the symbolic construction of algebra as a construction *in* or *through* symbols, i.e. as a construction with symbols as objects of intuition. I will refer to this as the "symbolic interpretation" of symbolic construction. The symbolic interpretation has several supporters in Kant scholarship,[7] but it has recently been confuted by Shabel (1998). According to Shabel, "symbolic constructions" are not kinds of constructions, that is, constructions of or out of symbols or characters. They are means to symbolize ostensive or geometrical constructions. Take the equation "$x = a/b$", which expresses a quantitative relationship between x, a, and b such that b: 1 = a: x. Now, this quantitative relationship is "exhibited in intuition" not because the equation itself, when it is written on paper, expresses that quantitative relationship. That quantitative relationship is "exhibited in intuition" because it allows the object x to be constructed geometrically as a line segment relative to the line segments a and b in accordance with the usual construction of the fourth proportional. The equation is not itself the construction; it is a means for the ostensive construction of a geometrical object. Shabel's may be called the "ostensive interpretation" of symbolic construction.

---

[6] For more recent work on Kant's philosophy of arithmetic see Carson (2020), Sutherland (2020), and Tait (2020). They do not directly discuss symbolic construction, however.

[7] "The 'symbolic construction' is essentially a construction with symbols as objects of intuition" (Parsons, 1992, p. 65); "the symbols are manipulated according to certain 'universal rules' and, I take it, the application of these rules is guided, or "controlled", by our intuition of the symbols (in the paradigm case, marks on paper) manipulated" (Brittan, 1992, p. 330). Cf. also Broad (1941), to which I return in Sect. 2.5 below.

Now, it is not my purpose here to argue in favor of either the symbolic or the ostensive interpretation of Kant's idea of symbolic construction. The purpose of this section was merely to show how important quantity is for Kant's general idea of construction. According to the ostensive interpretation, quantity is still relevant in algebra, in that algebra is merely an instrument to construct geometrical *quanta*. As we shall see, Peirce may be taken to accept some sort of symbolic interpretation. But this, I shall argue, is not in itself problematic for his adoption of Kant's notion of construction.

2.2 Peirce on Constructions

The best evidence that Peirce inherited the idea of mathematical construction from Kant is in a series of papers written by Peirce in the second half of the 1890s. I will start from an article published in 1898 in the *Educational Review* and titled "The Logic of Mathematics in Relation to Education". Here, Peirce begins by criticizing the traditional definition of mathematics as the science of quantity. The correct definition of mathematics, by contrast, was given by Benjamin Peirce, Charles' father, at the beginning of his *Linear Associative Algebra* (1870, p. 1): "mathematics is the science that draws necessary conclusions"; to which the son only added that since necessary conclusions can only be drawn from hypothetical states of things, mathematics must be the science that draws necessary (i.e. deductive) conclusions about hypotheses. Such a definition of mathematics is methodological, not substantial; it defines mathematics by its method (tracing deductive consequences from hypotheses) and not by its object. At this point, Peirce makes a direct reference to the *Disziplin*:

> Kant, in the *Critique of Pure Reason* (Methodology, chapter I, section 1), distinctly rejects the definition of mathematics as the science of quantity. What really distinguishes mathematics, according to him, is not the subject of which it treats, but its method, which consists in studying constructions, or diagrams. That such is its method is unquestionably correct; for, even in algebra, the great purpose which the symbolism subserves is to bring a skeleton representation of the relations concerned in the problem before the mind's eye in a schematic shape, which can be studied much as a geometrical figure is studied. (CP 3.556)

What Kant called a "construction" Peirce calls a "diagram":

> Kant is entirely right in saying that, in drawing those consequences, the mathematician uses what, in geometry, is called a "construction", or in general a diagram, or visual array of characters or lines. Such a construction is formed according to a precept furnished by the hypothesis. Being formed, the construction is submitted to the scrutiny of observation, and new relations are discovered among its parts, not stated in the precept by which it was formed, and are found, by a little mental experimentation, to be such that they will always be present in such a construction. Thus, the necessary reasoning of mathematics is performed by means of observation and experiment, and its necessary character is due simply to the circumstance that the subject of this observation and experiment is a diagram of our own creation, the conditions of whose being we know all about. (CP 3.560)

Mathematics is as observational and experimental as any positive science; only, the object of its observations and experiments is not a natural or empirical object; it is an ideal object, a purely hypothetical state of things, "the conditions of whose being we know all about", which guarantees that the conclusions we draw about it have the character of necessity. Mathematics is observational and experimental in that it constructs, observes and manipulates "diagrams". What does this mean?

It is important to notice at the outset that Peirce calls "mathematical" all deductive reasoning. What we call "formal" or "deductive" logic is for him a part of mathematics, indeed is a form of applied mathematics (mathematics applied to logic). So in the *Minute Logic* (1902) we read that "[t]here is a mathematical logic, just as there is a mathematical optics and a mathematical economics. Mathematical logic is formal logic. Formal logic, however developed, is mathematics" (CP 4.240). Logic as such is, of course, not to be identified with mathematics; logic in the broad sense includes inductive and abductive logic, and these are in no obvious sense mathematical (although they certainly have mathematical components, e.g. in the theory of probabilities, and mathematical roles, e.g. in the discovery of theorems). It is only formal deductive logic that Peirce classifies as applied mathematics. When he thus presents his doctrine of diagrammatic reasoning by the famous claim that "all necessary reasoning without exception is diagrammatic" (CP 5.156), both mathematics and formal logic are included under "necessary reasoning".[8]

Now, Kant, owing to the poor development that logic had reached at his time, took philosophical (deductive) knowledge to be discursive and mathematical knowledge to be constructive. This is an optical illusion due to the fact that very often the construction involved in deduction is simple and therefore neglected. Even simple syllogism involves a diagram:

> Why do the logicians like to state a syllogism by writing the major premiss on one line and the minor below it, with letters substituted for the subject and predicates? It is merely because the reasoner has to notice that relation between the parts of those premisses which such a diagram brings into prominence (CP 3.560).

> The traditional conception of syllogism [...] is that deductive reasoning is performed "symbolically" as Leibniz says, without any aid from "intuition," that is, from icons. But the truth of the matter is that in such reasoning the icon is the most vital element. (R 16 CSP 7)

Syllogistic is observational and experimental in that it performs observations and experiments on diagrams. All deductive thinking is diagrammatic in this sense, and no distinction should be made between "lower" forms of deduction, like syllogism, in which excessively simple diagrams are employed, and "higher" forms in which the diagrammatic component is evident and manifest.

It should be emphasized that in the second of the passages just quoted Peirce seems to take "intuition" to be equivalent to (what he calls) "icon". In saying that in syllogisms the icon is the most vital element, he is in fact saying that syllogistic

[8] Peirce's doctrine of diagrammatic reasoning has received a fairly generous bit of scholarly attention, especially in the last two decades; see Stjernfelt (2000, 2007, 2011), Hoffmann (2010), Pietarinen and Bellucci (2016).

reasoning is performed with the aid of intuition, i.e. is performed by constructing and manipulating diagrams. The terms "icon" and "diagram", though not perfectly synonymous, are roughly equivalent in this context. In other contexts Peirce is more precise. For example, in a note written in the *Logic Notebook* in 1906 we read: "The Diagram is an Interpretant of a Symbol in which the signification of the Symbol becomes a part of the object of the Icon" (R 339, p. 286r). This is tantamount to saying that a diagram constructs a concept (a symbol) in intuition (through icons), i.e. that it represents intuitively (iconically) the content of the concept.[9]

Mathematics being defined by its method—the employment of constructions or diagrams—and not by its object, the traditional definition of mathematics as the "science of quantity" should be abandoned. Yet, Peirce makes a further point that complicates things a bit. He says:

> But were I to drop this branch of the subject without saying one word more, my criticism of the old definition, "mathematics is the science of quantity", would not be quite just. It must be admitted that quantity is useful in almost every branch of mathematics. Jevons wrote a book entitled *Pure logic, the science of quality*, which expounded, with a certain modification, the logical algebra of Boole. But it is a mistake to regard that algebra as one in which there is no system of quantity. (CP 3.561)

Understanding Peirce's point in this passage requires that we situate his work in formal logic in the tradition to which it belongs, i.e. the Boolean tradition of the "algebra of logic". The idea that algebra is not necessarily connected with quantity lies at the foundation of this tradition. In his *Mathematical Analysis of Logic* (1847) Boole says that the theory of "symbolic algebra" does not depend on the quantitative interpretation of the symbols; indeed it does not depend on any interpretation of the symbols employed but only on the rules of their combinations.[10] In his address to the College of Cork for the academic year 1851–1852, Boole is again very clear that mathematics in the broad sense is "universal reasoning in symbolical forms",

---

[9] One should not be worried by the fact the, on Peircean principles, the sign "determines" the interpretant. In geometrical construction, the sign is the concept-symbol of a geometrical object, while the diagram-icon is its interpretant. To say that the concept-symbol of a geometrical object determines a diagram-icon of that geometrical object only means that the former determines the latter to represent that object. Thus, the latter represents the very same object that the former represents, only in an iconic-intuitive and not in a symbolic-conceptual manner. This sense of "determination" is fully compatible with the idea that the diagram "constructs" the object of the symbol, for "to determine" here simply means "determines to represent iconically". See Pietarinen and Bellucci (2016) for more on this. That icons in Peirce performs some of the functions that intuition does in Kant, while other functions are performed by indices, is one of the main points made in Gava (2014).

[10] "They who are acquainted with the present state of the theory of Symbolical Algebra, are aware, that the validity of the processes of analysis does not depend upon the interpretation of the symbols which are employed, but solely upon the laws of their combination. Every system of interpretation which does not affect the truth of the relations supposed, is equally admissible, and it is thus that the same process may, under one scheme of interpretation, represent the solution of a question on the properties of numbers, under another, that of a geometrical problem, and under a third, that of a problem of dynamics or optics" (Boole, 1847, p. 3).

and that mathematics as a science of quantity is just a particular development of it.[11] Logic and mathematics are just applications (interpretations) of a more general "qualitative algebra" or "mathematics in the broad sense". Taking "algebra" in this sense (as mathematics in the broad sense), the algebra of logic can be said to be an algebra of qualities as opposed to numerical or quantitative algebra. This idea emerges especially in some of Boole's followers. In the passage above Peirce refers to Jevons', 1864 book *Pure Logic, or the Algebra of Quality*, in which we read that "*logic is the algebra of kind or quality, the calculus of known and unknown qualities*, as algebra (more strictly speaking *universal arithmetic*, which does not recognise essentially negative quantities) is the calculus of known and unknown quantities" (1864, §15). Taking "algebra" in this generic sense (as universal reasoning in symbolic forms, or mathematics in the broad sense), one can say that numerical algebra, i.e. universal arithmetic, is the algebra of quantities in the same way as logic is the algebra of qualities. The same characterization of logic is given in MacFarlane's *Principles of the Algebra of Logic* (1879): "Logic, as the Algebra of Quality, is a *formal science*. It investigates the general properties of the symbol of Quality ..." (1879, §1.6).[12]

Now, in the passage above Peirce distances himself from this idea. He says that it is a mistake to regard Boole's algebra of logic as one in which there is no system of quantity. But in what sense does the algebra of logic involve quantity? Peirce's answer is that the algebra of logic is the algebra of the true and the false, i.e. is the algebra of a system of quantity that has two values: "the algebra of Boole is nothing but the algebra of that system of quantities which has but two values—the simplest conceivable system of quantity" (CP 3.561). The first part of chapter III of Peirce's 1902 book *Minute Logic*, devoted to deductive logic, was accordingly titled "The Simplest Possible Mathematical System". Here, we read:

> It is of the essence of mathematics that it should draw conclusions. There can be no reasoning where there was not previously room for doubt. Hence, the simplest mathematical hypothesis must allow the possibility of two answers to a question, and thus must involve at least two elements. Let us begin with the case in which it involves nothing more. We are to suppose, then, two independent units, or objects, which we may call V and F. What the nature of these units may be is of no consequence. (LoF 1, pp. 446–447)

Peirce is referring to the fact that Boole's algebra of logic may have a numerical interpretation restricted (by the famous index law: $x^2 = x$) to the values 1 and 0.

---

[11] "If it is asked whether out of these common principles of the reason we are able to deduce the actual expressions of its fundamental laws, I reply that this is possible, and that the results constitute the true basis of mathematics. I speak here not of the mathematics of number and quantity alone, but of mathematics in its larger, and I believe, truer sense, as universal reasoning expressed in symbolical forms, and conducted by laws, which have their ultimate abode in the human mind. That such a science exists is simply a fact, and while it has one development in the particular science of number and quantity, it has another in a perfect logic" (Boole cit. in Dooley, 2018, pp. 169–170)

[12] On this conception of the algebra of logic see Barone (1965).

Logic, treated algebraically, is therefore not just an algebra of qualities. Viewed from the point of view of quantity,[13] logic is, after all, a system of quantity.

Peirce's quantitative understanding of Boole's algebra of logic should help explaining in what sense he can take Kant's notion of construction and turn it into a doctrine of diagrammatic reasoning that holds of all deduction. For Kant only *quanta* can be constructed, i.e. exhibited in an intuition. Peirce may now agree with this idea and may argue that in the algebra of logic, logical concepts and relations are constructed, i.e. represented in diagrammatic form; and they can be constructed by means of that algebra precisely because that algebra has a quantitative interpretation.

However, Peirce goes further than this. In the 1898 "The Logic of Mathematics in Relation to Education" as well as in some drafts of a paper "On Quantity" probably dating from 1895–1896, he introduces the conception of a "scale of quantity". This conception is, I shall argue, Peirce's more considered (albeit implicit) answer to our problem.

2.3 Scales of Quantity

Our problem was the following. Does Peirce accept Kant's idea that only concepts of *quantity* can be constructed? If yes, since he takes Kant's notion of construction and turns it into the idea that all necessary (i.e. deductive) reasoning is diagrammatic, he should be committed to the idea that only concepts of quantity can be diagrammatized. As we have seen, in "The Logic of Mathematics in Relation to Education" Peirce argues that mathematics should not be defined as the science of quantity, but he also argues that Boole's algebra of logic has in fact a quantitative interpretation. He continues the discussion as follows:

> in solving a mathematical problem, we usually introduce some part or element into the construction which, when it has served our purpose, is removed. Of that nature is a scale of quantity, together with the apparatus by which it is transported unchanged from one part of the diagram to another, for the purpose of comparing those two parts. Something of this general description seems to be indispensable in mathematics. Take, for example, the Theorem of Pappus concerning ten rays in a plane. The demonstration of it which is now usual, that of von Staudt, introduces a third dimension; and the utility of that arises from the fact that a ray, or unlimited straight line, being the intersection of two planes, these planes show us exactly where the ray runs, while, as long as we confine ourselves to the consideration of a single plane, we have no easy method of describing precisely what the course of the ray is. Now this is not precisely a system of quantity; but it is closely analogous to such a system, and that it serves precisely the same purpose will appear when we remember that that same theorem can easily (though not so easily) be demonstrated by means of the bary-

[13] Cf. also the following passage: "Every mathematician understands that the use of an algebraical system of expression implies nothing in particular about the subject matter,—certainly, not, for instance, that it is of a quantitative character. […] The Boolian algebra deals with a system of quantity (if one conceives it from the point of view of quantity) having but two values." (R 812, 1894 = NEM 4, pp. 96–97).

> centric calculus. Although, then, it is not true that all mathematics is a science of quantity, yet it is true that all mathematics makes use of a scaffolding altogether *analogous* to a system of quantity (CP 3.561)

The notion on which the passage is based is that of a "scale of quantity". It is described as an element introduced into a demonstration for certain purposes that is removed afterwards. Peirce claims that something of this sort is necessary in all mathematical reasoning. His example is the "theorem of Pappus", actually Desargues's theorem, which Peirce most often calls the "Ten point theorem". Here Peirce refers to Karl von Staudt's demonstration of this theorem, which in brief consisted in seeing the two-dimensional diagram (Fig. 2.1) as three-dimensional, so that the three segments OX_2, OY_2, and OZ_2 may be viewed as a perspective representation in which the three rays form the top of a pyramid in which two triangles $X_1 Y_1 Z_1$ and $X_2 Y_2 Z_2$ are inscribed (Fig. 2.2).[14] The third dimension is not part of

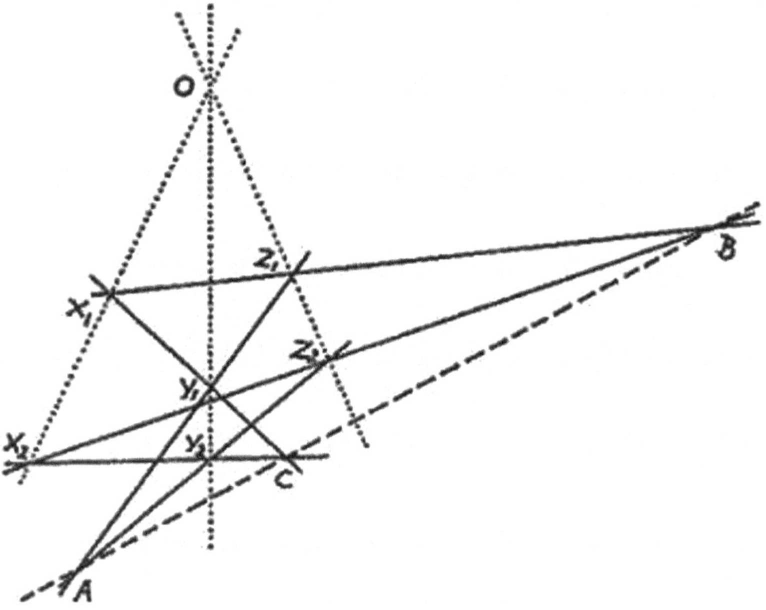

Fig. 2.1 Two-dimensional diagram for Desargues's theorem (from NEM 2, p. 212)

[14] See Hoffmann (2010) for a detailed analysis of Peirce's interpretation of Von Staudt's three-dimensional demonstration of Desargues's theorem. Later Peirce would use the theorem as an illustration of his idea of "theoric transformation" and the connected distinction of deductions into "theorematic" and "corollarial". In 1907 Peirce writes that "theoric" demonstrations consist "in the transformation of the problem, or its statement, due to viewing it from another point of view" (R 318 CSP 68); cf. R 201 and NEM 2, pp. 211–213; see also Stjernfelt (2014, pp. 263–292) and Pietarinen and Bellucci (2014, pp. 268–271). The mid-1890s notion of "scale of quantity" may well be an antecedent of the later notion of "theoric transformation", but I have no space to address this point here.

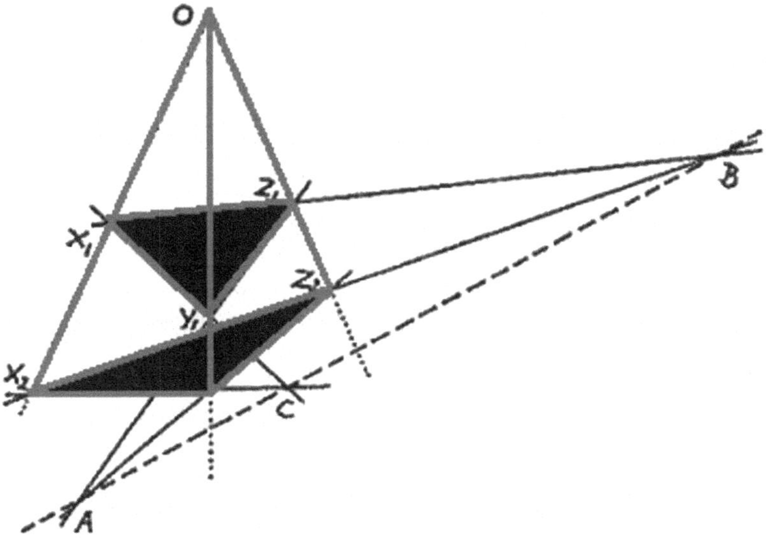

Fig. 2.2 Two-dimensional diagram for Desargues's theorem (from NEM 2, p. 212)

the statement of the premises, nor is it part of the statement of the theorem; it is only introduced in the demonstrative process as an aid to it which is removed after it has served the purpose for which it was introduced.

Further characterizations of the notion of a scale of quantity are in the drafts of "On Quantity" (R 14, R 16, R 17). In one of these we read:

> It is very useful in tracing effects of compounding relations to be provided with a lettered diagram of all possible combinations of the relations dealt with. [...] Such a diagram is a schema of the scale of quantity. Thus a *quantity* is nothing but one of the hecceities in the system of possible forms of combination of certain relations, whose characters except in regard to their forms of combination do not concern the scale of quantity. (R 14 = LoF 1, p. 81n23)

Take again Boole's algebra of logic. The two values 1 and 0, or the true and the false, are merely objects ("hecceities") of the system the algebra is about. They are "quantities" in the sense that the system refers, in this case, to a certain number of objects. These two obejcts are mere hecceities, i.e. mere singulars with no descriptive content. When we deal with this algebra as a non-interpreted system, the two objects have no other characters than those that are defined by the rules of their combination. The algebraic system itself (the "scale of quantity" or "diagram") is used as a means to trace the effects of those combinations without knowing what those objects or their characters are.

In another draft Peirce is more explicit:

> For the purposes of the present memoir, the chief element of mathematics upon which attention has to be fixed is the hypotheses. The general hypothesis of a branch of mathematics supposes a system of individual objects related to one another in a simple and general way. The hypothesis of each special problem supposes other objects whose relations to one another are to be determined by assigning to each one of them one of the individual objects of the general hypothesis, so that the system of these latter objects serves as a middle term

of comparison, for determining the relations of the objects of the special hypothesis to one another. For example, the objects of the general hypothesis may be the points, or indivisible places, of space. The objects of the special hypothesis may be particles, or indivisible things in space. Then the relations of these particles to one another are determined by ascertaining the point of space occupied by each particle. Thus the places are middle terms of comparison for determining the relations of the things. (R 16 CSP 8)

Here, Peirce imagines a "general hypothesis" that is a "middle term of comparison" for a "special hypothesis". Suppose a certain configuration of particles in space. This is what Peirce calls a "special hypothesis" and is the very object of mathematical investigation and reasoning. The particles are the "objects" ("hecceities" in the previous passage) of the special hypothesis. In order to study the special hypothesis we use a certain configuration of points of space. This is what Peirce calls the "general hypothesis", which is not the object of mathematical reasoning, but an instrument of it. By studying the relations of the objects of the general hypothesis—the relations of the points of space—one can determine the relations of the objects of the special hypothesis—the relations of the particles in space. The general hypothesis is thus a "scale of quantity" in that the relations of the points is a middle of comparison for determining the relations of the particles which is introduced in the course of the investigation and is removed afterwards.[15]

The draft continues as follows: "A multitude of individual objects in regular systematic relationship to one another, if it is employed as a middle term of comparison for determining the relations between other things, may be called a *scale of quantity*. Thus, the system of real numbers is used as such a middle term of comparison" (R 16 CSP 8). The example here is telling. When we say that the set of positive integers or the set of even integers have the same cardinality as the set of natural numbers (N), we use this latter as a middle term of comparison for determining the cardinality of the former. When we say that the power set of natural numbers has the same cardinality as the set of real numbers (R), we use this latter as a middle term of comparison for determining the cardinality of the former.

A third draft of "On Quantity" nicely summarizes all these aspects:

In order to comprehend the nature of quantity, it is necessary to consider what mathematics is and what its method of reasoning necessarily is. It will be found to relate to *hypotheses* and to reason about them by means of *diagrams*. The nature of these hypotheses and diagrams being examined, it will be found that a system of objects is required which are all alike, although distinguishable from one another; and between these is a regular system of relations. This system serves as a middle term of comparison to determine the relations of order between other objects and constitutes a *scale of quantity*. (R 17 CSP 1)

[15] The "removability" of the scale of quantity, although not always emphasized by Peirce, is crucial. It is in fact what specifically characterizes the experiments made in mathematical reasoning, where the object remains unlatered by its being treated by means of a scale of quantity. The same applies to what Peirce calles "theoric demonstrations", which consist in viewing the problem at hand from another point of view (see previous footnote). Changing the point of view on an object does not of course alter the object viewed. In empirical experiments nothing of this sort can happen; here the experiments are made on the object itself, which is thereby inevitably altered. I owe this point to an anonymous reader.

Peirce's final reference to "relations of order" may suggest that he has had such relations in mind all along. In 1881, he had published a paper containing a first axiomatization of natural numbers. In the course of the paper he defines a "system of quantity" (partially ordered set) and a "system of simple quantity" (totally ordered set) (CP 3.253–256). When in "On the Logic of Quantity" he speaks of determining "the relations of order between objects" he may have had in mind a case in which a "systems of quantity" is used to determine relations of order in a certain configuration of objects. In all such cases, a "systems of quantity" becomes a "scales of quantity".[16]

To sum up. In both "The Logic of Mathematics in Relation to Education" and in the drafts of "On Quantity" of 1895–1896, Peirce argues as follows. Mathematics does not necessarily deal with quantity. Yet in all mathematical (i.e. deductive) reasoning use must be made of something analogous to quantity, a "scale of quantity". This should now put us in a position to provide an answer to our original problem. How can logical objects, which are not quantitative, be constructed if only concepts of quantity can be constructed in Kant's sense? My proposal, which I will try to qualify further in the next section, is that logical objects can be constructed even if they are not quantitative because the construction (the diagram) is a quantitative middle term of comparison, i.e. is a scale of quantity.

2.4 Syllogism

We saw in Sect. 2.2 that in turning the Kantian doctrine of mathematical construction into a doctrine of diagrammatic reasoning, Peirce departs from Kant's too sharp distinction between philosophical-deductive knowledge and mathematical knowledge. According to Peirce, any such distinction is just the effect of an optical illusion due to the fact that very often the construction involved in deduction is too simple and is therefore neglected. His usual example is the syllogism: even in a syllogism (which for Kant counts as "discursive knowledge by means of concepts") there is an element of construction. My purpose in this section is to substantiate this idea by means of the notion of "scale of quantity". In brief, the idea is that the system of representation by means of which a syllogism is expressed is itself a scale of quantity, i.e. a middle term of comparison that determines the relations between certain objects (terms or classes).

Euler diagrams are a system of representation that is capable of representing syllogism with sufficient accuracy and which also constitutes a sort of algorithm by means of which the validity of syllogistic forms can be checked.[17] In the second

[16] On Peirce's 1881 axiomatization of arithmetic see Shields (2012).

[17] I will here leave aside the phenomenon of "over-specificity" in Euler diagrams: sometimes the information that is observable in an Euler diagram is not inferable from the information that was necessary to construct it; see Shimojima (1996a, 1996b).

section of the *Logical Tract No. 2*—a treatise on logical diagrams written in the summer or the early autumn of 1903—Peirce asks the following question: "What is it, then, that these diagrams are supposed to accomplis>sh? Is it to prove the validity of the syllogistic formula? That sounds rather ridiculous—as if anything could be more evident than a syllogism—yet that is not far from the opinion of Friedrich Albert Lange, a thinker of no ordinary force. Suppose we ask ourselves why it is that, if a circle P wholly encloses a circle M which itself wholly encloses a circle S, the circle P necessarily wholly encloses the circle S″ (CP 4.354). The reference to F.A. Lange is significant, because Lange had claimed, in his *Logische Studien* (Lange, 1877), that the validity of syllogism depends on the intuition of space relations.[18] Peirce admired Lange's book and accepted Lange's basic argument that intuitive images (*Anschauungsbilde*), i.e. diagrammatic constructions, are necessary in all deduction. Yet, he did not follow Lange in claiming that logical validity depends on geometrical evidence. Neither does logical validity depend on the holding of certain space relations, nor vice versa does the holding of certain space relations depend on logical validity. By a long and technical argument, Peirce reaches the conclusion that "as far as logical dependence goes, the validity of the syllogism and the property of the Eulerian diagram depend upon a common principle. They are analogous phenomena neither of which is, properly speaking, the cause or principle of the other" (CP 4.355). The common principle in question is the property of transitivity, which Peirce expresses in general terms as "At once *r* of and *r* of whatever is *r*'d by". Both the transitivity of universal predication ("at once predicates of and predicates of whatever is predicated by") and the transitivity of spatial inclusion ("at once includes and includes whatever is included by") are just two manifestations of transitivity "in itself", which thus grounds both logical and spatial inclusion. The same applies to exclusion: when a class excludes another, it excludes whatever the latter includes; thus exclusion "transitates" to whatever the excluded element includes. (The two principles—transitivity of inclusion, and transitivity of exclusion over inclusion—are what the medievals called the "dictum de omni et nullo").

Logical inclusion neither depends on spatial inclusion nor vice versa, yet the former may be constructed by means of the latter: when we represent the relations that hold between the three terms of a syllogism by means of the relations between spatial objects (the circles in Euler diagrams), we construct the former relations by means of the latter. Syllogistic relations are not quantitative. Yet, they can be represented by means of spatial, i.e. quantitative, relations. In other words, geometrical transitivity may be used to construct logical transitivity. In this sense, the system of Euler diagrams is a scale of quantity. Recall the example concerning particles in space and points of space that Peirce gives in R 16. It is easy to see how Euler diagrams and syllogistic relations may be susceptible of a similar analysis. Peirce imagines a "general hypothesis" that is a "middle term of comparison" for a

[18] See Bellucci (2013).

"special hypothesis". Suppose certain relations between classes are given. This is what Peirce calls a "special hypothesis" and is the very object of syllogistic investigation. The classes are the "objects" of the special hypothesis. In order to study the special hypothesis we use a certain configuration of geometrical objects, namely the circles. This is what Peirce calls the "general hypothesis", which is not the object of syllogistic, but an instrument of it. By studying the relations of the objects of the general hypothesis—the relations of the circles—one can determine the relations of the objects of the special hypothesis—the relations of the classes. The general hypothesis is thus a "scale of quantity" in that the relations of the circles is a middle of comparison for determining the relations of the classes which is introduced in the course of the investigation and is removed afterwards.

The same applies to the algebraical expression of syllogism, for example to Boole's algebraical analysis. According to Boole (1847), syllogistic reasoning is just a case of elimination of a variable from two equations. Elimination in ordinary (i.e. numerical) algebra is as follows. From two equations such as those in (2.1)

$$\begin{aligned} ay + b &= 0 \\ cy + d &= 0 \end{aligned} \tag{2.1}$$

the "y" may eliminated, which results in (2.2)

$$ad - bc = 0 \tag{2.2}$$

This rule of ordinary algebra may be applied to the algebraical expression of the premises of a syllogism, thereby obtaining the conclusion of the syllogism. So if the premises are "All Ys are Xs" and "All Zs are Ys", I can express these as the equations in (2.3)

$$\begin{aligned} y(1-x) &= 0 \\ z(1-y) &= 0 \end{aligned} \tag{2.3}$$

Applying the rule of elimination to (2.3) we get (2.4):

$$z(1-x) = 0 \tag{2.4}$$

which is the desired conclusion "All Zs are Xs" (Boole, 1847, pp. 31–34).

We saw in Sect. 2.1 that Peirce regarded Boole's algebra of logic as the algebra of the simplest system of quantities, namely the system with two values only. This was because Boole's algebra of logic has a numerical interpretation restricted to the values 1 and 0. Now, it is precisely the employment of the numerical algebra in logic that makes of it a "scale of quantity" for logic in Peirce's sense. Here is how Boole himself expresses this idea in *Laws of Thought*:

> [...] any system of propositions may be expressed by equations involving symbols x, y, z, which, whenever interpretation is possible, are subject to laws identical in form with the

> laws of a system of quantitative symbols, susceptible only of the values 0 and 1. But as the formal processes of reasoning depend only upon the laws of the symbols, and not upon the nature of their interpretation, we are permitted to treat the above symbols, x, y, z, as if they were quantitative symbols of the kind above described. *We may in fact lay aside the logical interpretation of the symbols in the given equation; convert them into quantitative symbols, susceptible only of the values 0 and 1; perform upon them as such all the requisite processes of solution; and finally restore to them their logical interpretation.* (Boole, 1854, pp. 69–70; emphasis in the original).

What Boole is saying here is precisely that in the conversion of the logical symbols into quantitative symbols susceptible only of the values 1 and 0, we treat the algebraical apparatus as a scale of quantity, i.e. as a middle term of comparison for the determination of certain relations between the logical symbols. The rule of elimination, which holds of numerical algebra generally, also holds for the algebra of 1 and 0, i.e. for the algebra of logic. But this is to say that numerical algebra is a scale of quantity for the algebra of logic, or more generally, that algebra is a scale of quantity for logic. Syllogism can be constructed algebraically in the sense that the apparatus of ordinary numerical algebra, which is quantitative, is used to represent logical relations and operations, which are not quantitative.

2.5 Back to Kant

In Sect. 2.1, I briefly discussed the distinction that Kant makes in the *Disziplin* between geometrical or ostensive construction and symbolic construction. It is important to notice that Peirce implicitly accepted this distinction: if Kant's construction becomes to Peirce a diagram, then there should be geometrical (ostensive) and algebraical (symbolic) diagrams. This is precisely what Peirce says on many occasions:

> In fact, there are two general classes of diagrams which are used to assist the mind in deduction; and there is no logical reason why either of these should not be sufficient. There are algebraical and geometrical diagrams. (RS 97 CSP 209, c. 1889)

> The diagrams in which the hypotheses are embodied are of two kinds. In the one kind the parts of the diagram are seen in the visual image to have the relations supposed. In the other kind of diagrams, the parts have shapes to which conventions or "rules" are attached, by means of which the supposed relations are attributed, or imputed, to the parts of the diagrams. Geometrical figures are diagrams of the inherential kind, while algebraical formulae are diagrams of the imputation[al] kind. (R 16 CSP 6, c. 1895–1896)

> It is necessary to set down, or to imagine, some individual and definite schema, or diagram;—in geometry, a figure composed of lines with letters attached; in algebra an array of letters of which some are repeated. (CP 4.233, 1902)

> We have to make choice, then, between a division of mathematics according to the matter of its hypotheses, or according to the forms of the schemata of which it avails itself. These latter are either geometrical or algebraical. Geometrical schemata are linear figures with letters attached; [...] The algebraical schemata are arrays of characters, sometimes in series,

sometimes in blocks, with which are associated certain rules of permissible transformation. (CP 4.246, 1902)[19]

I mentioned that Kant's idea of symbolic construction is susceptible of two distinct interpretations: a "symbolic" interpretation, according to which symbolic construction is construction in or through symbols, i.e. the symbols themselves are objects of intuition; and an "ostensive" interpretation, according to which symbolic construction is an indirect means for constructing a geometrical object, i.e. the symbols are not themselves objects of intuition; rather, they symbolize objects of intuition. Now, the passages just quoted, which testify to Peirce's implicit adherence to the Kantian distinction between ostensive and symbolic construction, also show that he leaned towards the "symbolic" interpretation of symbolic construction. Peirce certainly meant that the "array of letters" or string of symbols that algebra makes use of is itself the object of intuition, all the more so in that appeal to the interpretation of the characters is not necessary in algebraical manipulation and transformation. Peirce's position may be compared to C.D. Broad's, one of the first supporters of the "symbolic" interpretation of Kant's symbolic construction:

[…] if we look at the actual remarks of Kant's about algebra which I have quoted, we see that the only explicit statements which he makes might equally well have emanated from Leibniz himself, whose theory of mathematical reasoning is utterly different from that which is typical of Kant. All that his remarks amount to is that a good system of symbolism, which makes all our assumptions explicit and can be operated mechanically according to rules, is very useful for preventing us from falling into fallacies and for enabling us to solve complicated problems. This is exactly what Leibniz said. But the question arises: Why must this be confined to the treatment of extension and number? On Kant's earlier theory of mathematics, which makes it depend on the instantiation of concepts in the innate *intuita* of space and time, we can see why mathematical reasoning must be confined to the properties of figures and numbers. For these are the only concepts of which we can produce instantial constructions in imagination. But in his account of algebra he admits the merely symbolic and non-instantial construction of concepts. Of course these constructs themselves will be spatio-temporal, since the symbols will consist of perceived or imagined marks. But there seems no reason whatever why concepts which have nothing to do with extension or number should not be represented symbolically by spatio-temporal symbols. In fact everything that Kant says of algebra could be applied without change to symbolic logic; and yet he continued to hold that there is an absolutely fundamental difference between mathematics and formal logic. (Broad, 1941, p. 24)

The idea is the following. Symbols themselves are spatio-temporal objects, and qua spatio-temporal objects they are constructed a priori in intuition. And since symbols may stand for non-quantitative concepts, also the objects of non-quantitative concepts can be constructed.

[19] It has to be noticed that in one of the drafts of "On Quantity" (R 15) Peirce attempts to fit this distinction into a triad: "… a diagram, or visual image, whether composed of lines, like a geometrical figure, or an array of signs, like an algebraical formula, or of a mixed nature, like a graph" (NEM 4, p. 275). The resulting triad "geometrical figures, algebarical formulas, graphs" never emerges again in this set of writings or elsewhere. Thanks to Frederik Stjernfelt for drawing my attention to this passage. Cf. also Stjernfelt (2022, p. 156 and n143)

This, however, does not work. We saw that for Kant to construct a concept is to provide the a priori intuition that corresponds to the concept, and that only mathematical concepts can be constructed because these are concepts of quantity, i.e. concepts obtained by the composition of a homogeneous manifold of intuition. For Kant, to construct a mathematical concept is to construct the concept of a determined space or time, and the concept of a determined space or time is the synthesis of composition of a homogeneous manifold in intuition. Now, the "array of letters" or string of symbols that algebra employs are indeed spatio-temporal objects; but they are not concepts of a determined space or time; they are not obtained by the composition of a homogeneous manifold of intuition. As spatio-temporal objects, they are given in *empirical*, not in *pure* intution. When I draw a triangle on the blackboard, or when I draw a collection of five dots, I am bringing the homogeneous manifold of intuition to unity by means of concepts. But when I write the formula "$ax + b = 0$", there is no homogeneous manifold of intuition that is brought to unity. As Shabel (1998) has explained, the formula only stands for an object that can be constructed, but is not itself that construction. The formula is *given in empirical intuition*, it is not *constructed in pure intuition*. Spatio-temporal symbols are intuited, not constructed.

Be that as it may, Peirce's inclination towards what I have called the "symbolic" interpretation of Kant's symbolic construction does not invalidate his idea of a scale of quantity. The idea of a "scale of quantity" allows Peirce to maintain Kant's notion of construction and to extend it to all deductive reasoning, rather than limiting, as Kant did, to reasoning about quantities. The validity of this extension is independent of the way he understands symbolic constructions. No one would deny that symbolic constructions are given in empirical intuition (e.g. algebraical formulas are drawn on paper). In numerical algebra, they symbolize quantitative objects and relations; in logical algebra, they symbolize non-quantitative objects and relations. What is "constructive" about the algebra of logic is not that algebra constructs arrays of characters in the same way as geometry constructs figures in space. What is constructive about the algebra of logic is that a system of symbols and rules that has a quantitative interpretation is used to represent a system of objects and relations that are not quantitative. The constructive character of algebra does not consist in its constructing formulas, but in its using symbols of constructible objects to speak about non-quantitative, i.e. non constructible objects.

2.6 Conclusion

Kant holds that all mathematical knowledge is knowledge by the construction of concepts, and constructing a concept means exhibiting that concept a priori in intuition. Only mathematical concepts can be constructed because these are concepts of *quanta*, i.e. concepts obtained by the composition of a homogeneous manifold of intuition. Construction can be ostensive, when the concept itself is constructed, or symbolic, when only a formula standing for the construction is given.

Peirce transforms Kant's philosophy of mathematics into a theory of diagrammatic reasoning. In this sense, Peirce perfectly epitomizes Magnani's claim that the Kantian idea of construction become integrated into the analysis of deductive thinking (Magnani, 2001, pp. 208–209). For Peirce, all deductive (i.e. mathematical) reasoning proceeds by the construction and manipulation of diagrams. Peirce does not accept Kant's idea that only concepts of *quanta* and *quantitas* can be constructed; logical concepts and relations, which are not *prima facie* quantitative, can be diagrammatized, i.e. constructed in Kant's sense. What is quantitative in logical construction is not the object that is constructed, but the *means* of constructing it. In some papers written in the 1890s Peirce explains that in all deductive reasoning there is a "scale of quantity", i.e. an intermediate construction that is introduced in reasoning and removed afterwards. With respect to syllogistic logic, Euler diagrams and the Boolean algebra of logic are "scales of quantity", i.e. quantitative objects having quantitative relations that serve to the representation of non-quantitative objects having non-quantitative relations.[20]

References

Barone, F. (1965). *Logica formale e logica trascendentale. Vol. 2. L'algebra della logica*. Edizioni di filosofia.

Barwise, J., & Etchemendy, J. (1993). *The language of first-order logic*. CSLI Publications.

Barwise, J., & Etchemendy, J. (1994). *Hyperproof*. CSLI Publications.

Bellucci, F. (2013). Diagrammatic reasoning: Some notes on Charles S. Peirce and Friedrich A. Lange. *History and Philosophy of Logic, 34*, 293–305.

Bellucci, F., & Pietarinen, A.-V. (2016a). Existential graphs as an instrument of logical analysis: Part I. Alpha. *The Review of Symbolic Logic, 9*, 209–237.

Bellucci, F., & Pietarinen, A.-V. (2016b). From Mitchell to Carus. Fourteen years of logical graphs in the making. *Transactions of the Charles S. Peirce Society, 52*(4), 539–575.

Bellucci, F., & Pietarinen, A.-V. (2021). An analysis of existential graphs–part 2: Beta. *Synthese, 199*, 7705–7726.

Boole, G. (1847). *The mathematical analysis of logic*. Macmillan.

Boole, G. (1854). *An investigation of the laws of thought*. Walton and Maberly.

Brittan, G. J. (1992). Algebra and intuition. In C. J. Posy (Ed.), *Kant's philosophy of mathematics. Modern Essays* (pp. 315–340). Kluwer.

Broad, C. D. (1941). Kant's theory of mathematical and philosophical reasoning. *Proceedings of the Aristotelian Society (New Series), 42*, 1–24.

Carson, E. (2020). Arithmetic and the conditions of possible experience. In C. J. Posy & O. Rechter (Eds.), *Kant's philosophy of mathematics. Vol. 1: The critical philosophy and its roots* (pp. 231–247). Cambridge University Press.

Dooley, B. (Ed.). (2018). *The continued exercise of reason. Public addresses by George Boole*. The MIT Press.

[20] I presented a part of this research in an online workshop on "Peirce and the History of Logic" organized by Lu Jiang at the Department of Philosophy, Sun Yat-sen University, Guangzhou, China, in May 2024. Thanks to Lu, Ahti-Veikko Pietarinen and the audience for constructive comments. I should also thank Frederik Stjernfelt, who read an earlier version of this paper, and two anonymous readers, for valuable feedback. Mistakes and shortcomings are my responsibility only.

Friedman, M. (1990). Kant on concepts and intuitions in the mathematical sciences. *Synthese, 84*, 213–257.
Gava, G. (2014). What is wrong with intuitions? An assessment of a Peircean Criticism of Kant. *Transactions of the Charles S. Peirce Society, 50*(3), 340–359.
Hoffmann, M. H. G. (2010). "Theoric Transformations" and a new classification of abductive inferences. *Transactions of the Charles S. Peirce Society, 46*(4), 570–590.
Jevons, S. (1864). *Pure logic, or the logic of quality apart from quantity*. Stanford.
Kant, I. (1781). *Kritik der reinen Vernunft*. 2nd ed. 1787. Riga: Hartnoch; Eng. ed. *Critique of Pure Reason*. Transl. by P. Guyer and A. W. Wood (The Cambridge Edition of the Works of Immanuel Kant). Cambridge University Press.
Kant, I. (2002). *Theoretical philosophy after 1781*. Ed. by H. Allison, P. Heath. Transl. by G. Hatfield, M. Friedman, H. Allison, P. Heath. (The Cambridge Edition of the Works of Immanuel Kant). Cambridge University Press.
Kant, I. (2005). *Notes and fragments*. Ed. by P. Guyer. Transl. by C. Bowman, P. Guyer, F. Rauscher. (The Cambridge Edition of the Works of Immanuel Kant). Cambridge University Press.
Lange, F. A. (1877). *Logische Studien*. Baedecker.
MacFarlane, A. (1879). *Principles of the algebra of logic*. David Douglas.
Magnani, L. (2001). *Philosophy and geometry. Theoretical and historical issues*. Kluwer.
Melnick, A. (1973). *Kant's analogies of experience*. The University of Chicago Press.
Parsons, C. (1992). Kant's philosophy of arithmetic. In C. J. Posy (Ed.), *Kant's philosophy of mathematics. Modern Essays* (pp. 43–68). Kluwer.
Peirce, B. (1870). *Linear associative algebra*. Washington City.
Peirce, C. S. (1932–1958). In C. Hartshorne, P. Weiss, & A. W. Burks (Eds.), *Collected papers of Charles Sanders Peirce* (Vol. 8 vols). Harvard University Press.
Peirce, C. S. (1976). *The new elements of mathematics by Charles S. Peirce*. 4 vols. Edited by C. Eisele. The Hague: Mouton.
Peirce, C. S. (2019–2024). *Logic of the future. Peirce's writings on existential graphs*. 3 vols. in 5 books. Edited by A.-V. Pietarinen. Berlin: De Gruyter.
Pietarinen, A.-V. (2011). Existential graphs: What the diagrammatic logic of cognition might look like. *History and Philosophy of Logic, 32*(3), 265–281.
Pietarinen, A.-V. (2015). Exploring the beta quadrant. *Synthese, 192*(4), 941–970.
Pietarinen, A.-V., & Bellucci, F. (2014). New light on Peirce's conceptions of retroduction, deduction and scientific reasoning. *International Studies in the Philosophy of Science, 28*(4), 353–373.
Pietarinen, A.-V., & Bellucci, F. (2016). The iconic moment: Towards a Peircean theory of scientific imagination and abductive reasoning. In O. Pombo & A. Nepomuceno (Eds.), *Epistemology, knowledge and the impact of interaction* (pp. 463–481). Springer.
Roberts, D. D. (1973). *The existential graphs of Charles S. Peirce*. Mouton.
Robin, R. (1967). *Annotated catalogue of the papers of Charles S. Peirce*. University of Massachusetts Press.
Robin, R. (1971). The Peirce Papers: A supplementary catalogue. *Transactions of the C. S. Peirce Society, 7*(1), 37–57.
Segner, J. A. (1739). *Elementa arithmeticae et geometriae*. Cunon.
Sgarbi, M. (2010). Matematica e filosofia trascendentale in Kant. Note a margine di una fonte dimenticata della *Kritik der reinen Vernunft*. *Philosophical Readings, 2*(1), 209–224.
Shabel, L. (1998). Kant on the "symbolic construction" of mathematical concepts. *Studies in the History and Philosophy of Science, 29*(4), 589–621.
Shields, P. (2012). *Peirce on the Logic of Number*. Docent Press; or. ed. 1978.
Shimojima, A. (1996a). *On the efficacy of representation*. Ph.D. thesis, Indiana University.
Shimojima, A. (1996b). Operational constraint in diagrammatic reasoning. In G. Allwein & J. Barwise (Eds.), *Logical reasoning with diagrams* (pp. 27–48). Oxford University Press.
Stjernfelt, F. (2000). Diagrams as centerpiece of a Peircean epistemology. *Transactions of the Charles S. Peirce Society, 36*(3), 357–384.

Stjernfelt, F. (2007). *Diagrammatology. An investigation on the borderlines of phenomenology, ontology, and semiotics*. Springer.

Stjernfelt, F. (2011). On operational and optimal iconicity in Peirce's diagrammatology. *Semiotica, 186*(1), 395–419.

Stjernfelt, F. (2014). *Natural propositions*. Docent Press.

Stjernfelt, F. (2022). *Sheets, diagrams, and realism in Peirce*. De Gruyter.

Sutherland, D. (2020). Kant's philosophy of arithmetic: An outline of a new approach. In C. J. Posy & O. Rechter (Eds.), *Kant's philosophy of mathematics. Vol. 1: The critical philosophy and its roots* (pp. 248–266). Cambridge University Press.

Sutherland, D. (2021). *Kant's mathematical world*. Cambridge University Press.

Tait, W. W. (2020). Kant on "number". In C. J. Posy & O. Rechter (Eds.), *Kant's philosophy of mathematics. Vol. 1: The critical philosophy and its roots* (pp. 267–291). Cambridge University Press.

Thompson, M. (1992). Singular terms and intuitions in Kant's epistemology. In C. J. Posy (Ed.), *Kant's philosophy of mathematics. Modern Essays* (pp. 81–101). Kluwer.

Young, J. M. (1992). Construction, schematism and imagination. In C. J. Posy (Ed.), *Kant's philosophy of mathematics. Modern Essays* (pp. 159–176). Kluwer.

Chapter 3
Idealizations, Scientific Models and Fictions

Marco Buzzoni

Abstract Lorenzo Magnani has pointed out the danger of considering scientific models not only as useful tools for explaining facts or discovering new entities, laws and theories, but also as fictions, surrogates, parables, etc. It is true that underlying many current positions is the inability to clearly distinguish between the epistemic and non-epistemic use of idealizations and/or fictions in the scientific enterprise. However, an account that primarily interprets idealization not at the level of theoretical modelling and thought experiment, but at the level of real experimentation and, more generally, human agency, can acknowledge the fictional character of empirical-scientific models and thought experiments while maintaining a firm distinction between empirical-scientific and fictional work.

Keywords Abduction · Fictions · Idealization · Models · Thought experiments

3.1 Introduction

Especially in two articles (Magnani, 2012, 2013), Lorenzo Magnani pointed out the danger of unilaterally insisting on the proximity of fictional narratives to scientific models. He argued that scientific models must be clearly distinguished from fictional narratives, since only the latter explain facts or discover new entities, laws or theories. In this claim there is much that I agree with, and the main purpose of my paper is to restate it, though with some modifications and supplements.

Section 3.2 will offer a brief sketch of Magnani's main theses. As I shall try to show, he is right that many current positions are based on the inability to clearly distinguish between epistemic and nonepistemic uses of idealizations and/or fictions, but an adequate account of the idealization process should not deny a specific

M. Buzzoni (✉)
Department of Humanistic Studies, University of Macerata, Macerata, Italy
e-mail: marco.buzzoni@unimc.it

S. Arfini (ed.), *Scientific Cognition, Semiotics, and Computational Agents: Essays in Honor of Lorenzo Magnani - Volume 2*, Synthese Library 506,
https://doi.org/10.1007/978-3-031-96688-0_3

epistemic character to the works of fictional narratives (and, more generally, to artwork). The main thesis defended in Sect. 3.3 will be that the counterfactual assumptions on which any empirical-scientific idealization depends make it possible in principle for scientific 'fictions' to transform themselves (*in principle* completely) into real chains of empirical events, that is, into scientific experiments; they are the condition of the possibility of the production of increasingly good technical-practical realisations of (empirical-scientific) fictions, be they models, thought experiments, or computer simulations. For this reason, they simultaneously clarify the explanatory capacity of scientific idealizations and their principled distinction from artistic fictions. In the artistic-fictional case the demand upon the *in principle complete* technical-experimental realizability of scientific counterfactuals does not make sense and must be abandoned. What takes its place does not remove all cognitive value from fictional narratives, for two distinct reasons.

First, fictional narratives cannot do without some real—not only physical, but also social and cultural—constraints, on pain of becoming incomprehensible (this is an element of truth of any cognitive interpretation of fictional narratives); what constitutes fictive narratives is not the denial of all empirical knowledge, but only the fact that they suspend some (*but not all*) everyday and scientific causal links.

Secondly, both scientific and artistic fictions arise from the capacity of human reason to construct counterfactual scenarios, but counterfactual assumptions in empirical-scientific "fictions" are justified insofar as they play the role of making possible, in principle, reproducible real chains of empirical events (i.e., scientific experiments): as in part already seen by Hugo Dingler, they are the condition of the possibility of the production of increasingly good technical-practical realisations and predictions of (empirical) thought experiments and models. On the contrary, artistic fictions exploit the freedom, typical of human imagination, to suspend some everyday and/or scientific causal links, in order to induce us to relive in the first person thoughts/feelings that, on the one hand, are *de facto* always connected to a life worth living, but, on the other hand, are in principle distinct from them, since they always remain, no matter how minimal, a represented or symbolic reality.

Of course, it is beyond the scope of this paper to elucidate the nature of artistic fictions, but the last section, assuming the concepts of idealization and counterfactuality already outlined in the paper, will at least sketch out a working hypothesis in need of further investigation but compatible with both the principled distinction and the similarity of empirical science and artistic fictions.

3.2 "Scientific Models Are Not Fictions"

As is well-known, the fundamental philosophical problem associated with the concept of "Fiktion" was posed by Hans Vaihinger in the preface to his most famous work, *Die Philosophie des Als Ob* (1922), with a question similar to that later asked by Kuhn in relation to thought experiments: "The question asked in this book should therefore be formulated clearly and sharply right at the beginning: How is it that we achieve the true [*Richtiges*] with deliberately false ideas?"[1]

Despite the great popularity that Hans Vaihinger enjoyed at the time, and despite his constant references to both empirical and mathematical science, his rediscovery in the philosophy of science is relatively recent (on Vaihinger see for example Boniolo, 1997; Fine, 2009; Kobow, 2014). His ideas in recent decades have contributed to a fast-growing literature on the subject, from different perspectives, devoted to the role of fiction and idealization not only in the construction of models, but also of thought experiments, computer simulations, and even real-world experiments.

Although the function of idealization in science has not been a central theme of logical empiricism, there are a few notable exceptions. For example, Carl Gustav Hempel already discussed the connection between the use of idealized (or "ideal-typical") concepts and the *ceteris paribus* clause in the testing of empirical theories (cf. Hempel, 1965, p. 169–170; on this point, see Buzzoni, 2024). With reference to the status of scientific laws, an important contribution came from Scriven (1961). Alluding to the problems generated by the presence of idealizations in scientific laws right from the title ("The key property of physical laws—inaccuracy"), Michael Scriven pointed out that "[t]he most interesting fact about laws of nature is that they are virtually all known to be in error", since their drive to explain produces "idealization" or "oversimplification" (1961, p. 98; in a similar context, cf. also Scheffler, 1963, pp. 182–222).

In the philosophy of science, however, the notion of idealization became central from the 1960s onwards. Taking up the concept of model (and with it that of analogy and metaphor) developed by Norman Robert Campbell (1920), authors such as Mary Hesse (1963), Rom Harré (cf. e.g. 1970) and, with the important grafting of Weber's doctrine of the ideal type, Leszek Nowak (1971, 1980, 1992) took up the idea of the model not as an abstract logical-mathematical structure, but as a concrete or "iconic" representation (Harré, 1970, pp. 36 ff.). But it is especially through the contributions of Nancy Cartwright (1983) and Ronald Giere (1988) that the theme of the nature and role of idealization in science became central. Nancy Cartwright, in particular, defended a thesis (that scientific theories "lie") that will further the rediscovery of Hans Vaihinger and, with it, the interpretation of both models and thought experiments in the terms of the concept of "Fiktion."

As far as more recent literature is concerned, we should first mention a few collective volumes that, given the variety of positions expressed, can be considered to a large extent representative of the current debate: Jones and Cartwright (eds.) (2005), Brzeziński et al. (eds.) (2007), Suárez (ed.) (2009b), Woods (ed.) (2010), Frigg and Hunter (eds.), (2010), and Magnani and Bertolotti (eds.) (2017). In addition to these books, however, there are many interesting individual contributions that discuss, in the last analysis, the problem raised by Vaihinger, though from

[1] Vaihinger 1922, p. XII; my translation, since there is no exact correspondence with the English edition. For Kuhn, see 1964, p. 308: "How, then […] can a thought experiment lead to new knowledge or to new understanding of nature? […] What, if anything, can scientists hope to learn from thought experiments?"

different points of view and in synergy with some focal aspects of the current philosophical debate. Apart from the authors of the volumes just mentioned, see for example: Barr (1971, 1974), Laymon (1985, 1991, 1995), Musgrave (1981), McMullin (1985), Niiniluoto (1986, 2007), Dilworth (1989), Bokulich and Liu (1999), Liu (1999), Teller (2004), Mäki (2005, 2009, 2020), Weisberg (2007), Portides (2007, 2021), Buzzoni (2004, 2008), Godfrey-Smith (2009), Kuorikoski and Lehtinen (2009), Frigg (2010), Toon (2010), Woods and Rosales (2010), Arcangeli (2010), Bokulich (2011), Bueno and French (2011), Contessa (2010), Mizrahi (2012), Norton (2012), Morrison (2015), Currie (2016), Appiah (2017), Potochnik (2017), Reiss (2018), Wheeler (2018), Nguyen (2021).

Mauricio Suárez, presenting the essays he collected, introduced a distinction that, although somewhat simplifying, is of some use in orienting oneself in the current debate. He distinguished between a "wide" and a "narrow fictionalism": according to the former, "fictive assumptions and representations are fictions", while according to the latter (in which he includes the papers by Morrison, Winsberg, Teller, and Giere, but to which many of the above-mentioned contributions should be added) fictive representations are a case of idealization or abstraction. For the latter authors, "fiction" is a term that does not fit with realistic representations and assumptions:

> *Wide fictionalism* takes it that idealization and abstraction are subspecies of fiction, and consequently that most modeling practice involves fiction and fictionalizing most of the time. *Narrow fictionalism* by contrast takes it that fictions are not idealizations or abstractions, and thus fictionalizing plays a less ubiquitous role in modeling. Perhaps not surprisingly, the promoters of *wide fictionalism* tend toward instrumentalism in the epistemology of science, whereas the defenders of *narrow fictionalism* are friendlier toward scientific realism. (Suarez, 2009a, p. 13)

In the following we shall discuss in more details two important essays by Lorenzo Magnani (2012, 2013) that are related to this debate and can be considered as representative of what Suarez has called narrow functionalism: expressed in its most laconic form, their main thesis is that "scientific models are not fictions."

According to Magnani, if one were to say that models are fictions, one would be correct, but running the risk of insisting too much on a similarity at the expense of very relevant differences: "to consider models as fictions would destroy the well regarded distinction between science and science fiction. This attitude can present cultural dangers: is science just a matter of fictions?" (Magnani, 2012, p. 6; cf. also Magnani, 2012, p. 7 and 2013, p. 904) Magnani rightly insists on the need to distinguish epistemic and non-epistemic models. Both are the product of abduction, but "one thing is to abduce fictions, like in the case of creations in literature, another thing is to abduce models in empirical science" (Magnani, 2013, p. 900).

The epistemic models of the empirical sciences, which range from the simple perceptual to the more complex models of the developed sciences (cf. Magnani, 2012, p. 10), are characterized by what he calls manipulative abduction, i.e. an abduction that is connected to experimental doing:

> in my perspective models cannot be considered neither abstract (in the traditional ambiguous sense) nor fictional: scientists do not have any intention to propose fictions, instead they

> provide models as tools that reshape a generic cognitive niche as an epistemological niche to the aim of performing a genuine struggle for representing the external world. (Magnani, 2012, p. 14)

Although fictional narratives can sometimes provide interesting insights,

> to say that scientific and literary fictions are both "good" fictions is a bit of a theoretical oversimplification, because it is science that created, beyond literature and poetry, *new* kinds of models committed to a specific production of truth, constitutively aiming at not being fictional.[...] I confess I cannot see how we can speak of the ideal pendulum in the same way we speak of Anna Karenina: it seems to me that we are running the risk of inadvertently opening the gates of epistemology to a kind of relativistic post-modernism *à la mode*, even if fictionalists seem to avoid this possible confusion by producing—often useful—taxonomies about the slight differences between fictions in science and in other cognitive practices. (Magnani, 2012, p. 16)

Echoing Latour's (and Kuhn's) distinction between a static and a dynamic aspect of science, and further reiterating his point, he writes:

> Scientific models [...] play fundamental 'rational' knowledge enhancing roles: in a static perspective (e.g. when inserted in a textbook) scientific models can appear fictional to the epistemologist, but their fictional character disappears if a dynamic perspective is adopted. Abduction in scientific model-based reasoning is not a suspicious process of guessing fictions. (Magnani, 2013, p. 882)

Now, my agreement with these considerations by Magnani is very large, ranging from the basic manipulative-operational perspective on empirical science to the general methodological caveat of not neglecting relevant differences when relevant similarities exist. This large agreement, despite some terminological differences, extends to the fundamental centrality of the abductive procedure, even if there is an important trend in the history of induction (more precisely, a minority one, but defended by important authors) that brings induction very close to what Peirce and Magnani would call abduction (cf. Buzzoni, 2023).

However, with respect to some points there is a significant divergence. According to Magnani, idealizing assumptions in science are not constitutive, as happens in narrative fiction, but "evidentially inert". This applies to the more general principles of science, whose nature is that of "conventions" which, even though they escape empirical falsification, are "knowledge enhancing". They would be "evidentially inert fruits of abduction—at least from the point of view of their impossible falsification—but nevertheless knowledge enhancing." (Magnani, 2013, p. 898) Even if "evidentially inert" in themselves, the abduced models of scientific reasoning (and of human perception)

> cannot be considered as neither mere fictions, simple surrogates or make-believe, nor they are unproblematic idealizations. [...] The abduction of these models in science is epistemically productive, models are just inert in the perspective of a *direct* empirical significance but they play a 'causal' role in generating it: scientific models can be empirically false, but they are not fictions, instead they are knowledge enhancing devices, which play an important role in reaching empirically fecund knowledge. (Magnani, 2013, p. 899)

Now, two points raise some concerns. First, the epistemological and methodological status of the theories or scientific models of high level of generality is puzzling.

Certainly this is an idea that reappears in many philosophers of science—from Poincaré to Hacking and Cartwright, to name but a few examples. But in the first place, any scientific statement with a part of content that cannot in principle be traced back to experience is, strictly speaking, non-empirical or metaphysical (in the most negative sense of lacking reasons in its favour), and can only be accepted as scientific, i.e. intersubjectively testable, to the extent that, even if its empirical truth or falsity is not well established at the moment, we do not exclude a possible future ascertainment of it (for further details on this controversial point, see Buzzoni, 1995, 2008).

Secondly, Magnani himself acknowledges that "we cannot surely imagine that literature does not provide knowledge of some kind." (Magnani, 2013, pp. 899–900) But even if we leave aside the status of these principles or models of high generality, the fact remains that, while what the failure of a scientific 'fiction' consists of is relatively clear, namely some kind of experimental refutation—but we will see that some clarifications are necessary here as well—, it is not at all clear what Magnani means by "aesthetic failure" (Magnani, 2013, p. 900). And this (at least from a cognitivist point of view), leaves the possibility open of a defense of the epistemic value of artistic fictions in a sense that, again, would lead to their epistemological and methodological confusion with scientific 'fictions'.

Clearly, from the point of view of a philosophy of science, the first point is the fundamental one and our main focus will be directed towards its clarification. However, as already mentioned in the introduction, this clarification will also suggest a tentative answer regarding the epistemic and aesthetic value of artistic fictions. Even if this answer—if nothing else, because of the limits set to this paper—cannot receive an adequate philosophical justification, it will have the advantage of being suggested by (and being consistent with) the different way in which, as we shall see, scientific laws (and the empirical idealizations underlying them) are used in science and artistic fiction. In this way, the difference in principle between scientific imagination and artistic imagination, on which Magnani has rightly insisted, will be both reaffirmed and, I hope, better established.

3.3 Idealization and Counterfactuals in Scientific Thought and Practice

The great majority of authors distinguish scientific 'fictions'—be they theoretical models, computer simulations, thought experiments, etc.—from actual experiments on the basis that only the first require at least one hypothetical or counterfactual assumption, possibly together with other assumptions held to be descriptively true (for a possible exception, however, see Brown, 2004, p. 25, according to whom "idealization is neither necessary nor sufficient" in order to have a thought experiment). It seems impossible to construct a theoretical model or a thought experiment simply on the basis of a supposition which we believe to be factually true (such as, say, that the Earth rotates on its axis or is flat), for if this statement corresponded in all respects to the reality that we can ascertain from experience, we would be describing a concrete state of affairs rather than constructing a theoretical model or formulating a thought experiment (which as such have no perfect counterpart in reality). To construct a theoretical model or to formulate a thought experiment, we would have to, at least implicitly, counterfactually assume, say, that the Earth is perfectly round or that it does not move at all (cf. Norton, 1996, p. 357).

From this arises the fundamental problem with all conceptual 'fictions' used in empirical science already addressed by Vaihinger. Idealizations in theoretical models or thought experiments are the product of counterfactual theoretical assumptions that introduce premises which, if taken literally, are false, and this seems to rule out their empirical testability. In real experiments, on the contrary, even a correct conclusion drawn from false premises would count as an error: in real experiments, at most, we can neglect certain effects that are considered secondary with respect to others (cf. e.g. Hon, 1998, pp. 233–4 and 237–238). To the extent that the 'fictions' with which we will be concerned intend to speak about the reality of experience—regardless of whether they are theoretical models, computer simulations or thought experiments[2]—we must face Vaihinger's problem.

Addressing this problem, by means of his interpretation of Hans Vaihinger's account of "fictions", Arthur Fine ultimately reaffirmed his own instrumentalism. According to him, in the last analysis, we should not pay too much attention to the fact that certain fictions are contradictory (which, according to Vaihinger, is one of the hallmarks of his "fictions"), but we should evaluate their actual and practical usefulness:

> What Vaihinger seems to be suggesting is that one ought to reject these questions about contradictions. Instead of pursuing these "how is it possible that" questions, simply pay detailed attention to reflective practice and notice how fictions are actually used. (Fine, 2009, p. 29)

Now, if in practice this can sometimes be sensible (although it is not always the most far-sighted strategy), not so in philosophy, where we want not a rule of thumb of only practical value, but an analytical solution of the problem. In other words, if we want a genuinely philosophical solution, we cannot content ourselves with refuting Zeno by walking back and forth while he argues against the existence of movement. We cannot evade the question of whether, and if so to what extent and in what sense, we can succeed (both theoretically and practically) with our fictions.

The solution I propose is based on the technical-experimental meaning of the counterfactuals underlying idealizational assumptions in science. As we shall see, it

[2] For the legitimacy of treating thought experiments, theoretical models, and computer simulations as one concept, I must refer to Buzzoni (2004, 2008, 2016). Similarly, but for different reasons, Mäki (2005) argued that, as the title of the papers states, "Models are experiments, experiments are models," since there are "strong structural similarities between material experimentation and theoretical modelling. The structure of experimentation, involving controls and isolation, is the same, while what is different is the way these controls and isolations are effected: by way of thinking and assuming, and by way of material or causal manipulation" (Mäki, 2005, p. 308; one of the most recent restatements of a similar thesis is Karadimas, 2024).

is their technical-experimental or pragmatic-operational meaning that clarifies under what conditions it is legitimate to introduce counterfactual assumptions and idealizations that we know to be 'false'.

In our endeavour to understand the nature and function of counterfactual assumptions, we shall start by analysing the process of idealization which, according to Max Weber, is characteristic of scientific discourse.[3]

Weber famously defied the traditional mistrust of "what if" questions in history when he argued that, in order to understand the causal-probabilistic importance of an event such as the battle of Marathon for the development of Greek culture and Western civilisation, a historian must start by asking (explicitly or implicitly) what would have happened if the Persians had won rather than the Greeks. The historian must answer this question *by abstracting from what really happened and constructing an ideal, or possible, course of events.* If, relying on general knowledge of a nomological kind (Weber 1906/1949, p. 287, Engl. transl., pp. 185–186), the historian is led to believe that, had the antecedent been different, the subsequent course of events would have been different too, he can conclude that the antecedent is an "adequate" cause of the subsequent events. For example, factual knowledge at our disposal regarding both Persian politics in conquered territories and the objective conditions in Greece at the time, leads us to believe that a Persian victory at Marathon would probably have steered the development of Greek civilization in a theocratic direction.

The decisive point of Weber's analysis is the connection between counterfactual conditionals and causal imputation, and more precisely the fact that the understanding of causal connections is possible only by the use of counterfactual conditionals. As Weber aptly puts it, "[i]n order to penetrate to the real causal interrelationships, *we construct unreal ones.*" (Weber, 1906, p. 287, Engl. transl., pp. 185–186) When Weber proposes the counterfactual possibility of a Persian victory at Marathon, his intention is not just to construct a mere 'fiction' that clarifies certain conceptual connections or removes certain ambiguities; rather, he believes that this 'fiction' (which is actually a historical theoretical model or historical thought experiment) is also *an essential tool for explaining reality, that is, for understanding particular causal connections.*

This, however, is still not enough to solve the problem we started with, even if it leads us to formulate it in a way that brings us closer to the solution. Weber, who suggested that in order to understand real causal connections, we must construct counterfactual scenarios, did not give a clear answer as to the basis for the empirical legitimacy of this type of construction: how can we give empirical content to ideal entities (Vaihinger's "fictions") that are the product of counterfactual assumptions?

[3] Cf. Weber, 1904/1949, p. 192, Engl. transl., p. 92. A number of authors (including, above all, the excellent works of Mäki already mentioned) used Weber's ideas in the context of the problem we are dealing with. Apart from other minor shortcomings, from my point of view their main weakness is the lack of awareness of the intrinsic connection between science and technique, which implies that the predictive capacity connected to the use of ideal types is still understood as a mere theoretical performance, not necessarily linked to its technical-practical realisations.

To answer this question, we need to clarify the connection between, on the one hand, theoretical models or thought experiments and, on the other hand, real-world experiments. To this end, it is crucial to apply an idea of Hugo Dingler concerning the nature of real-world experiments to the connection between experiments and theoretical 'fictions', albeit freed from some of the more dubious philosophical assumptions to which he had linked it. Even though Dingler's analysis, limited as it was to real experiments, did not grasp their connection with thought experiments or theoretical models, his account of the function of idealizations (and counterfactual assumptions) in real-world experiments is decisive for understanding their nature and connection to thought experiments (cf., e.g., Dingler, 1938, p. 146).

Like Weber, Dingler believed that the construction of ideal models was a necessary premise of causal imputations; unlike Weber, however, Dingler (who, I repeat, only took in account real experiments) explicitly addressed the problem of the technical-operational foundation of idealizations. He argued that the use of "ideal forms" is one side of a process. The other, apparently opposite, side is the concrete technical-operational construction in reality—in Dingler's words, the "realization" *(Realisierung)*—of those ideal forms. In order to perform reproducible measurements of physical magnitudes, our measuring instruments must be made, and tested, without relying on other instruments, since this would involve an infinite regress. According to Dingler, the independent criteria that allow us to judge whether instruments have been built correctly, are certain ideal "elementary forms" supplied by Euclidean geometry and Newtonian dynamics; they are the necessary conditions for initiating processes of technical "realization" in which real forms increasingly approximate ideal ones (cf. Dingler, 1938, p. 59).

The example that best illustrates this point is perhaps the procedure where three plates were used to produce plane surfaces without resorting to a sample plane. For Dingler, a plane is a surface whose opposite sides are indistinguishable; to know whether a surface is a plane "it is enough to have, as well as the surface's two sides S_1 and S_2, also a testing surface S_3. If the testing surface fits the surface's two sides, then they too fit each other, and the surface is a plane" (Dingler, 1928, p. 59). In fact, plane surfaces can be technically produced by a procedure which realizes this definition: two steel plates are smoothed by grinding them on each other with the aid of an abrasive substance, until they fit perfectly whatever their position. A plane is, per se, an ideal geometrical form (so that its concept is already the result of a counterfactual reasoning) with no perfect empirical exemplification; and yet for this very reason it can function as the *criterion* of a process whereby increasingly plane real surfaces are "realized" in a properly technical-operational sense.[4]

[4]Cf. Dingler, 1930, p. 39. After Dingler, the thesis that idealizations in the empirical sciences depend for their legitimacy on their capacity to lead to correct predictions has been upheld by various authors, and especially by Nowak (cf., e.g., 1971, 1980, 1992), Barr (1974) and Laymon (1985, pp. 155–158, 1991 or 1995). Like Dingler, all these authors hold that the formulation of theories about the causes that may interfere with an experiment is an essential stage of an indefinite process of "approximation" (cf. e.g. Laymon, 1991, p. 173, and Barr, 1974, p. 59). There are however two fundamental differences. First of all, these authors neglect almost completely the intrinsic

Now, once we get rid of flawed and in the last analysis platonic notions of "ideal forms", it is only natural to extend Dangler's view of the relationship between "ideal forms" (resulting from counterfactual assumptions) and their use in real experiments *to the relationship between thought experiments or theoretical models on the one hand and real-world experimental practices on the other.* It follows that even in thought experiments and theoretical models idealizations and counterfactual assumptions are closely connected with the possibility of our technical-practical intervention in the world.

This amounts to an interpretation in technical-operational terms (or, perhaps better, in terms of technical-experimental agency) of counterfactual assertions and is the decisive step in my argument to meet Vaihinger's (and Kuhn's) already mentioned challenge. Much has been written about counterfactual assumptions, starting with Rescher (1964) and Lewis (1973a, 1973b). However, here we are not so much interested in the logical structure of counterfactual assumptions as in the epistemological problem they raise. And from this point of view, it is their use in the context of the design and execution of human actions that is methodically primary, not their logical analysis (what Vaihinger had perhaps, but in any case only partly, intuited by saying that fictions are "purposive").

The very important function performed by counterfactual assumptions in empirical-experimental discourse lies in the fact that they indicate operational procedures for achieving certain goals (which in turn justifies corresponding theoretical assertions). On reflection, "If the Persians had defeated the Greeks at Marathon, the history of the West would have been very different" is a proposition that not only explains the actual course of Western history, emphasizing the importance of ancient Greece, but also provides, at least in principle, a method for modifying the actual course of history in analogous situations. Likewise, the proposition "if at sea level the temperature were raised to 100 °C, water would boil" both explains why water is not boiling now, and also provides a method for making water boil. "If you opened the shutters, you would see a magnificent sunset" both explains why we are not seeing a magnificent sunset now, and also provides a method for seeing one.

This is the ultimate reason why idealizations that omit certain variables—and so lead to statements that are empirically 'false' from certain points of view—may have empirical meaning, that is, if the technical-experimental, reproducible, but counterfactual conditions under which specific omissions take place are clearly spelt out and explicitly used in the courses of action directly suggested by the corresponding counterfactuals. Once this is done, it becomes clear that, from the

connection between science and technique upheld by Dingler; accordingly, for them the ability to predict correctly becomes a theoretical performance, unrelated to technical realisations. Moreover, these authors fail to grasp the reflexive-transcendental relationship made up of both unity and distinction between thought experiments or theoretical models on the one side and real-world experiments on the other; as a consequence, they distinguish them also with regard to the logical intension of their concepts, which is in my opinion impossible. For an adequate justification of the latter thesis, which would require more space than is available, I must refer to Buzzoni 2004 and 2008, of which it is one of the key theses.

epistemic point of view (that is, here, regardless of the personal abilities of those who apply the theory), idealization, far from jeopardising the connection between truth and technical applicability, presupposes it, and can do so because at the foundation of every idealization lies, in a more or less explicit sense, a corresponding counterfactual, which makes possible a course of action and the realization of particular practical aims.

In the empirical sciences idealization is a matter of practical-technical relevance: it is illegitimate to idealize (or to overlook) variables in such a way that it would be in principle impossible to reproduce an experiment, and therefore also impossible to formulate correct predictions and to intervene practically in an effective manner. Contrary to Vaihinger's distinctions between "fictions" and "hypotheses", *all* empirical fictions or idealizations have at bottom the character of hypotheses, *and only the technical-experimental criterion can decide whether they are relevant, irrelevant or even wrong*. The construction of classical models requires that we make many idealizing assumptions, such as perfectly smooth, frictionless surfaces, non-deformable bodies, inextensible strings, negligible masses and weights etc.; they are ultimately justified because they allow us to predict more accurately nature's behaviour under certain experimental conditions, and in that way extend our mastery over nature.

To sum up, the presence of counterfactual assumptions in thought experiments or theoretical models prevents neither their realization nor the possibility of obtaining concrete empirical information from them which extends our mastery of reality. On the contrary, from a pragmatic-experimental point of view, counterfactual assumptions are the condition that makes it possible in principle for thought experiments or theoretical models to transform themselves into real chains of empirical events; they are the condition of the possibility of the production of increasingly good technical-practical realizations and predictions of (empirical) thought experiments and models. This closely links scientific 'fictions', such as empirical thought experiments or models, to real experiments, since in both cases good idealizations, unlike bad ones, generate counterfactuals that allow predictions that fit what we can ascertain experimentally and realize technically.

3.4 Scientific Idealizations and Artistic Fictions

So far, we have tried to clarify the role played in empirical science by idealizing assumptions, made possible by their intrinsic connection to the practical-experimental nature of counterfactual reasoning. Thereby we are in a better position to inquire in what particular sense it is possible to recognize an epistemic or cognitive function to artistic fictions without in any way reducing them to scientific ones.

As far as the cognitive value of fictional narratives is concerned, at the risk of some over-simplification, we may divide the critics into two main groups and, according to some suggestions in the relevant literature (see for example all Davies,

2010 and Egan, 2016), we may classify them as pro-cognitivist or as anti-cognitivist. One group argues that fictional narratives convey knowledge either in principle similar to that of empirical or philosophical knowledge (see e.g. Elgin, 2007; Davies, 2010) or have a *sui generis* cognitive capacity (see e.g. Young, 2001). The other denies any cognitive value to art: artistic fictions wouldn't be intended to contribute to knowledge, do not have the means to do so, and can convey contradictory knowledge without losing their aesthetic value; when art promotes knowledge, this knowledge must be accompanied by external evidence already justified by other means: "even when such knowledge shapes the work of art, it is not [...] the work of art that confirms." (Stolnitz, 1992, p. 198; in this sense, see also Dohrn, 2016 and Egan, 2016) Between these poles there is a variety of intermediate views (for example Davies, 2007, or Putnam, 1976, who attributes to art a conceptual knowledge or a "knowledge of a possibility": cf. pp. 89–90).

In what follows, I will attempt to interpret and perhaps shed some new light on this debate, addressing the problem of the cognitive scope of artistic fictions starting from the conclusions already reached about the nature of counterfactuals and idealizations in science. From this point of view, the fundamental difference between artistic fictions, on the one hand, and scientific models or thought experiments, on the other, is that in the scientific case there is always implicit, from the beginning, a request (based on the counterfactuality intimately connected with our interventions in our surrounding world) for the *in principle complete* technical-experimental realizability of any conceptual content. On the contrary, in a fictional narrative, despite certain empirical constraints (to which we will return in a moment), this demand does not make sense and is abandoned.

Let us assume, for example, that the situation envisaged in Einstein's lift experiment is *in principle and forever*, and not only for factual reasons and temporarily, unrealizable. Well, this thought experiment would still retain a close relationship to an empirical thought experiment, because there are known objects from common experience that are manipulated, dropped or thrown according to known laws, etc. But strictly speaking it would no longer be a scientific thought experiment, but a (very short) fictional story. By discarding the claim that the technical-experimental translation must in principle be complete with regard to all conceptual content, one leaves the realm of scientific discourse and enter the realm of fantasy (although this is of course still not enough to guarantee the literary or artistic value of the result).

Up to now we have concentrated on what distinguishes, from an epistemological point of view, empirical-scientific models or thought experiments, on the one hand, and fictional narratives, on the other. However, again from an epistemological point of view, what they have in common is just as, if not more, important for understanding their mutual nexus of unity and distinction. On reflection, that artistic fictions are a product of an imagination that exhibits a certain freedom from the law is only a half-truth, which risks hiding the other half. The other half, which should deserve greater attention, is that no imagination, not even that of the wildest science fiction or the strangest delirium, can suppress *all* known causal nexus, on pain of its total unintelligibility (cf. Buzzoni, 2004, pp. 267–268; for some cursory hints, see also

Currie, 2016, pp. 302–303).[5] Even science fiction stories—which continually break certain laws of nature, for instance by hypothesizing the existence of spaceships travelling at speeds faster than the light—actually have a structure similar to that of scientific models or thought experiments, insofar as part of the consequences of their hypothetical statements must be in principle controllable. Some ingredients of fantastic inventions must necessarily be well known and corroborated by everyday or scientific experience. What the protagonist does in Defoe's Robinson Crusoe, for instance, is to a large extent easily controllable from an empirical point of view, since it is often only a question of establishing whether, with the tools and in the situation in which he found himself, it is possible to technically reproduce what is described in the novel.

Here lies the most important relation of unity and distinction between empirical or scientific and artistic fictions. It is only because we are normally able to realize violations of well-known causal regularities that we are also able to distinguish and recognize a conceptual content as fantastic and, therefore, to recognize a piece of language as fiction. In other words, one could not even identify an artistic fiction as such if one did not presuppose the existence of empirical everyday and/or scientific regularities, while at the same time suspending some of them and letting others subsist.[6]

Thus, on the one hand, artistic fictions cannot do without some real constraints, on pain of becoming incomprehensible: they will necessarily use certain items from the physical sciences, the human sciences, other literary works and direct experience drawn from everyday life. But on the other hand, artistic fictions, abandoning the demand of a technical-practical realization, in principle complete, of their conceptual contents, exploit the freedom, typical of human imagination, to suspend some (but not all) everyday and scientific causal links.[7]

---

[5] Interesting empirical studies exist in connection with this point. See for example Margoni et al. (2024) for a review on the cognitive ability to realise a violation of empirical expectations already in the first three years of life. From another point of view, but still in relation to the point I am trying to make, Bartsch et al. (2024) showed that, without the point of reference of known laws or empirical regularities, one could not even use fictitious entertainment as a source of knowledge.

[6] An anonymous reviewer remarked that it does not seem necessary to assume that artistic fictions must always suspend certain everyday or scientific causal links, since there are 'realist' fictions (such as, for instance, those of Dickens or Austin) that try to represent things, as far as causality is concerned, as close as possible to real life. But even fictions that attempt describe things as close to real life as possible cannot help but deviate from reality in at least some respects (think, for instance, of the fact that they take place in a time and/or space other than our common intersubjective space-time continuum, which is nonetheless a violation of the causal structure of the world in which we live). Moreover, this deviation from reality, which can be exploited in various ways by both the author of the work (for her/his artistic purposes) and the interpreter (for his/her efforts of understanding), is a necessary condition for the typically symbolic meaning of artistic fictions. For more details on this point, see Buzzoni (2025).

[7] The need to distinguish in order to unite this dual aspect of the imagination—which in one sense can, but in another cannot, prescind from experience and its regularities—is intimately connected with the distinction drawn by Stuart (2021) between, on the one hand, imagination as an ability (imagination_0), and, on the other hand, two kinds of imagination as different particular manifestations or "uses" of it (imagination_1 and imagination_2).

Summarising what has been argued so far, it is only this dual aspect of preservation and suspension of known empirical laws that characterizes the relation of unity and distinction between the use of counterfactuals (and the idealizations connected to them) in, respectively, scientific and artistic fictions.

This immediately raises the question "Why does the artist drop some causal links and not others?" "What is the criterion of this choice?"

With this question we move from a predominantly epistemological to a predominantly aesthetic consideration, and in this regard it is obvious that it cannot be the purpose of this paper to clarify the nature of artistic fictions. However, assuming the concepts of idealization and counterfactuality already outlined in the paper, I will sketch out a working hypothesis, certainly in need of further investigation, but compatible with both the distinction and the likeness of scientific and artistic fictions defended up to this point.

In a context very similar to ours, Hans Vaihinger already made a suggestion which, in one of its many forms, is widely accepted in the philosophy of art and which we shall take here, at least in its most general sense, as the basis for our comparison between scientific and fictional idealizations. On the one hand, according to this author, the free creative play of psychical activity, which expresses itself "in arbitrary combinations and alternations of elements existing in the world of fact", is something that the fictions of science as well as those of myth and art have in common. But on the other hand, he introduced two different terms to distinguish accurately these fields: according to his example, "Pegasus is a figment [*Figment*], atom, a fiction [*Fiktion*]" (Vaihinger, 1922/1924, p. 129, Engl. Transl., p. 81). Like the scientific fictions, figments "are not an end in themselves but a means for the attainment of higher ends." But figments "serve the purpose of awakening within us certain uplifting or otherwise important feelings." (Vaihinger, 1922/1924, Engl. Transl., p. 82).

Now, with this mention of the role of feeling in aesthetic experience, Vaihinger touched on a point on which many philosophical theories of art converge (the most notable exception being perhaps Goodman: see e.g. 1976, p. 264). Using his suggestion in its most general sense for our purposes, we can say that, whereas both scientific and artistic fictions arise from the capacity of human reason to construct counterfactual scenarios always connected to potential courses of action (which is the sense in which there is a cognitive dimension common to science and art), the crucial difference lies in the different purposes to which the corresponding constructions are subordinated. In the case of artistic fictions, the goal is to make thought-feelings resonate, at the same time, in our minds and bodies, whether it be the indignation (and perhaps social-political reflections) aroused by a Zola novel about the plight of socially disadvantaged people, the existential despair experienced by reading a Giacomo Leopardi's lyric, or the anguish, despair and bewilderment we see and feel in Munk's Howl.

What is missing in artistic fictions, compared to science, is the possibility of demonstrating the existence of some feature of reality by simply making it interact experimentally with our embodied organic reality. What there is more in art than in science is the fact that a painting, a sculpture, a piece of music, are indeed in front of us in their materiality in a manner similar to that of an experimental set-up, but that what is in 'movement' is not something empirical and, so to speak, external—as can sometimes be also found in contemporary art—, but is the first person re-enactment of a thought-feeling (or feeling-thought, since thought and feeling are things that, de facto, can never be entirely separated) that can in principle be evoked and relived in the body-mind of the interpreter.

Following the indications left by the author in his/her work of art, the interpreter expects to find, at the end of his experience, an ideal meaning—universal in the sense that in principle it can be re-appropriated in the first person by anyone. Although here we find an important similarity with science, it goes hand in hand with an equally important difference. The reproducibility of scientific technique is in the service of an impersonal truth, which refers to a reality that is as it is, regardless of our desires or goals. In art, on the other hand, technique is at the service of a re-enactement that in one sense has a foundation in a technically reproducible external world (as a statue, painting, etc.), but in another sense is made possible only by a re-appropriation in the first person of a content that is emotionally, ethically and existentially relevant.

This last observation allows us to address another important difference in principle between scientific and artistic fictions, based upon the intimate relationship between art and morality. The artistic representation of feelings is always intimately connected with an at least implicit demand for an evaluative stance. Artistic fictions are meant to induce us to re-enact thoughts and/or feelings, but they cannot do so without presenting us with content toward which we cannot help but take, explicitly or implicitly, an evaluative stance.

In other words, the aesthetic value of an artistic fiction, although it must *in principle* be distinguished from its moral value, is *in fact* never entirely separable from the latter. The artist's effort to drop this or that physical, chemical, psychological, sociological constraint (and so on, on the basis of our best empirical knowledge) responds to the need to foreground an emotional-ethical-existential value that can be experienced in the first person and that can range from a strong social, religious, political commitment to the limit of pure entertainment (which is however also a value in human life). Art for art's sake is a more or less attainable ideal, and can even be accompanied by a strong moral and civic tension (think of the use of art to protest against those who would reduce every aspect of human life to politics or any other particular aspect of our existence), but it is never completely attainable: if taken to its ultimate end, it would be an aesthetic pleasure detached from any represented content, which means, on reflection, no longer a properly human pleasure, but an affection that would be produced in us due to a natural agent completely disconnected from our cultural dimension as human beings.

But, as already mentioned, it was obviously not within the scope of this paper to address the aesthetic problem. The reflections of this section only were meant to sketch out a possible interpretation of artistic fictions at the same time coherent with and suggested by our epistemological analysis of the role played by counterfactuals and idealizations in empirical-scientific reasoning.

3.5 Conclusion

As we have seen, while Lorenzo Magnani concedes that even fictional narratives can teach us something about human reality and life, he intends to make a clear distinction—as far as their epistemic status is concerned—between scientific models and literary models: the latter, unlike the former, could be reduced to mere entertainment, with no epistemic relevance. But despite his just demand upon this distinction (of which this paper was intended to be a restatement, refinement and supplement), I have raised some doubts about two particular points, which are closely connected: the first and most important concerned the way in which idealizations can contribute to the empirical and technological confirmation of scientific 'fictions' (such as thought experiments and theoretical models); the second concerned the explanation of the way in which artistic fictions, despite their non-scientific character, are able to teach us something about life and reality.

We have argued that counterfactual conditions, which underlie idealized scientific laws or principles, are also an essential ingredient of artistic fictions. With regard to the difference in principle between scientific fictions and fictional narratives, counterfactual assumptions in scientific fictions are justified insofar as they play the role of making possible, in principle, reproducible real chains of empirical events (i.e., scientific experiments): as in part already seen by Hugo Dingler, they are the condition of the possibility of the production of increasingly good technical-practical realizations and predictions of (empirical) thought experiments and models. On the contrary, artistic fictions, abandoning the demand of a technical-practical realization, in principle complete, of their conceptual contents, exploit the freedom, typical of human imagination, to suspend some (*but not all*) everyday and scientific causal links, in order to induce us to relive in the first person thoughts/feelings that, on the one hand, are *de facto* always connected to a life worth living, but, on the other hand, are in principle distinct from them, since they always remain, no matter how minimal, a represented or symbolic reality. But this is another story, to which I hope to return.

References

Appiah, K. A. (2017). *As if: Idealization and ideals*. Harvard University Press.

Arcangeli, M. (2010). Imagination in thought experimentation: Sketching a cognitive approach to thought experiments. In Magnani, Carnielli, & Pizzi (Eds.), *Model-based reasoning in science and technology. abduction, logic, and computational discovery* (pp. 572–587). Springer.

Barr, W. F. (1971). A syntactic and semantic analysis of idealizations in science. *Philosophy of Science, 38*, 258–272.

Barr, W. F. (1974). A pragmatic analysis of idealizations in physics. *Philosophy of Science, 41*(1), 48–64.

Bartsch, A., Mares, M. L., Schindler, J., Kühn, J., & Krack, I. (2024). Trust but verify? A social epistemology framework of knowledge acquisition and verification practices for fictional entertainment. *Human Communication Research, 50*(2), 194–207.

Bokulich, A. (2011). How scientific models can explain. *Synthese, 1*, 33–45.

Bokulich, A., & Liu, C. (1999). Approximation, idealization, and laws of nature. *Synthese, 118*, 229–256.
Boniolo, G. (1997). On a unified theory of models and thought experiments in natural sciences. *International Studies in the Philosophy of Science, 11*(2), 121–142.
Brown, J. R. (2004). Why thought experiments transcend empiricism. In C. Hitchcock (Ed.), *Contemporary debates in philosophy of science* (pp. 23–43). Blackwell.
Brzeziński, J., & Nowak, L. (Eds.). (1992). *Idealization III. Approximation and truth*. Rodopi.
Brzeziński, J., Klawiter, A., Kuipers, T. A. F., Lastowski, K., Paprzycka, K., & Przybysz, P. (Eds.). (2007). *The courage of doing philosophy*. Essays Presented to Leszek Nowak.
Bueno, O., & French, S. (2011). How theories represent. *The British Journal of Philosophy of Science, 62*, 857–894.
Buzzoni, M. (1995). *Scienza e tecnica. Teoria ed esperienza nelle scienze della natura*. Studium.
Buzzoni, M. (2004). *Esperimento ed esperimento mentale*. Angeli.
Buzzoni, M. (2008). *Thought experiment in the natural sciences. An operational and reflective-transcendental conception*. Königshausen+Neumann.
Buzzoni, M. (2016). Thought experiments and computer simulations. In L. Magnani & C. Casadio (Eds.), *Model-based reasoning in science and technology. Logical, epistemological, and cognitive issues* (pp. 57–78). Springer.
Buzzoni, M. (2023). Inductive exemplification in real and thought experiments. *Distinctio: Journal of Intersubjective Studies, 2*(2), 17–43.
Buzzoni, M. (2024). Carl Gustav Hempel. Thought experiments between methodological monism and the discovery/justification dichotomy. *Epistemology & Philosophy of Science, 61*(1), 202–222.
Buzzoni, M. (2025). *Empirical-scientific and fictional thought experiments: A comparison*. Global Philosophy (collection "Science and the Arts"), 35, 1–18.
Campbell, N. R. (1920). *Physics. The elements*. Cambridge University Press.
Cartwright, N. (1983). *How the laws of physics lie*. Oxford University Press.
Contessa, G. (2010). Scientific models and fictional objects. *Synthese, 172*, 215–229.
Currie, G. (2016). Models as fictions, fictions as models. *The Monist, 99*(3), 296–310.
Davies, D. (2007). Thought experiments and fictional narratives. *Croatian Journal of Philosophy, 7*, 29–45.
Davies, D. (2010). Learning through fictional narratives in art and science. In: Frigg R. and Hunter MC. (eds.), 51–70.
Dilworth, C. (1989). Idealization and the abstractive-theoretical model of scientific explanation. *Poznań Studies in the Philosophy of the Sciences and the Humanities, 16*, 167–181.
Dingler, H. (1928). *Das experiment. Sein Wesen und seine Geschichte*. Reinhardt.
Dingler, H. (1930). *Das system. Das philosophisch-rationale Grundproblem und die exakte Methode der Philosophie*. Reinhardt.
Dingler, H. (1938). *Die Methode der Physik*. Reinhardt.
Dohrn, D. (2016). Fiction and thought experiment – A case study. *Teorema: Revista Internacional de Filosofia, 35*(3), 185–219.
Egan, D. (2016). Literature and thought experiments. *The Journal of Aesthetics and Art Criticism, 74*, 139–150.
Elgin, C. Z. (2007). The laboratory of the mind. In W. Huemer, J. Gibson, & L. Pocci (Eds.), *A sense of the world: Essays on fiction, narrative, and knowledge* (pp. 43–54). Routledge.
Fine, A. (2009). Fictionalism. In Suárez, M. (Ed.), *Fictions in science* (pp. 19–36).
Frigg, R. (2010). Models and fiction. *Synthese, 172*, 251–268.
Frigg, R., & Hunter, M. C. (Eds.). (2010). *Beyond mimesis and nominalism: Representation in art and science*. Springer.
Giere, N. (1988). *Explaining science: A cognitive approach*. University of Chicago Press.
Godfrey-Smith, P. (2009). Models and fictions in science. *Philosophical Studies, 143*, 101–116.
Goodman, N. (1976). *Languages of art: An approach to a theory of symbols* (2nd ed.). Hackett Publishing Company.
Harré, R. (1970). *The principles of scientific thinking*. Macmillan.

Hempel, C. G. (1965). Typological methods in the natural and the social sciences. In *Aspects of scientific explanation and other essays in the philosophy of science* (pp. 155–171). The Free Press.
Hesse, M. (1963). *Models and analogies in science*. Univ. of Notre Dame Pr..
Hon, G. (1998). 'If This Be Error': Probing experiment with error. In M. Heidelberger & F. Steinle (Eds.), *Experimental essays—Versuche zum Experiment* (pp. 227–248). Nomos.
Horowitz, T., & Massey, G. J. (Eds.). (1991). *Thought experiments in science and philosophy*. Rowman & Littlefield.
Jones, M., & Cartwright, N. (Eds.). (2005). *Idealization XII: Correcting the model, idealization and abstraction in the sciences*. Rodopi.
Karadimas, P. (2024). Thought experiments and the pragmatic nature of explanation. *Foundations of Science, 29*, 257–280.
Kobow, B. (2014). How to do things with fictions: Reconsidering Vaihinger for a philosophy of social sciences. *Philosophy of the Social Sciences, 44*(2), 201–222.
Kuhn, Th. S. 1964. A function for thought experiments. In: R. Taton et I.B. Cohen (dir.), *L'aventure de la science, Mélanges Alexandre Koyré* (vol. II, pp. 307–337). Hermann.
Kuorikoski, J., & Lehtinen, A. (2009). Incredible worlds, credible results. *Erkenntnis, 70*, 119–131.
Laymon, R. (1985). Idealisation and the testing of theories by experimentation. In P. Achinstein & O. Hannaway (Eds.), *Observation, experiment, and hypothesis in modern physical science* (pp. 147–173). MIT Press.
Laymon, R. (1991). Thought experiments by Stevin, Mach and Gouy: Thought experiments as ideal limits and as semantic domains. In: Horowitz and Massey (eds.), 167–191.
Laymon, R. (1995). Experimentation and the legitimacy of idealisation. *Philosophical Studies, 77*, 353–375.
Lewis, D. (1973a). *Counterfactuals*. Blackwell.
Lewis, D. (1973b). Causation. *Journal of Philosophy, 70*, 556–567.
Liu, C. (1999). Approximation, idealization, and laws of nature. *Synthese, 118*, 229–256.
Magnani, L. (2012). Scientific models are not fictions. In L. Magnani & P. Li (Eds.), *Model-based science as epistemic warfare* (pp. 1–38). Springer.
Magnani, L. (2013). Is abduction ignorance-preserving? Conventions, models and fictions in science. *Logic Journal of the IGPL, 21*(6), 882–914.
Magnani, L., & Bertolotti, T. (Eds.). (2017). *Handbook of model-based science*. Springer.
Mäki, U. (2005). Models are experiments, experiments are models. *Journal of Economic Methodology, 12*(2), 303–315.
Mäki, U. (2020). Puzzled by idealizations and understanding their functions. *Philosophy of the Social Sciences, 50*(3), 215–237.
Margoni, F., Surian, L., & Baillargeon, R. (2024). The violation-of-expectation paradigm: A conceptual overview. *Psychological Review, 131*(3), 716–748.
McMullin, E. (1985). Galilean idealisation. *Studies in History and Philosophy of Science, 16*, 247–273.
Mizrahi, M. (2012). Idealizations and scientific understanding. *Philosophical Studies, 160*, 237–252.
Morrison, M. (2015). *Reconstructing reality*. Oxford University Press.
Musgrave, A. (1981). Unreal assumptions' in economic theory: The F-twist untwisted. *Kyklos, 34*, 377–387.
Nguyen, J. (2021). Do fictions explain? *Synthese, 199*(1), 3219–3244.
Niiniluoto, I. (1986). Theories, approximations, idealizations. In R. Barcan Marcus, G. J. W. Dorn, & P. Weingartner (Eds.), *Logic, methodology and philosophy of science VII, 255–289*. North-Holland.
Niiniluoto, I. (2007). Idealization, counterfactuals, and truthlikeness. In: Brzezhiski et al. (eds.), 103–122.
Norton, J. (1996). Are thought experiments just what you thought? *Canadian Journal of Philosophy, 26*, 333–366.
Norton, J. D. (2012). Approximation and idealization: Why the difference matters. *Philosophy of Science, 79*(2), 207–232.

Nowak, L. (1971). *Foundations of marxian methodology of science*. National Scientific Publishing House. [Polish Title: U podstaw marksowskiej metodologii nauki].
Nowak, L. (1980). *The structure of idealization. Towards a systematic interpretation of the marxian idea of science*. Reidel.
Nowak, L. (1992). The idealizational approach to science: A survey. In: Brzeziński and Nowak (eds.), 1992, 9–63.
Portides, D. P. (2007). The relation between idealization and approximation in scientific model construction. *Science & Education, 16*, 699–724.
Portides, D. (2021). Idealization and abstraction in scientific modeling. *Synthese, 198*(Suppl 24), 5873–5895.
Potochnik, A. (2017). *Idealization and the aims of science*. University of Chicago Press.
Putnam, H. (1976). Literature, science, and reflection. In H. Putnam (Ed.), *Meaning and the moral sciences* (pp. 83–94). Routledge & Kegan Paul.
Reiss, J. (2018). Thought experiments and idealizations. In M. T. Stuart, Y. Fehige, & J. R. Brown (Eds.), *Routledge companion to thought experiments* (pp. 469–483). Routledge.
Rescher, N. (1964). *Hypothetical reasoning*. North-Holland.
Scheffler, I. (1963). *The anatomy of inquiry: Philosophical studies in the theory of science*. Knopf.
Scriven, M. (1961). The key property of physical laws – Inaccuracy. In H. Feigl & G. Maxwell (Eds.), *Current issues in the philosophy of science* (pp. 91–101). Holt.
Stolnitz, J. (1992). On the cognitive triviality of art. *British Journal of Aesthetics, 32*, 191–200.
Stuart, M. T. (2021). Towards a dual process epistemology of imagination. *Synthese, 198*(2), 1329–1350.
Suárez, M. (2009a). Fictions in scientific practice. In M. Suárez (Ed.), *Fictions in science: Philosophical essays on modeling and idealization* (pp. 3–15). Routledge.
Suárez, M. (Ed.). (2009b). *Fictions in science: Philosophical Essays on Modeling and Idealization*. Routledge.
Teller, P. (2004). The law-idealization. *Philosophy of Science, 71*(5), 730–741.
Toon, A. (2010). The ontology of theoretical modelling: Models. *Synthese, 172*, 301–315.
Vaihinger, H. (1922/1935). *Die Philosophie des Als Ob*. Meiner, Leipzig (7. u. 8. Aufl.); Engl. Transl. by C.K. Ogden (based upon an edition abbreviated by the author), The Philosophy of 'As if'. Kegan Paul, London 1935 (2nd edition).
Weber, M. (1904/1949). *Die "Objektivität" sozialwissenschaftlicher und sozialpolitischer Erkenntnis*. Archiv für Sozialwissenschaft und Sozialpolitik 19, 22–87, repr. in Gesammelte Aufsätze zur Wissenschaftslehre, Mohr, Tübingen, 1922; 5th ed. 1982 (quotations are from this edition), 146–214. Engl. Transl. by E.A. Schils and H.A. Finch, "Objectivity" in Social Science and Social Policy. In: M. Weber, The Methodology of the Social Sciences, The Free Press, Glencoe, Illinois, 1949, 50–112.
Weber, M. (1906/1949). *Kritische Studien auf dem Gebiet der kulturwissenschaftlichen Logik*. Archiv für Sozialwissenschaft und Sozialpolitik 22, 143–207, repr. in Gesammelte Aufsätze zur Wissenschaftslehre, Mohr, Tübingen, 1922; 5th ed. 1982 (quotations are from this edition), 215–290. Engl. Transl. by E.A. Schils and H.A. Finch, Critical Studies in the Logic of the Cultural Sciences. In: M. Weber, The Methodology of the Social Sciences, The Free Press, Glencoe, Illinois, 1949, 113–188.
Weisberg, M. (2007). Three kinds of idealizations. *Journal of Philosophy, 104*(12), 639–659.
Wheeler, B. (2018). *Idealization and the laws of nature*. Springer.
Woods, J. (Ed.). (2010). *Fictions and models: New essays*. Philosophia Verlag.
Woods, J., & Rosales, A. (2010). Virtuous distortion. Abstraction and idealization in model-based science. In: Magnani et al (eds.), *Model-based reasoning in science and technology*, 3–30.
Young, J. O. (2001). *Art and knowledge*. Routledge.

Chapter 4
Rethinking Cognition: Morphological Info-computation and the Embodied Paradigm in Life and Artificial Intelligence

Gordana Dodig-Crnkovic

Abstract This study aims to place Lorenzo Magnani's Eco-Cognitive Computationalism within the broader context of current work on information, computation, and cognition. Traditionally, cognition was believed to be exclusive to humans and a result of brain activity. However, recent studies reveal it as a fundamental characteristic of all life forms, ranging from single cells to complex multicellular organisms and their networks. Yet, the literature and general understanding of cognition still largely remain human-brain-focused, leading to conceptual gaps and incoherency. This paper presents a variety of computational (information processing) approaches, including an info-computational approach to cognition, where natural structures represent information and dynamical processes on natural structures are regarded as computation, relative to an observing cognizing agent. We model cognition as a web of concurrent morphological computations, driven by processes of self-assembly, self-organization, and autopoiesis across physical, chemical, and biological domains. We examine recent findings linking morphological computation, morphogenesis, agency, basal cognition, extended evolutionary synthesis, and active inference. We establish a connection to Magnani's Eco-Cognitive Computationalism and the idea of computational domestication of ignorant entities. Novel theoretical and applied insights question the boundaries of conventional computational models of cognition. The traditional models prioritize symbolic processing and often neglect the inherent constraints and potentialities in the physical embodiment of agents on different levels of organization. Gaining a better info-computational grasp of cognitive embodiment is crucial for the advancement of fields such as biology, evolutionary studies, artificial intelligence, robotics, medicine, and more.

G. Dodig-Crnkovic (✉)
Department of Computer Science and Engineering, Chalmers University of Technology, Gothenburg, Sweden

Division of Computer Science and Software Engineering, School of Innovation, Design and Engineering, Mälardalen University, Lidingö, Sweden
e-mail: gordana.dodig-crnkovic@chalmers.se

S. Arfini (ed.), *Scientific Cognition, Semiotics, and Computational Agents: Essays in Honor of Lorenzo Magnani - Volume 2*, Synthese Library 506,
https://doi.org/10.1007/978-3-031-96688-0_4

Keywords Computationalism · Eco-cognitive computationalism · Info-computationalism · Computing nature · Physical computing · Morphological computing · Information · Computation · Cognition

4.1 Lorenzo Magnani's Eco-cognitive Computationalism and the Idea of Computational Domestication of Ignorant Entities

Lorenzo Magnani's Eco-Cognitive Computationalism (Magnani, 2018, 2021, 2022b; Arfini & Magnani, 2022) is a philosophical framework that explores the interplay between cognitive processes and their environmental contexts, particularly emphasizing the role of computational tools in enhancing human cognitive abilities. The concept of "computational domestication of ignorant entities" within this framework can be interpreted as the use of computational methods and models to harness and transform less understood or unorganized aspects of the environment (which he refers to as "ignorant entities") into structured, intelligible forms that can be more easily manipulated and understood by human cognition.

This approach suggests that computational tools not only extend our cognitive capabilities but also reshape our cognitive environment, making it more open to human understanding and control. By "domesticating" these ignorant entities through computational means, we effectively restructure our cognitive landscape, enhancing our ability to solve complex problems and generate new knowledge.

Magnani's theory touches on themes common in the philosophy of cognitive science, such as the *extended mind thesis* and *situated cognition*, but places a unique emphasis on the *transformative power of computational technology* in these processes. This perspective is particularly relevant in the context of today's data-driven science and technology, where computational tools play a critical role in organizing, analyzing, and interpreting large datasets, effectively bringing clarity and insights from what might initially appear as chaotic or overwhelming information.

In the contemporary landscape of cognitive science and artificial intelligence, Lorenzo Magnani's book "Eco-Cognitive Computationalism: Cognitive Domestication of Ignorant Entities" presents a groundbreaking perspective. The concept of eco-cognitive computationalism offers a novel approach to understanding computation as an active process in physical entities, which are *transformed into cognitive mediators*. The main themes and concepts presented in Magnani's work emphasize the dynamic interplay between internal processes, the body, and the environment in the emergence of information, computation, and cognition. Eco-cognitive computationalism views computation as an active process in distributed physical entities that are transformed to encode and decode data. This approach *moves* beyond the traditional notion of digital computation, encompassing a broader range of computational activities performed by physical devices, including the brain. Magnani argues that this perspective avoids reducing the analysis of computation to a rigid, one-sided view, instead proposing an intellectual framework that respects the historical and dynamic nature of concepts of information, computation, and cognition (Magnani, 2022b).

One of the central themes in Magnani's approach is the evolutionary emergence of information and cognition as the result of dynamic interactions *between internal processes, the body, and the external environment*. Magnani emphasizes that understanding computation requires acknowledging its context within the ecology of cognition, which includes both organic and artefactual agents (Magnani, 2022b). Turing's insights into the emergence of cognition and computation are fundamental in this discussion. Turing's concept of *unorganized machines, which can be educated* through interaction with their environment, illustrates the dynamic nature of cognitive development. This analogy between human brains and computational machines underscores the *coevolution of information, computation, and cognition* in both natural and artificial systems (Turing, 1948).

Magnani extends the discussion of eco-cognitive computationalism to include embodied and distributed cognition. He argues that cognitive processes are not confined to the brain but are distributed across the body and external environment. This view aligns with recent cognitive science theories that emphasize the role of the body and environment in shaping cognitive tasks, (Clark, 1989, 1997, 2008, 2016). The creation of *mimetic bodies*, which enhance computational capabilities through *morphological features*, exemplifies this approach. These bodies simplify cognitive and motor tasks, highlighting the *interdependence of internal and external resources* in cognitive processes.

Magnani (2022a, b) details the cognitive strategies employed by deep learning machines, such as AlphaGo. Magnani distinguishes between locked and unlocked strategies, examining how these approaches impact the creativity and adaptability of AI systems. Locked strategies, which are limited in their eco-cognitive openness, contrast with the more flexible and creative reasoning observed in human cognition.

The final chapter in (Magnani, 2022b) explores the *domestication* of physical and biological bodies through computational means. Magnani introduces the concept of *overcomputationalism*, which critiques the tendency to overextend computational applications to all aspects of life. He emphasizes the need to *protect and preserve the unique qualities of ignorant entities*, advocating for a balance between computational and human-centered approaches. The future of eco-cognitive settings, Magnani argues, should be tailored to enhance both computational efficiency and human well-being.

Magnani's Eco-cognitive computationalism presents a comprehensive and dynamic framework for understanding computation, information, and cognition. By integrating insights from cognitive science, artificial intelligence, and philosophy, Magnani presents a holistic view of computational systems as integral to both natural and artificial environments. His emphasis on the dynamic and distributed nature of cognition and computation challenges traditional notions and opens new perspectives for research and application in these fields.

4.2 Computation in a Computing Universe. Nature as a Network of Morphological Info-computational Processes (ICON) for Cognitive Agents

Magnani's Eco-cognitive computationalism with cognitive domestication of ignorant entities can be related to a broader computational perspective of Info-computationalism, which addresses nature in computational terms. The info-computational approach (ICON) presents a view of nature as a network of information-processing agents organized in a dynamic hierarchy of levels. This framework aims to unify diverse understandings of natural, formal, technical, behavioral, and social phenomena through the perspective of information and computation. The idea that, for us, information is the fabric of reality goes back to Wheeler's "it from bit" (Wheeler, 1990). Our reality is a result of the processing of information of multimodal signals coming through our senses combined with information processes in our bodies that unfold on several levels of organization/scale. Among prominent works on informational reality are (Baeyer von, 2004; Lloyd, 2006; Seife, 2006; Vedral, 2010; Davies & Gregersen, 2010; Burgin, 2010). A world exists on its own with human agents in it and we discover its structures and processes, patterns and laws, by interactions and via self-structuring (self-organization) of information obtained through morphological/physical computation processes (Pfeifer & Iida, 2005; Pfeifer & Gomez, 2009; Dodig-Crnkovic, 2013). In this analysis, a natural philosophy framework is presented that provides a foundation for connecting domain-specific knowledge obtained from a variety of sciences and other knowledge fields into an interdisciplinary/transdisciplinary perspective.

Unlike old-day stand-alone calculating machines, represented by the abstract logical Turing Machine Model, modern networked and concurrently executing computers are used for worldwide communication, information processing, cyber-physical control, and data-, information-, and knowledge management. They serve as cognitive tools of the extended mind (Clark & Chalmers, 1998), facilitating social interactions, decision-making, and control of physical processes. This interconnected concurrent nature of computation aligns with the actor model of computation (Hewitt, 2007, 2012), emphasizing distributed, reactive, agent-based, and parallel execution. Such systems operate continuously, focusing on behavior, response to changes, and adaptability rather than predetermined algorithm termination.

4.2.1 History of the Idea of the Computing Universe

Konrad Zuse first suggested that the physical behavior of the universe could be computed on a basic level, using cellular automata (Zuse, 1970). A similar, pan-computationalist view is supported by various scientists (Wheeler, 1990, 1994; Fredkin, 1990; Wolfram, 2002; Dodig-Crnkovic, 2012) who see natural phenomena as results of computational processes. This perspective aligns with the idea that the universe computes its next state from its current state, with interactions and information exchanges driving its evolution (Chaitin, 2006).

4.2.2 Natural, Physical, and Morphological Computation

Natural computation encompasses a range of computational systems inspired by nature, including neural networks, genetic algorithms, and quantum computing. This field promoted a generalized model of computation beyond traditional Turing machines, recognizing the complexity of biological and cognitive systems. Morphological computation, a key concept, explains how physical interactions lead to the development of complex structures and processes in living systems (Dodig-Crnkovic, 2013, 2017; Dodig-Crnkovic & Lowe, 2017; Pfeifer et al., 2006; Hauser et al., 2014; Ghazi-Zahedi et al., 2017; Pfeifer & Iida, 2005).

Natural info-computation provides a unified framework for understanding cognitive systems through morphological info-computation across various levels of organization. It advocates for a new philosophy of nature, integrating information and computation to explain diverse phenomena. By generalizing computational models, this approach aims to encompass the full range of natural processes, from physics to cognition, highlighting the essential role of agents in these systems. The framework necessitates generalizing computational models beyond the traditional Turing machine model, requiring agent-based concurrent resource-sensitive models to cover phenomena from physics to cognition.

In this approach, nature computes on a variety of levels of organization where informational structures (as observed by diverse cognizing agents at different levels) dynamically change through interactive processes of natural computation. Natural computing (Rozenberg & Kari, 2008; Rozenberg et al., 2012) encompasses a wide range of areas, including quantum computing, chemical computing, biological computing, evolutionary algorithms, swarm intelligence, reservoir computing, thermodynamic computing, and cognitive and intelligent computing. Each of these areas draws on principles and phenomena observed in the natural world to inform and improve computational techniques.

Natural computing (unconventional computing) raises philosophical questions about the nature of computation and intelligence, as it challenges traditional notions about where and how computation can occur. Natural computing continues to grow as a field of research that bridges the gap between natural sciences and computer science, enabling new solutions to complex problems by using the power of natural processes. It exemplifies a profound shift toward computing that is not confined to silicon but is increasingly interwoven with biological and quantum systems, (Denning, 2003, 2007, 2009, 2010; Paun et al., 2017; Rozenberg, 2008; Rozenberg & Kari, 2008).

4.2.3 The Necessity of an Agent/Observer/Actor for the Generation of Knowledge

Information, a concept widely used and ambiguous, is essential for understanding both everyday applications and formal definitions in various fields of research. Current understanding of matter/energy also reflects ambiguity, as observable matter constitutes only a small percentage of the universe, with the rest attributed to dark matter and dark energy. This indicates a need to re-examine our understanding of the physical world. What is missing in the traditional picture is the model of a physical observer, such as proposed in second-order cybernetics. As (Goyal, 2012) explains, an observer in classical physics is a highly idealized omnipotent agent looking at the world from nowhere (a "God's eyes" perspective). Or, as (Fields, 2012) expresses it, a "Galilean observer" is a bare "point of view", able to observe and conceptualize the world without affecting it or being affected. There is a hope to gain more clarity on the structures and dynamics of the universe by connecting information and matter/energy. Information is defined as differences that make a difference (Bateson, 1972), or "Information expresses the fact that a system is in a certain configuration that is correlated to the configuration of another system. Any physical system may contain information about another physical system." (Hewitt, 2007) In other words, information is *relational*, which is fundamental.

4.2.4 Relational Info-computation as a Reality for an Agent

> What we call reality arises in the last analysis from the posing of yes-no questions and the registering of equipment-evoked responses; in short, all things physical are information-theoretic in origin and this is a participatory universe. (Wheeler, 1990)

We can find the relational view of information in Floridi's work (Floridi, 2003, 2008a, 2008b) in the form of Informational Structural Realism (ISR). This relational view in physics is formulated by Carlo Rovelli (2015, 2018). Even Stephen Wolfram's "Project to Find the Fundamental Theory of Physics" (Wolfram, 2020) is based on the relational view, as well as the approach of (Chaitin, 2006, 2018).

Barry Cooper (2015) addresses the issue of informational structures, citing (Floridi, 2011):

> Structural objects work epistemologically like constraining affordances: they allow or invite certain epistemic constructs (they are affordances for inforgs like us, who elaborate them) and resist or impede some others (they are constraints for the inforgs), depending on the interaction with, and the nature of, the informational organisms that process them. They are exploitable by a theory, at a given LoA, as input of adequate queries to produce information (the model) as output. (p. 340)

and

> all that physics—or science in general—tells us about, and hence all that we can know, on the basis of that physics, is structure. (p. 343)

This question of informational structures that are not "flat" is directly connected to models of computation as elaborated in the article "From the closed classical algorithmic universe to an open world of algorithmic constellations" (Burgin & Dodig-Crnkovic, 2012), indicating fundamental connections between information and computation as well as the importance of morphological (structural and material) aspects of both.

Matter and energy are higher-order concepts derived from interactions between observers and the universe, including mutual interactions among observers. This does not imply that reality is merely an agreed-upon fiction. The world exists independently, with human observers/agents discovering its structures and processes through interactions and self-structuring of information. The reality for a cognitive agent is built upon external information and the agent's information processing architecture, leading to diverse realities for different types of agents (von Uexküll, 1957). Every subsystem in the universe interprets every other subsystem in a way dependent on their information processing capacities, (Friston et al., 2010; Friston & Kiebel, 2009; Isomura et al., 2023).

All living beings have the ability to adapt to the environment and learn so as to increase their chances of survival. The life itself can be understood as a process of cognition (Maturana & Varela, 1980; Stewart, 1996). Building on Maturana and Varela's understanding of cognition the author proposed an info-computational constructive framework (Dodig-Crnkovic, 2011a, b) to explain how increasingly complex structures develop as a result of information processing in nature. These information processes can be modeled as natural morphological computation, in the network of networks of physical processes where constant exchange (communication) of data (signals) among agents establishes new structures on physical, chemical, and biological levels. Instead of symbol manipulation, morphological computation is a result of physical object interactions. Natural computation (Rozenberg et al., 2012) can be used to explain the emergent phenomena by complexification of information through processes of signals exchange at different levels of organization or scale. The new framework is used as a tool for studying cognitive systems such as living organisms or artifactual cognitive agents on different levels of complexity.

For example, only recently we learned about the essential role of viruses and bacteria for the development of life on Earth, (Witzany, 2012). Agents exchanged among organisms can be bacteria, viruses, or pieces of DNA that can actively alter the organism that receives them, as they continue to act within an organism. All those processes of communication along with self-organization (autopoiesis) unfold in thermodynamically open living systems, and they need energy to persist, in the regime on the edge of chaos (Prigogine & Stengers, 1984; Prigogine, 1980).

4.2.5 Morphological Info-computation in Cognitive Systems

Over billions of years, nature has developed information processing networks in living organisms, enabling them to effectively cope with environmental complexity. The evolution from inanimate to animate matter, though largely unknown, suggests that self-assembly and self-organization of increasingly complex structures led to the first molecular machines and cells, eventually forming multicellular organisms. Communication processes and self-organization unfold in thermodynamically open systems, highlighting the significance of energy in maintaining these systems. Morphological computation, a result of physical object interactions, provides a framework for understanding the emergent phenomena in living systems (Dodig-Crnkovic, 2017).

4.2.6 Evolution in the Framework of Info-computationalism/ Computational Naturalism (ICON)

Evolution in the framework of ICON is understood as the dynamic process of creating, transferring, and transforming informational structures across multiple scales and levels of organization in nature. This perspective views nature as a hierarchical network of informational agents engaged in computational processes, where information processing (computation) occurs. Morphological computation considers computation as the physical process of structural change in informational structures. Evolution is driven by these dynamic processes, where informational agents (such as cells or organisms) interact with their environment and with each other, creating new informational structures. Lower-level informational processes give rise to higher-level emergent phenomena through self-organization, with top-down control structures like the nervous system and the brain, regulating bodily functions.

The evolution of information is driven by multi-level interactions and information transfers between hierarchical levels, aligning with the principles of the extended evolutionary synthesis (EES) (Laland et al., 2015; Müller, 2017; Noble, 2002, 2012, 2016a, 2016b, 2020; Corning et al., 2023; Torday et al., 2020; Torday & Miller, 2020).

The info-computationalism/computational naturalism/ICON framework provides a perspective on the evolution of complex systems by viewing them as hierarchical networks of informational agents engaged in morphological computation. Nature is organized into a dynamic hierarchy of levels, with informational processes (computation) occurring at each level, from elementary particles to biological systems. This hierarchical structure allows for the emergence of complex systems through interactions and information transfer across levels. The physical form (morphology) of informational agents and their interactions with the environment shape the computational processes underlying their behavior and intelligence. These interactions across genetic, epigenetic, behavioral, and cultural processes contribute to the emergence and evolution of complex systems.

By grounding cognition in morphological computation and embodied dynamics, the ICON framework provides a theoretical basis for rethinking and developing new approaches to artificial intelligence (AI) that better capture the embodied and situated nature of intelligence as observed in biological systems.

Traditional AI has typically focused on disembodied, symbolic reasoning systems, but the ICON framework suggests that truly intelligent systems should incorporate principles of embodied cognition, where the physical form and environmental interactions play a central role in shaping computational processes and cognitive behavior. Morphological computation could inspire the development of AI systems based on the computational capabilities of physical structures and their interactions with the environment, rather than relying solely on abstract algorithms and symbolic representations. This could lead to more robust, adaptive, and context-aware AI systems that better mimic the intelligence observed in natural systems.

Furthermore, the ICON emphasis on hierarchical organization, self-organization, and multi-level interactions suggests the idea of developing AI systems that can exhibit emergent behavior and adapt to complex environments through bottom-up and top-down interactions across multiple scales and levels of organization.

4.3 The Relationships Between Magnani's Eco-cognitive Computationalism and Other Views of Cognition Based on Physical Computation

The concepts of information, computation, and cognition are central to various branches of philosophy, cognitive science, and computer science. Lorenzo Magnani, Gordana Dodig-Crnkovic, Gualtiero Piccinini, Oron Shagrir, Marcin Miłkowski, Nir Fresco and Michael Levin —each provide different perspectives on these topics.

Lorenzo Magnani's Eco-Cognitive Computationalism (Magnani, 2018, 2021, 2022b; Arfini & Magnani, 2022) and Gordana Dodig Crnkovic's Info-Computationalism both study the integration of computation with cognitive and informational processes, but they focus on different aspects of this integration and have distinct foundational perspectives. Magnani is particularly interested in the concept of distributed cognition, which emphasizes that human cognition extends beyond the individual and into the environment and artifacts used. This approach views information as something not solely processed internally but also distributed across and manipulated by external media (Magnani, 2025).

Eco-cognitive computationalism, emphasizes the role of computational tools in reshaping human cognitive processes and environments. While it recognizes the importance of computational methods in extending and enhancing human cognitive capacities, *it is more focused on the interaction between humans and their cognitive tools in specific environments. Magnani's approach is particularly interested in how computational tools can domesticate complex, poorly understood, or "ignorant" entities, making them more accessible and manipulable by human cognition.*

Info-Computationalism (ICON), (Dodig-Crnkovic, 2007, 2008; Ehresmann, 2014), in contrast, views nature itself as computational, where information processing is fundamental to the physical, biological, and cognitive processes that constitute reality for a cognizing agent. This framework posits that all natural phenomena can be understood in terms of information processing and that knowledge generation is a result of computational processes in the natural world. The universe, from this perspective, is seen as a network of networks of computational processes where different levels of physical, biological, and cognitive phenomena interact through computational dynamics. ICON views information as a fundamental building block of reality for an agent, a perspective based on Informational Structural Realism (ISR) (Floridi, 2008a). In this interpretation, information, and computation are seen as essential for understanding natural and artificial processes, implying that computation is a broader phenomenon than what happens in digital computers. Cognition is intrinsically connected to the processing of information integrating this into the broader ecosystem, positioning cognition as an information-processing activity at all levels of biological and artificial systems (what Levin calls a "multiscale competency") (Levin, 2023). This approach introduces an integrative dimension to the study of cognition, with cognitive processes being more widespread in biology than previously thought. It supports a view similar to that of Magnani's distributed cognition, where cognitive processes are not confined to the brain but are distributed throughout the body and interact closely with the environment. The relation between these two theories can be seen in how they both extend the traditional view of cognition and incorporate computational thinking into the fabric of understanding and interacting with the world. However, while Eco-cognitive computationalism focuses on the enhancement of human cognition and the domestication of complexity through computational tools, Info-computationalism proposes a foundational theory where computation and information processing manifest in all natural processes.

Thus, while both frameworks elaborate the role of computation in understanding and interacting with the world, Eco-cognitive computationalism is applied and focused on the enhancement of human cognitive environments through specific computational interventions, whereas Info-computationalism provides a foundational perspective on nature as it appears to us, including psychological and social aspects. Both perspectives enrich our understanding of the computational nature of cognition and reality but from complementary angles.

Gualtiero Piccinini's approach offers another important dimension to consider regarding varieties of computationalism (Piccinini, 2012, 2020; Piccinini & Scarantino, 2011; Piccinini & Shagrir, 2014). Piccinini focuses on the philosophy of mind and the foundations of cognitive science, specifically the nature of computation in cognitive systems. He proposes a mechanistic view of computation, where a physical system is computational if it manipulates data through a set of functional transitions as specified by its architecture. Piccinini distinguishes between different kinds of computation based on how they are implemented physically and the kinds of processes they involve, such as digital versus analog. His theory is comprehensive in its definition and classification of computational processes, insisting on clear, functional, and mechanistic explanations to demarcate what counts as computation and what does not. Piccinini's emphasis on the physical basis of computation aligns well with Magnani's Eco-cognitive computationalism as both see computational tools as extensions and enhancements of human cognitive capabilities in tangible, real-world settings. Where Piccinini provides a framework for understanding how computational processes operate mechanistically within cognitive systems, Magnani looks at the outcomes of these processes—specifically, how they can domesticate "ignorant" entities to augment human cognitive capacities and modify cognitive environments.

Oron Shagrir's views on computation (Shagrir, 2010, 2022) further enrich the discourse, by addressing topics of physical computation, computational models, computational complexity, and hypercomputation (physically realizable computational models that can solve problems beyond the capabilities of Turing machines, including analog computation, quantum computation, and exploiting physical phenomena like chaos). Shagrir highlights the importance of mapping computational descriptions onto physical phenomena, asserting that a physical system is computational if there is a coherent way to map computational states onto the states of the physical system. His approach extends the understanding of computation beyond the strictly mechanistic framework of Piccinini, proposing a more inclusive approach that can accommodate different types of computations based on their functional realization in physical substrates. Shagrir extends the computational theory of mind by exploring how various types of computations can explain different cognitive abilities. He discusses information in the context of how it is represented and processed by the brain, viewing computation as a model for explaining how cognitive processes are implemented neurally. Shagrir provides a comprehensive and rigorous analysis of the physical foundations of computation, bridging the gap between theoretical computer science and the physical realizability of computational processes. Shagrir's way of approaching physical systems computationally complements Magnani's emphasis on using computational tools to extend human cognitive processes. In both frameworks, the physical instantiation of computation plays a central role.

Marcin Miłkowski (Miłkowski, 2013b) presents a rigorous analysis of explanation in cognitive science through his mechanistic theory of computation. He explores various computational explanations in cognitive science through four diverse case studies: classical computational simulation, connectionist modeling, computational neuroscience, and radical embodied robotics. Miłkowski outlines different computational processes, emphasizing that computation in cognitive science is broader than traditional symbolic representations. He argues for a mechanistic account of computation that includes digital, analog, and even quantum computation, highlighting the importance of information processing rather than mere symbol manipulation. He defends the view that computation in cognitive science is a physical process that underlies cognitive phenomena. Miłkowski points out the importance of a clear definition of computation that avoids pancomputationalism (the idea that everything computes). He advocates for the mechanistic framework as the most suitable for elucidating computational explanations, although he acknowledges the value of other explanatory frameworks. His approach to cognition is also mechanistic, viewing cognitive processes as computational operations defined by their causal structure and organization.

Michael Levin provides deep insights into cognition, particularly in non-neural biological systems (Baluška & Levin, 2016; Bongard & Levin, 2023; Fields & Levin, 2020; Levin, 2019, 2020, 2022, 2023; Levin & Pezzulo, 2016; Manicka & Levin, 2019, 2022; Mcmillen & Levin, 2024; Pezzulo & Levin, 2015). His work significantly expands the conventional understanding of cognition by studying how cells and non-neural organisms process information and exhibit behavior that suggests a form of basal intelligence. Drawing parallels between his work and the concepts of information, computation, and cognition, can provide a broadened perspective on these concepts. Levin's research primarily focuses on the bioelectric processes that underlie the decision-making and behavior in cells and tissues, independent of the presence of a nervous system. His studies have shown that cells and groups of cells can process information and make complex decisions, which are critical for morphogenesis, regeneration, and cancer suppression. This suggests a form of distributed cognition that occurs at cellular and tissue levels. In the context of information, Levin's work suggests that all living systems, not just neural systems, engage in processing environmental signals (information) to maintain homeostasis and to guide development and regeneration. This is akin to the information-centric view of ICON, where information is a fundamental component of all natural processes. Regarding computation, Levin demonstrates that non-neural and even single-cell organisms perform complex, goal-directed tasks by using bioelectrical signaling pathways to process inputs and drive outputs, effectively computing responses to their environment. Levin's research implies that biological computation can be seen in many forms and across various biological entities, supporting a broader, more inclusive definition of computation. Regarding cognition, Levin's work challenges the traditional brain-centered approaches. He proposes that cells and tissues exhibit a form of cognition, evidenced by their ability to process information, learn, make decisions, and remember states—behaviors typically reserved for organisms with brains. This aligns well with Karl Friston's Free Energy Principle, where organisms act to minimize surprise by forming predictive models of their environment. Levin's studies suggest that even single cells engage in predictive modeling and decision-making processes to minimize free energy, thus displaying a rudimentary form of cognition.

Levin's work shows that information processing and computation are basic life functions that occur at multiple levels of biological organization, thereby enriching our understanding of cognition in a broad biological context.

4.4 Magnani on the Interrelatedness of Information, Computation, and Cognition

Magnani (2025) explores the interrelated concepts of information, computation, and cognition, highlighting their dynamic and evolving nature in the framework of Eco-cognitive computationalism. Magnani views these concepts from the perspective of ecological interactions between the brain, body, and environment, emphasizing their *coevolutionary aspects*. Information, computation, and cognition are conceptually deeply intertwined and influence each other's evolution.

Magnani argues against viewing these concepts as static or isolated, suggesting instead that they dynamically interact and evolve within cultural, technological, and biological contexts.

Eco-cognitive computationalism suggests that understanding computation should extend beyond traditional digital computation to include a variety of physical systems and processes. Computation is seen as active in various entities that have been transformed to encode and decode data, thus participating in cognitive processes. Semiosis (the process of signification or meaning-making) is fundamental to the development of information and cognition. Magnani integrates insights from processes of meaning-making from the catastrophe theory of René Thom—topology, morphogenesis, and structuralism (Aubin, 2020) in conjunction with Peircean semiotics to explain how biological and physical phenomena emerge as salient forms carrying information. Magnani discusses how changes in technology and cultural practices impact the meanings and functions of information, computation, and cognition. These changes facilitate the emergence of new forms of cognitive processes and information handling, reflecting a coevolutionary development between humans and their technological environments.

Magnani explores practical instances, such as how signs and tools externalize and distribute cognitive processes in the environment, a concept linked to extended cognition theories (Clark & Chalmers, 1998). He makes an important observation that *the meanings of information, computation, and cognition are not fixed but are subject to continual evolution and reinterpretation*. This emphasizes the need for an integrated approach that considers the *historical, ecological, and semiotic aspects* of these concepts to fully understand their roles and impacts in both natural and artificial systems. These concepts are not merely technical or theoretical but are deeply embedded in ecological, semiotic, and cognitive frameworks.

4.5 Karl Friston's Work Related to Information, Computation, and Cognition

Karl Friston's work, with his formulation of the Free Energy Principle (FEP), provides a perspective that relates to the concepts of information, computation, and cognition. Friston's theory is influential in neuroscience, psychology, and cognitive sciences, and it offers a unifying framework that can be related to the views of the other scholars mentioned earlier.

The Free Energy Principle suggests that all living systems (from single-cell organisms to human brains) aim to minimize their free energy. In this context, free energy is a mathematical concept borrowed from statistical physics but *represents an upper bound on the surprise or unpredictability of sensory data given a model of the world.* Minimizing free energy can be understood as *minimizing uncertainty or prediction error about the sensory inputs received from the world.*

In terms of information, Friston's approach can be seen as treating *cognition as fundamentally an information-processing problem.* The brain continuously infers and updates its internal models of the world to minimize the informational discrepancy (or "surprise") between its predictions and the actual sensory input. This aligns with views like those of (Dodig-Crnkovic, 2012, 2014a, 2014b), who also see information as essential to understanding of cognitive processes.

Computation in Friston's theory is about the *neural mechanisms* that the brain employs to achieve this minimization of free energy. It involves complex computational strategies such as *predictive coding, where the brain uses hierarchical models to predict sensory inputs and adjust its beliefs based on prediction errors.*

The FEP effectively offers a theory of how cognition functions: it is the process by which the brain models the world to minimize the surprise associated with new sensory inputs. Cognition, therefore, is about reducing uncertainty through perpetual inference and learning. Predictive coding is a central aspect of Karl Friston's work, and a theory primarily concerned with how the brain processes *sensory* information. The key idea is that the brain is constantly generating and updating a mental model of the environment. This model is used to predict sensory input before it arrives. The brain then compares the actual sensory input with its predictions and adjusts its model based on the differences. The process is *hierarchical.* Higher levels of the neural network make predictions about what lower levels will perceive. Lower levels send back the prediction error (the difference between the expected sensory input and the actual sensory input) to higher levels. This feedback leads to the updating of the predictive model. At each level of this hierarchy, the goal is to minimize prediction error. *Minimizing prediction error can be equated to minimizing free energy.* Essentially, by minimizing the surprise or the prediction error, the brain is trying to improve its model of the world to better predict future sensory inputs.

Predictive coding has implications for learning and adaptation. As the brain continuously adjusts its model to minimize prediction error, it learns about the environment. This learning is the brain's way of becoming more efficient at predicting the world, which leads to better decision-making and behavior.

Neurobiologically, predictive coding is supported by the structure and function of the brain's cortex. Neurons in higher cortical layers are suggested to represent the predictions, while lower layers represent the prediction errors. These layers interact dynamically to update and refine predictions through synaptic weight adjustments, reflective of learning. Predictive coding has implications not only for neuroscience but also for psychology and cognitive sciences, explaining phenomena such as perception, motor control, and even psychiatric disorders. Predictive coding offers a comprehensive and computationally detailed account of how the brain might function as a prediction and error correction machine, within the Free Energy Principle.

Friston's framework thus provides a biological and theoretical foundation for discussing cognition as a process rooted in the principles of information theory and computational neuroscience. It bridges the gap between abstract computational theories and concrete biological mechanisms, offering a robust model that explains both the adaptiveness and the efficiency of cognitive processes in terms of energy minimization strategies.

Friston's work can be seen as both complementary to the interpretations of information, computation, and cognition discussed by Magnani, Dodig-Crnkovic, Piccinini, Shagrir, Miłkowski, and Levin. Each brings a unique perspective, but together, they enhance our understanding of the complex interplay between the brain, its computational processes, and the information it handles. Magnani focuses on how cognition extends beyond the brain to include interactions with the world. Piccinini emphasizes the mechanistic nature of computation in explaining cognitive functions. Shagrir elaborates on computational theories to account for various cognitive capabilities. Miłkowski provides a precise, mechanistic understanding of how computational processes underpin cognition. From the biological point of view, Levin expands the conventional understanding of cognition by studying how cells and non-neural organisms process information and exhibit behavior that suggests basal cognition and intelligence. Dodig-Crnkovic integrates information and computation as structure and process aspects into a naturalist framework. Their work collectively enriches the discourse of these foundational concepts, offering diverse yet complementary perspectives that advance our understanding of the mind and its processes.

4.6 Future Developments, Cognitive Computing, and AI

Eco-cognitive computationalism, proposed by Magnani, studies the role of the environment and external representations in shaping cognitive processes. It suggests that cognition is not only an internal process but involves interactions with the external world, including the manipulation of physical objects and representations. This perspective aligns with the embodied and situated cognition paradigm, which recognizes the importance of an agent's physical form and environmental interactions in shaping cognitive behavior. Similarly, the ICON (info-computationalism or computational naturalism) framework, discussed before, highlights the concept of morphological computation, where the physical structure and embodied interactions of agents shape their computational processes.

In the context of AI development, particularly for large language models (LLMs), eco-cognitive computationalism could inform the design of more grounded and context-aware systems. LLMs are trained on vast amounts of textual data, which means disconnected from the physical world of embodied experiences.

By incorporating principles from Eco-cognitive computationalism, LLMs could be designed to integrate external representations, such as images, diagrams, or physical objects, into their cognitive processes. This would involve multimodal learning, where the model learns to associate textual information with visual or other sensory inputs, potentially leading to more grounded and contextually relevant language understanding and generation.

Furthermore, Eco-cognitive computationalism emphasizes the role of external representations in cognitive offloading and distributed cognition. This could inspire the development of interactive AI systems that leverage external representations and tools to augment their cognitive capabilities. The Eco-cognitive perspective could inform the development of AI systems that can better adapt to and interact with dynamic environments, as it recognizes the importance of environmental factors in shaping cognitive processes. The principles of embodied and situated cognition, as well as the integration of external representations and environmental interactions, could potentially contribute to the development of more grounded, context-aware, and adaptive AI systems.

Similarly, the application of ICON framework to AI systems, with the principles of embodied and situated cognition, hierarchical organization, self-organization, information dynamics, morphological computation, and constructive approach offers insights and perspectives for rethinking the development of AI systems, including LLMs, to better capture the embodied and adaptive nature of intelligence observed in natural systems. Hierarchical Organization and Self-Organization could inform the development of architectures that incorporate hierarchical representations and self-organizing mechanisms, enabling the emergence of higher-level cognitive capabilities from lower-level interactions. Such architectures could potentially exhibit more flexible and adaptive behavior, akin to the self-organization observed in natural systems. The ICON framework adopts a constructivist approach, where meaning and semantics emerge from the interactions between agents and their environments, rather than being predefined or imposed externally. For LLMs, this could inspire the development of models that learn to construct meaning and semantics through interactions with multimodal data and environments, rather than relying solely on predefined symbolic representations. Such models could potentially exhibit more flexible and context-aware language understanding and generation capabilities. The concept of morphological computation could inform the design of LLMs that leverage physical or virtual embodiments to shape their computational processes. Natural computation is known for its energy efficiency and energy consumption and resource efficiency may be the decisive factors for the future of AI.

4.7 Computationalism, Its Varieties, Its Critics and Its Raison D'être

The current article presents a new approach to cognition, framed in terms of information and computation within the broader context of computing nature, specifically focusing on morphological (physical, embodied) computation. It situates Lorenzo Magnani's Eco-cognitive computationalism within this wider discourse. Given the persistent critiques against computationalism since its inception, one might wonder how anyone could venture to build upon the "quicksand" of computationalism—particularly in its pancomputationalist form.

Computationalism in different forms has been criticized by so many, that it is impossible to address all those criticisms. Gualtiero Piccinini e.g. provides a systematic critique of limited, unlimited, and ontic Pancomputationalism (Piccinini, 2017) while the new book by Piccinini and Anderson provides a critique of Ontic Pancomputationalism (Anderson & Piccinini, 2024). Piccinini advocates his own mechanistic version of computationalism with the requirement of "robust mapping". Under this interpretation not even neurons compute as they do a variety of other life-sustaining tasks which do not directly contribute to cognitive computing as described in Neural Computation and the Computational Theory of Cognition by the same author, (Piccinini & Bahar, 2013).

Nir Fresco in Physical Computation and Cognitive Science (Fresco, 2014), and Marcin Miłkowski (2013a, b), navigate carefully avoiding Scylla and Charybdis of pancomputational grounding, staying with the proposals that always leave space for *processes in nature that are not computational.*

Dominic Horsman, Viv Kendon and Susan Stepney propose a concrete schema to explicitly define conditions under which physical system computes in (Horsman et al., 2017). Similarly, the recent work done by physicists proposes "a category theory-based framework to redefine physical computing in light of advancements in quantum computing and non-standard computing systems" (Dehghani & Caterina, 2024) asking what physical systems can be considered computers and what constraints enable physical systems to compute.

The Church-Turing-Deutsch principle, posits computation as inherently physical (Deutsch, 1985).

On the other hand, Giuseppe Longo compares (certain applications of) computationalism to scientism and mysticism (Longo, 2020). He critiques symbolic, discrete computation but embraces geometry (and by extension, continuous computation) as crucial for understanding embodied cognition (Longo, 2024). Longo suggests that cognition cannot be reduced to symbolic manipulation but instead involves continuous, analog processes that are central to how humans perceive and act in the world.

There are even authors defending new directions of computationalism, like Matthias Scheutz (2002) who in an edited collection of essays shows how computationalism evolved from the simplistic idea that the mind is computable on the symbol manipulating Turing-Machine, to concurrent forms of embodied computation beyond the abstract Turing Machine model.

One recurrent theme is that the assertion "*all of nature continuously computes*" is a vacuous statement. Similar holds for the statement that "*all of nature is an informational structure*", a position known as Informational Structural Realism (ISR).

In the dialogue article (Dodig-Crnkovic & Miłkowski, 2023) I replied to Marcin Miłkowski's objections against computing nature/natural computationalism. Likewise, in the earlier dialogue article, Vincent Muller presented similar objections to pancomputationalism (Dodig-Crnkovic & Müller, 2011). The main stumbling block is that the assertion: "*nature is an info-computational phenomenon for an observer*" evokes fears of a "grey goo" vision where no distinctions can be made in that info-computational soup. However, I defend the position that natural computationalism is not trivial but a very productive approach where our present computing machinery is used to leverage discovery processes in nature organized in variety of levels of abstraction, with the help of a more general idea of natural computation as envisaged in (Rozenberg et al., 2012) and (Rozenberg & Kari, 2008).

In 2008 Luciano Floridi wrote 'A Defense of Informational Structural Realism' (Floridi, 2008a) article defending ISR as a viable and coherent philosophical position that reconciles the strengths of both Epistemic Structural Realism (ESR) and Ontic Structural Realism (OSR), arguing that the world can be understood as a totality of informational objects interacting within a structural framework. Info-computationalism adds to this "pan-informational" structure its "pan-computational" dynamics.

In conclusion, the diversity of interpretations—ranging from mechanistic computationalism to natural computationalism—demonstrates the flexibility and relevance of this framework. The ongoing dialogue between critics and proponents shows that computationalism's raison d'être lies in its ability to provide new perspectives on cognition, nature, and the role of information. While critiques are numerous and necessary to refine the theory, the continued development of embodied, physical, and natural computational paradigms ensures that computationalism will remain a productive and evolving area of inquiry.

References

Anderson, N. G., & Piccinini, G. (2024). *The physical signature of computation*. Oxford University Press. https://doi.org/10.1093/9780191872075.001.0001

Arfini, S., & Magnani, L. (2022). *Embodied, extended, ignorant minds: New studies on the nature of not-knowing/edited by Selene Arfini, Lorenzo Magnani*. Springer.

Aubin, D. (2020). Forms of explanation in the catastrophe theory of René Thom: Topology, morphogenesis, and structuralism. In *Growing Explanations*. https://doi.org/10.1515/9780822390084-005

Baeyer von, H. (2004). *Information: the new language of science*. Harvard University Press.

Baluška, F., & Levin, M. (2016). On having no head: Cognition throughout biological systems. *Frontiers in Psychology, 7*, 902. https://doi.org/10.3389/fpsyg.2016.00902

Bateson, G. (1972). *Steps to an ecology of mind*. The University of Chicago Press.

Bongard, J., & Levin, M. (2023). There's plenty of room right here: Biological systems as evolved, overloaded, multi-scale machines. *Biomimetics, 8*(1), 110. https://doi.org/10.3390/biomimetics8010110

Burgin, M. (2010). *Theory of information: fundamentality, diversity and unification*. World Scientific Pub.

Burgin, M., & Dodig-Crnkovic, G. (2012). From the closed classical algorithmic universe to an open world of algorithmic constellations. In G. Dodig-Crnkovic & R. Giovagnoli (Eds.), *Computing nature* (pp. 1–24). Springer. https://doi.org/10.1007/978-3-642-37225-4_16

Chaitin, G. (2006). Epistemology as information theory. Alan Turing lecture given at E-CAP 2005. *Collapse, 1*, 27–51.

Chaitin, G. (2018). Building the world out of information and computation: Is god a programmer, not a mathematician? BT. In S. Wuppuluri & F. A. Doria (Eds.), *The map and the territory: Exploring the foundations of science, thought and reality* (pp. 431–438). Springer. https://doi.org/10.1007/978-3-319-72478-2_24

Clark, A. (1989). *Microcognition: Philosophy, cognitive science, and parallel distributed processing*. MIT Press.

Clark, A. (1997). *Being there: Putting brain, body and world together again*. Oxford University Press.

Clark, A. (2008). *Supersizing the mind embodiment, action, and cognitive extension*. Oxford University Press.

Clark, A. (2016). *Surfing uncertainty: Prediction, action, and the embodied mind*. Oxford University Press.

Clark, A., & Chalmers, D. J. (1998). The extended mind. *Analysis, 58*(1), 7–19.

Cooper, S. B. (2015). "Real" information is not flat – And why it matters. *Journal of Experimental and Theoretical Artificial Intelligence, 27*(1). https://doi.org/10.1080/0952813X.2014.940694

Corning, P. A., Kauffman, S. A., Noble, D., Shapiro, J. A., Vane-Wright, R. I., & Pross, A. (2023). *Evolution "on purpose": Teleonomy in living systems*. The MIT Press. https://doi.org/10.7551/mitpress/14642.001.0001

Davies, P., & Gregersen, N. H. (2010). *Information and the nature of reality from physics to metaphysics* (pp. 1–398). Cambridge University Press.

Dehghani, N., & Caterina, G. (2024). Physical computing: A category theoretic perspective on physical computation and system compositionality. *Journal of Physics: Complexity, 5*(3), 035005. https://doi.org/10.1088/2632-072X/ad6051

Denning, P. (2003). Great principles of computing. *ACM Communications, 46*(11), 15–20.

Denning, P. (2007). Computing is a natural science. *Communications of the ACM, 50*(7), 13–18. http://cs.gmu.edu/cne/pjd/PUBS/CACMcols/cacmJul07.pdf

Denning, P. (2009). Computing: The fourth great domain of science. *Communications of the ACM, 52*. https://doi.org/10.1145/1562164.1562176

Denning, P. (2010). What is computation?: Editor's Introduction. *Ubiquity*, October, 1–2.

Deutsch, D. (1985). Quantum theory, the Church-Turing Principle and the universal quantum computer. *Proceedings of the Royal Society of London A, 400*, 97–117.

Dodig-Crnkovic, G. (2007). Epistemology naturalized: The info-computationalist approach. *APA Newsletter on Philosophy and Computers, 06*(2), 9–13.

Dodig-Crnkovic, G. (2008). Semantics of information as interactive computation a brief, sketchy picture of infocomputationalism. In M. M. W. Neuser & T. Roth-Berghofer (Eds.), *Philosophy* (pp. 1–48). Springer, DFKI Technical Reports. 91.

Dodig-Crnkovic, G. (2011a). Info-computational philosophy of nature: An informational universe with computational dynamics. In C. P. and T. T. Sørensen Bent (Ed.), *Festschrift for Søren Brier* (pp. 97–129). CBS University Press.

Dodig-Crnkovic, G. (2011b). Significance of models of computation, from turing model to natural computation [Article]. *Minds and Machines, 21*(2), 301–322. https://doi.org/10.1007/s11023-011-9235-1

Dodig-Crnkovic, G. (2012). Info-computationalism and morphological computing of informational structure. In P. L. Simeonov, L. S. Smith, & A. C. Ehresmann (Eds.), *Integral Biomathics* (Issue Sloman 1996, pp. 97–104). https://doi.org/10.1007/978-3-642-28111-2_10

Dodig-Crnkovic, G. (2013). The Info-computational nature of morphological computing. In V. Müller (Ed.), *Philosophy and theory of artificial intelligence. Studies in applied philosophy, epistemology and rational ethics* (Vol. 5, pp. 59–68). https://doi.org/10.1007/978-3-642-31674-6_5

Dodig-Crnkovic, G. (2014a). Info-computational constructivism and cognition. *Constructivist Foundations, 9*(2), 223–231. http://search.ebscohost.com.lib.kaplan.edu/login.aspx?direct=true&db=a9h&AN=96034017&site=eds-live

Dodig-Crnkovic, G. (2014b). Modeling life as cognitive info-computation. In A. Beckmann, E. Csuhaj-Varjú, & K. Meer (Eds.), *Computability in Europe 2014. LNCS* (pp. 153–162). Springer.

Dodig-Crnkovic, G. (2017). Nature as a network of morphological infocomputational processes for cognitive agents. *The European Physical Journal Special Topics, 226*(2), 181–195. https://doi.org/10.1140/epjst/e2016-60362-9

Dodig-Crnkovic, G., & Lowe, R. (2017). Morphological computing and cognitive agency. *Proceedings, 1*(3), 185–188. http://www.es.mdh.se/publications/4959-

Dodig-Crnkovic, G., & Miłkowski, M. (2023). Discussion on the relationship between computation, information, cognition, and their embodiment. *Entropy, 25*(2), 310. https://doi.org/10.3390/E25020310

Dodig-Crnkovic, G., & Müller, V. (2011). A dialogue concerning two world systems: Info-computational vs. mechanistic. In G. Dodig-Crnkovic & M. Burgin (Eds.), *Information and computation 2011*. World Scientific Publishing. Series in Information Studies. https://doi.org/10.1142/9789814295482_0006

Ehresmann, A. C. (2014). A mathematical model for info-computationalism. *Constructivist Foundations, 9*(2), 235–237.

Fields, C. (2012). If physics is an information science, what is an observer? *Information, 3*(1), 92–123.

Fields, C., & Levin, M. (2020). Integrating evolutionary and developmental thinking into a scale-free biology. *BioEssays, 42*(8), e1900228. https://doi.org/10.1002/bies.201900228

Floridi, L. (2003). Informational realism. In J. Weckert & Y. Al-Saggaf (Eds.), *Selected papers from conference on Computers and philosophy – Volume 37 (CRPIT '03)* (pp. 7–12). Australian Computer Society.

Floridi, L. (2008a). A defense of informational structural realism. *Synthese, 161*(2), 219–253.

Floridi, L. (2008b). The method of levels of abstraction. *Minds and Machines, 18*(3), 303–329. https://doi.org/10.1007/s11023-008-9113-7

Floridi, L. (2011). *The philosophy of information*. Oxford University Press.

Fredkin, E. (1990). Digital mechanics: An information process based on reversible universal cellular automata. *Physica D, 45*, 254–270.

Fresco, N. (2014). *Physical computation and cognitive science*. Springer.

Friston, K., & Kiebel, S. (2009). Predictive coding under the free-energy principle. *Philosophical Transactions of the Royal Society B: Biological Sciences, 364*(1521), 1211–1221. https://doi.org/10.1098/rstb.2008.0300

Friston, K. J., Daunizeau, J., Kilner, J., & Kiebel, S. J. (2010). Action and behavior: A free-energy formulation. *Biological Cybernetics, 102*, 227–260. https://doi.org/10.1007/s00422-010-0364-z

Ghazi-Zahedi, K., Langer, C., & Ay, N. (2017). Morphological computation: Synergy of body and brain. *Entropy, 19*(9), 456. https://doi.org/10.3390/e19090456

Goyal, P. (2012). Information physics—Towards a new conception of physical reality. *Information, 3*, 567–594.

Hauser, H., Füchslin, R. M., & Pfeifer, R. (2014). *Opinions and outlooks on morphological computation* (e-book). https://tinyurl.com/4adt9j75

Hewitt, C. (2007). What is commitment? Physical, organizational, and social. In P. Noriega, J. Vazquez-Salceda, G. Boella, O. Boissier, & V. Dign (Eds.), *Coordination, organizations, institutions, and norms in agent systems II* (pp. 293–307). Springer.

Hewitt, C. (2012). What is computation? Actor model versus Turing's model. In H. Zenil (Ed.), *A computable universe, understanding computation & exploring nature as computation* (pp. 159–187). World Scientific Publishing Company/Imperial College Press. https://doi.org/10.1142/9789814374309_0009

Horsman, D., Kendon, V., & Stepney, S. (2017). The natural science of computing. *Communications of the ACM, 60*(8). https://doi.org/10.1145/3107924

Isomura, T., Kotani, K., Jimbo, Y., & Friston, K. J. (2023). Experimental validation of the free-energy principle with in vitro neural networks. *Nature Communications, 14*(1), 4547. https://doi.org/10.1038/s41467-023-40141-z

Laland, K. N., Uller, T., Feldman, M. W., Sterelny, K., Müller, G. B., Moczek, A., Jablonka, E., & Odling-Smee, J. (2015). The extended evolutionary synthesis: Its structure, assumptions and predictions. *Proceedings of the Royal Society B: Biological Sciences, 282*, 1–14. https://doi.org/10.1098/rspb.2015.1019

Levin, M. (2019). Modeling somatic computation with non-neural bioelectric networks. *Scientific Reports, 9*, 18612. https://doi.org/10.1038/s41598-019-54859-8

Levin, M. (2020). Life, death, and self: Fundamental questions of primitive cognition in biological systems. *Biochemistry and Cell Biology, 98*(5), 654–664.

Levin, M. (2022). Technological approach to mind everywhere: An experimentally-grounded framework for understanding diverse bodies and minds. *Frontiers in Systems Neuroscience, 16*, 768201. https://doi.org/10.3389/fnsys.2022.768201

Levin, M. (2023). Darwin's agential materials: Evolutionary implications of multiscale competency in developmental biology. *Cellular and Molecular Life Sciences, 80*(6), 1–33. https://doi.org/10.1007/S00018-023-04790-Z

Levin, M., & Pezzulo, G. (2016). Top-down models in biology: Explanation and control of complex living systems above the molecular level. *Journal of the Royal Society Interface, 13*(124), 20160555. https://doi.org/10.1098/rsif.2016.0555

Lloyd, S. (2006). *Programming the universe: A quantum computer scientist takes on the cosmos*. Knopf.

Longo, G. (2020). *Information, science and democracy, for an ethics of scientific knowledge* (pp. 63–79). https://doi.org/10.1007/978-3-030-44018-3_5

Longo, G. (2024). *Geometry and cognition from the foundations of mathematics to theories of knowledge and cognition* (pp. 263–273). https://doi.org/10.1007/978-3-031-51993-2_15

Magnani, L. (2018). Eco-cognitive computationalism: From mimetic minds to morphology-based enhancement of mimetic bodies. In *Entropy* (Vol. 20, Issue 6, p. 430). https://doi.org/10.3390/e20060430

Magnani, L. (2021). Computational domestication of ignorant entities. *Synthese, 198*(8), 7503–7532. https://doi.org/10.1007/s11229-020-02530-5

Magnani, L. (2022a). AlphaGo, locked strategies, and eco-cognitive openness. In *Cognitive systems monographs* (Vol. 43). https://doi.org/10.1007/978-3-030-81447-2_3

Magnani, L. (2022b). *Eco-cognitive computationalism – Cognitive domestication of ignorant entities* (Vol. 43). Springer. https://doi.org/10.1007/978-3-030-81447-2

Magnani, L. (2025). Information, computation, and cognition. Their emergence and disappearance, and the ever-changing entanglement of their meaning. In M.J. Schroeder and & W. Hofkirchner (Eds.), *Understanding information and its role as a tool. In memory of mark Burgin.*

Manicka, S., & Levin, M. (2019). Modeling somatic computation with non-neural bioelectric networks. *Scientific Reports, 9*, 18612. https://doi.org/10.1038/s41598-019-54859-8

Manicka, S., & Levin, M. (2022). Minimal developmental computation: A causal network approach to understand morphogenetic pattern formation. *Entropy, 24*(1), 107. https://doi.org/10.3390/e24010107

Maturana, H., & Varela, F. (1980). *Autopoiesis and cognition: The realization of the living*. D. Reidel Pub. Co.

McMillen, P. T., & Levin, M. (2024). Collective intelligence: A unifying concept for integrating biology across scales and substrates. *Communications Biology, 7*, 378. https://doi.org/10.1038/s42003-024-06037-4

Miłkowski, M. (2013a). Computational theory of mind. In *Internet encyclopedia of philosophy*.

Miłkowski, M. (2013b). *Explaining the computational mind*. MIT Press.

Müller, G. B. (2017). Why an extended evolutionary synthesis is necessary. *Interface Focus, 7*, 201700152 0170015.

Noble, D. (2002). The rise of computational biology. *Nature Reviews Molecular Cell Biology, 3*(6), 459–463. https://doi.org/10.1038/nrm810

Noble, D. (2012). A theory of biological relativity: No privileged level of causation. *Interface Focus, 2*(1), 55–64. https://doi.org/10.1098/RSFS.2011.0067

Noble, D. (2016a). *Dance to the tune of life: Biological relativity*. https://doi.org/10.1017/9781316771488

Noble, D. (2016b). The music of life: Biology beyond the genome. In *Lavoisierfr*.

Noble, D. (2020). Charles Darwin, Jean-Baptiste Lamarck, and 21st century arguments on the fundamentals of biology. Progress in Biophysics and Molecular Biology, 153, 1–4. https://doi.org/10.1016/j.pbiomolbio.2020.02.005

Paun, G., Petre, I., Rozenberg, G., & Salomaa, A. (2017). At the intersection of computer science with biology, chemistry and physics – In Memory of Solomon Marcus. *Theoretical Computer Science, 701*, 1–234.

Pezzulo, G., & Levin, M. (2015). Re-membering the body: Applications of computational neuroscience to the top-down control of regeneration of limbs and other complex organs. *Integrative Biology, 7*(12), 1487–1517. https://doi.org/10.1039/C5IB00221D

Pfeifer, R., & Gomez, G. (2009). Morphological computation – Connecting brain, body, and environment. In K. B. Sendhoff, O. Sporns, E. Körner, H. Ritter, & K. Doya (Eds.), *Creating brain-like intelligence: From basic principles to complex intelligent systems* (pp. 66–83). Springer.

Pfeifer, R., & Iida, F. (2005). Morphological computation: Connecting body, brain and environment. *Japanese Scientific Monthly, 58*(2), 48–54.

Pfeifer, R., Iida, F., & Gomez, G. (2006). Morphological computation for adaptive behavior and cognition. *International Congress Series, 1291*, 22–29.

Piccinini, G. (2012). Computation in physical systems. In *The Stanford encyclopedia of philosophy* (Fall 2012).

Piccinini, G. (2017). Computation in physical systems. In *The Stanford encyclopedia of philosophy* (Fall 2017, pp. 27–47).

Piccinini, G. (2020). *Neurocognitive mechanisms: Explaining biological cognition*. Oxford scholarship online.

Piccinini, G., & Bahar, S. (2013). Neural computation and the computational theory of cognition. *Cognitive Science, 37*(3), 453–488. https://doi.org/10.1111/cogs.12012

Piccinini, G., & Scarantino, A. (2011). Information processing, computation, and cognition. *Journal of Biological Physics, 37*(1), 1–38. https://doi.org/10.1007/s10867-010-9195-3

Piccinini, G., & Shagrir, O. (2014). Foundations of computational neuroscience. *Current Opinion in Neurobiology, 25*, 25–30.

Prigogine, I. (1980). *From being to becoming: Time and complexity in the physical sciences*. W.H. Freeman.

Prigogine, I., & Stengers, I. (1984). *Order out of Chaos: Man's new dialogue with nature*. Flamingo.

Rovelli, C. (2015). *Relative information at the foundation of physics*. https://doi.org/10.1007/978-3-319-12946-4_7

Rovelli, C. (2018). *Meaning and Intentionality = Information + Evolution*. https://doi.org/10.1007/978-3-319-75726-1_3

Rozenberg, G. (2008). Computer science, informatics and natural computing—personal reflections. In *New computational paradigms: Changing conceptions of what is computable* (pp. 373–379). Springer.

Rozenberg, G., & Kari, L. (2008). The many facets of natural computing. *Communications of the ACM, 51*, 72–83. https://doi.org/10.1145/1400181.1400200

Rozenberg, G., Bäck, T., & Kok, J. N. (2012). *Handbook of natural computing*. Springer. https://doi.org/10.1007/978-3-540-92910-9

Scheutz, M. (2002). *Computationalism new directions*. MIT Press.

Seife, C. (2006). *Decoding the universe: How the new science of information is explaining everything in the cosmos, from our brains to black holes*. Viking.

Shagrir, O. (2010). Brains as analog-model computers. *Studies in History and Philosophy of Science Part A, 41*(3), 271–279. https://doi.org/10.1016/j.shpsa.2010.07.007

Shagrir, O. (2022). *The nature of physical computation*. Oxford University Press. https://doi.org/10.1093/oso/9780197552384.001.0001

Stewart, J. (1996). Cognition = life: Implications for higher-level cognition. *Behavioral Processes, 35*, 311–326.

Torday, J., & Miller, W. (2020). *Cellular-molecular mechanisms in epigenetic evolutionary biology*. Springer. https://doi.org/10.1007/978-3-030-38133-2

Torday, J., Miller, W., Torday, J., & Miller, W. (2020). Four domains: Cognition-based evolution. In *Cellular-molecular mechanisms in epigenetic evolutionary biology*. https://doi.org/10.1007/978-3-030-38133-2_13

Turing, A. M. (1948). Intelligent machinery. In D. C. Ince (Ed.), *Mechanical intelligence* (pp. 107–127). Elsevier.

von Uexküll, J. (1957). A stroll through the worlds of animals and men. In C. Schiller (Ed.), *Instinctive behavior* (pp. 5–80). International Universities Press.

Vedral, V. (2010). *Decoding reality: The universe as quantum information*. Oxford University Press.

Wheeler, J. A. (1990). Information, physics, quantum: The search for links. In W. Zurek (Ed.), *Complexity, entropy, and the physics of information*. Addison-Wesley.

Wheeler, J. A. (1994). *It from bit, at home in the universe*. American Institute of Physics.

Witzany, G. (2012). Viruses: Essential agents of life. In *Viruses: Essential Agents of Life*. Springer Netherlands. https://doi.org/10.1007/978-94-007-4899-6

Wolfram, S. (2002). *A new kind of science*. Wolfram Media. https://www.wolframscience.com/nks/

Wolfram, S. (2020). *A project to find the fundamental theory of physics*. Wolfram Media.

Zuse, K. (1970). *Calculating space. Translation of "Rechnender Raum"*. MIT Technical Translation.

Chapter 5
The Nature of Social Facts. Collective Intentionality, Status Function and the Role of an "Eco-cognitive" Perspective

Raffaela Giovagnoli

Abstract My contribution concerns the ontological state of social facts as objects that fulfill social functions and those social facts that become institutional facts. Fundamental insights stem from John Searle's philosophy of society presented in Making the Social World as an extension of the book The Construction of Social Reality. Classically, Searle distinguishes between brute physical facts and mental facts caused by the brain/world relationship. Mental facts can be unconscious or conscious. In the latter case, conscious mental states are intentional and can occur individually or collectively. Social facts depend on collective intentionality; for example, the value of a five-dollar bill is a social fact that depends on our agreement that a five-dollar bill is worth something. Social facts acquire a certain social status through the assignment of a status function (y counts as x in the context c) and require collective intentionality.

Despite the assumption of a background of abilities and skills described in the book Intentionality, research on the nature of social facts may require an analysis of the mental processes carried out by a plausible model of extended mind, that investigates the human manipulation of the environment at the basis of the same status function. A promising vision shows the important relationship between abduction and extension, disembodiment and distribution of the mind to grasp the role of habits, rituals and symbols in human cognition. Lorenzo Magnani's work, presented in Abduction, Reason and Science and in other important writings, introduces a fundamental "eco-cognitive" perspective that is based on the notion of "manipulative abduction"; the relationship between internal and external representations underlines the role of the environment in human cognition. Extending, disembodying and distributing the mind means that the human mind can delegate to external representations important cognitive functions already present in our environment.

R. Giovagnoli (✉)
Pontifical Lateran University, Vatican City, Italy
e-mail: giovagnoli@pul.it

S. Arfini (ed.), *Scientific Cognition, Semiotics, and Computational Agents: Essays in Honor of Lorenzo Magnani - Volume 2*, Synthese Library 506,
https://doi.org/10.1007/978-3-031-96688-0_5

5.1 Introduction

The concept of "social fact" is implicitly present in the thought of several classical sociologists from Saint-Simon to Marx, who strongly emphasized the objective autonomy of the social being. Along this line, the sociologist Emile Durkheim formally introduced the term "social fact" into modern sociology. He indicated four criteria to define a social fact. From a research perspective, social facts are characterized, in the first place, by their remarkable to exist from individual consciousness as an autonomous reality. We can consider examples of social facts: legal and moral rules, dogmas and religious organizations, political and literary schools, professional societies, financial systems and some forms of unorganized social life. Secondly, social facts possess, by their intrinsic nature, an imperative and coercive power by virtue of which they impose themselves on individuals. For example, the rules of law can be ignored or violated; however they work to prevent an action, if still possible, so as to cancel it and restore it in the standard form, otherwise they act as sanctions. A similar pressure of varying intensity is exerted by moral norms, customs, public opinion and conventions.

Third, a social fact must be considered as a "thing" that as such offers and imposes itself on the observer. Social facts must be studied from an external point of view detached from conscious individuals who represent them in their minds. Finally, society is a system that has its own specific nature and social facts are created by other social facts; the specific nature of society resides in forms of association that force individuals towards a specific elaboration from which a new form of existence is created. We can observe a sort of "emergence" of social facts that are obviously linked to psychological and physical phenomena, but the level of what emerges from them has an autonomous and distinct reality.

Along this line of thought, important authors of North American pragmatism such as Stephen Cohen and Arthur Bentley elaborate a notion of social fact essentially linked to communicative and linguistic processes. The French sociologist Adolphe Coste used the term social fact very rigorously. Unlike Durkheim, Coste separated social facts from ideological facts. In this context, social facts are governments, economics, religious practices, some forms of solidarity (such as territorial communities, social classes) and ideological facts are philosophy, the arts, political ideologies, literature, theology. Social facts are the object of sociology and ideological facts should have been investigated by a new science, namely "Ideology". This dichotomy recalls Marx's distinction between "structure" and "superstructure", between "civilization" and "culture" or between "real factors" and "ideal factors".

The field that investigates social facts from a philosophical point of view is "social ontology" which presents a lively debate in contemporary thought. However, we can find arguments about the nature of the social world in classical philosophical thought as part of metaphysical investigations. Social ontology studies concern entities in the world that derive from social interaction. Social ontology investigates a wide range of social entities such as money, sex and gender, social groups and group cognition. But, we can include language, norms, family, law, societies, art and artifacts. To schematize, there are several theories on the constitution of social entities that move from contrasting perspectives. Some theories argue that entities are constructed by people's psychological states; others believe that they are constructed by human actions or practices.

We start our discussion of social facts with a brief reference to studies on collective intentionality in the field of social ontology, because they are at the center of important theories. Collective intentionality is our shared direction towards an object and is fundamental for the constitution of the social world (Jankovic & Ludwig, 2018, p.1):

> It subsumes centrally the intentionality of collectives and intentional phenomena involving groups of agents in contexts in which they are acting or oriented toward action as a group.

We can trace the notion of collective intentions back to Plato's *Republic*, in which he speaks of the cooperative determination of social order and laws by the members of a society. Hobbes and Locke consider this form of collective action essential for the social contract. We can find a brilliant analysis of collective intentions in the contemporary thought of the philosopher of science Wilfrid Sellars.

In the ongoing debate on collective intentionality, Raimo Tuomela and Kaarlo Miller played a fundamental role in the discussion of the conditions to be met to make a collective act. Each member of the group must intend to do his part; he must believe that he can do his part and the other members also intend to do their part. Finally, each member believes that the other members can also play their part. Tuomela and Miller's accounts are considered "individualistic" because collective intentionality originates from the individual intentions of the group members. Michael Bratman also maintained an individualist position and presents an account of collective intentionality in terms of "shared cooperative activity". Just as an individual can plan to act alone, that same individual can also act according to a plan shared by a group. The shared cooperative activity could also involve sub-plans compatible with the common shared program and do not involve conflicts between the individual sub-plans.

Some important theories are based on the thesis of the "irreducibility" of We-intentionality, that is to say intentionality is not reducible to I-intentionality. The social phenomenologist Alfred Schütz studied the constitution of the social world starting from social action based on a culturally shaped and temporally segmented common lifeworld. In the contemporary debate on collective intentionality, John Searle and Margaret Gilbert introduced the most relevant theories for the argument of "irreducibility". Gilbert investigated coordination in the conversation as an example of joint activity that cannot be reduced to individual intentionality. There is a joint commitment that characterize collective intentionality that affects human beings who possess a sort of "natural" willingness to cooperate in the social world, reminiscent of Rousseau's notion of collective will. John Searle starts from a similar notion of collective will in which the individual does his part in a collective enterprise and has his own individual mental content only because there is a shared goal they want to achieve together. Individual content is in a sense "parasitic" in terms of collective intentionality.

5.2 Searle on Social Facts: We-intentionality, Status Function and Institutional Facts

First, we consider Searle's perspective on the constitution of the social world as the set of social facts in general and, secondly, we discuss the nature of institutional facts that are fundamental to the social order.

The main question about the constitution of social reality is the following (Searle, 2010, p. IX):

> How can there be an objective world of money, property, marriage, governments, elections, football games, cocktail parties and las courts in a world that consists entirely of physical particles in fields of force, and in which some of these particles are organized into systems that are conscious biological beasts, such as ourselves?

The Construction of Social Reality could be considered as a proposal of a special theory. In particular, Searle extended research on the constitutive rule "X counts as Y in C" as a status function indicating that a person or object has a certain social status in the appropriate context. This is sufficient to explain the constitution of the social world. In *Making the Social World* we find also a deep analysis of the role of language in the process of constitution of the social world that leads to the centrality of the speech act of *Declaration* of the status function and other forms that derive from this basic notion. First, we observe that cases such as a tribe that considers a stone line as a border (without having a general constitutive rule) is only a step before introducing new institutional facts such as the election of a leader in this particular territory. Second, in response to the critical remarks raised by Barry Smith, Searle considers "independent Y terms" that occur when a status function is created without there being an existing person to be imposed this status function. In the case of a corporation, there is nothing corresponding to it but, because of the existence of people and norms and the corresponding deontic powers, we attribute existence to it. Similarly, in the case of electronic money we simply have an electronic representation of money. Thirdly, there are institutional facts that do not require collective recognition, but are unexpected effects or consequences of institutional events as in the case of an unexpected economic recession.

To grasp the process of constituting social reality as a set of social events, we must sketch the relationship between collective intentionality and status function. However, "constitutive rules" come into play when we use ordinary language through classes of speech acts. In the case of the constitution of social reality, the speech act of declaration is fundamental. Searle outlines that human beings along with many other social animals have collective intentionality (Searle, 2010, p. 43). It is a complex notion that we cannot reduce to I-intentionality, that is to say we cannot replace "We" with "I" even if all intentionality is in our brains. One person's intentionality can only vary on the actions that can be caused personally and in cooperative behavior we observe a sort of intentionality that goes beyond the range of one's causality. Since there are many forms of collective behavior in which we should cooperate in action, the content of what one is doing must be different from what others are doing to achieve what we are both trying to do together. There are important conditions that must be met to clearly characterize the CI (Searle, 2010, pp. 44–46):

- We must have clear distinction between prior intentions and intentions-in-actions
- We need to make it absolutely clear that the conditions of satisfaction of both prior intentions and intentions-in-action are causally self-referential
- All intentionality, whether collective or individual, has to exist inside individual's heads
- In cases of CI, each individual's intentionality cannot require that they know what the intentionality on the part of other is. In complex forms of teamwork or collective behavior, one typically does not know what the others are doing in detail
- The mutual knowledge about the intention that occurs between two people, when each intends, and each knows that thex other intends what they intend to do.

According to Searle, the fact that intentionality exists in individual brains does not imply that the content that exists in the individual brain cannot exist in a plural grammatical form. He gives the example of cleaning the yard together. In this case, I have the thought "we are cleaning the yard together" and you have the thought "we are cleaning the yard together" and when we decide to cooperate in a joint activity each of us does his part, that is to say he has the content suitable to start his action as I clean one part you clean the other.

Let's remember the famous case of the "business school" in which students venture into economic activities after graduation in two different contexts that exemplify the fact that acts involving people with mutual knowledge without cooperation cannot involve collective intentionality. Cooperation implies the existence of common knowledge or common belief, but common knowledge or belief together with individual intentions to achieve a common goal is not in itself sufficient for true cooperation.

Collective intentionality should move the individual bodies but it is not reducible to the I-intentionality. There are previous intentions before the start of an action and intentions in action that are components of actions (Searle, 2010, p. 51). Searle explains it better with the sample of raising my arm in which the prior intention makes me perform the action of raising my arm and the intention in action makes my arm rise. In reality there are two different explanations of a performance: by means of doing something else. We shoot a gun by pulling a trigger; but we vote by raising our arm. Pulling the trigger makes the gun fire, but raising the arm does not make the vote take place. In the joint activity, if our individual contribution causes some additional effect, then we have a causal relationship. Otherwise, if our efforts constitute a certain intentional result, we have a constitutive relationship. The case of preparing a sauce together can be considered as an example of causal by means relationship. In the case of performing a duet, in which I play the piano part and you play the violin part, our playing does not cause the performance but constitutes it. In the joint activity it is assumed that each agent must believe that the other member of the collective is doing his part.

It implies that there is a shared knowledge implicit in collective behavior. But it is not part of the content of the intentionality of a person who refers to things that she can cause. To engage in collective behavior one has to believe that others are collaborating with her. Their cooperation will consist of having intentions in action that specify the same goal, but they do not have to specify the same means for the goal. At the same time, we don't need to know what the contribution of others is. A person can only control their performance as in the case of the duet. They must assume that the other person plays the violin; the intention refers to things that the agent can achieve and cannot involve references to the actions of others (Searle, 2010, p. 56).

According to Searle, collective intentionality may have a "strong" or "weak" form. Collective recognition is weaker than true cooperation, but it is especially important for the functioning of institutional facts. A piece of paper is recognized and accepted as money when we accept the general institution of money and the institution of commerce. Institutional structures require collective recognition by participants to the institution to function, but particular transactions within the institution require genuine cooperation. The couple planning a wedding must recognize and accept the institution of marriage before the wedding. The current wedding ceremony is an example of genuine cooperation; we return to this point in the final session of this document.

The most important contribution of *Making the Social World* is the general theory of institutional and institutional facts, that is to say the results of a peculiar relationship between language and social reality (Searle, 2010, p. 91):

> Institutional facts range all the way from the informality of friendship to the extreme legal complexities of international corporation. The simplest test for whether a phenomenon or fact is genuinely institutional is to ask, Does its existence imply deontic powers such as those of rights, duties, obligations, requirements, and authorizations? There are (uncodified) rights and obligations of citizenship and employment. There are deontologies without institutional facts (I am for example under moral obligation to help people who are in desperate need of immediate help and whom I am able to help), but there are no institutional facts without some form of deontology. These institutional deontic powers have a common logical structure that I am to reveal in this chapter.

Probably the best way to isolate the general notion of social fact and not only to codified institutions (governmental institutions, economic sport, general purpose, structural institutions such as monetary marriage, government, private property) is to refer to types of what Searle calls non-linguistic institutional reality: informal unstructured institutions (mostly uncoded) (friendship, family, love affairs, parties); general forms of human activity that are not institutions (science, religion, recreation, literature, sex, nutrition) and professional activities that are not institutions but contain institutions (law, medicine, academia, theater, carpentry, retail trade).

The test to establish whether a noun appoints an institution is whether under that description the named object acquires deontic powers. For example, the Catholic Church is an institution, religion is not.

All institutional facts are created by the same logical operation: the creation of a reality representing it as existing:

> We make it the case by declaration that Y status function exists.

Status functions have simple forms like in the case of the tribe establishing a boundary to more complex cases like corporations. Declarations assign deontic powers to person and objects but without collective recognition they cannot be implemented in society. So, in addition to power creation operator (PCO) we need a collective or acceptance operator that enables that continued existence of the SF:

> We collectively recognize or accept (S has power (S does A)).
>
> A further step is to include both SF and PCO within the scope of collective recognition:
>
> We collectively recognize or accept (There exists Y in C, and because SRY (S has power (S does A)).

The formula means that we collectively recognize that a Y status exists in context C, and because a human subject S stands in suitable relationships R to the status function Y in C, we further recognize that S has the power to do A, the acts determined by the Y status function.

As noted above, we have different forms of collective recognition that is fundamental to the functioning of the institution. Collective intentionality in the form of full-fledged cooperation occurs when we actually create institutional or acquire a new status, such as the couple through marriage. But collective recognition occurs when we regularly follow codified procedures in everyday life such as using money to pay for anything. Searle also recognizes cases where we passively acknowledge decisions made by a political group or party.

The driving force of the constitution of social reality is the collective which is pre-linguistic, that is, it allows the very possibility of cooperation and coordination in linguistic situations. Intentionality is the basic notion for understanding the human representation of reality where the notion of representation is "functional" and not "ontological". This fact means that we do not have internal images of the world. We simply have unconscious and conscious mental states (convictions, desires, intentions and emotions) with different directions of fit as in the case of speech acts. The functioning of our intentional life is based on a network of intentional states; for example, I can't plan to go to the movies unless I have a lot of other beliefs and desires (like a belief about movies as forms of entertainment, the fact that I have to pay for a ticket etc.). In addition, Searle introduces the notion of a set of assumptions for the application of intentionality, including a set of skills. There are facts that we take for granted like how we can drive our car and we will travel on the surface of the earth. Most of what we take for granted, and our abilities, do not seem to be well described in terms of intentional states. Searle thinks we need a set of skills, abilities to apply those intentional states and this set is what he calls "background" as that know-how that allows us to carry out our intentions and apply our intentional states in general (Searle, 1983).

5.3 The Eco-Cognitive Perspective and Its Relevance for the Constitution of Social Facts

As we have seen, Searle introduced the notion of a background of ability and ability that shows our direct relationship with the outside world, so we only discover them through perception and action. He describes the mental states and speech acts with their corresponding directions of fit, that indicate their conditions of satisfaction. Individual and collective intentionality are based on the functioning of this complex apparatus and, for the constitution of social facts, we need status functions to attribute deontic states to objects and people. Both Searle and Magnani present a form of externalism. In the case of the use of a verb such as "open" we can have a literal or metaphorical meaning as we can "open a bottle" but only in a metaphorical sense can we "open a discussion" (Searle, 1983). Along this pragmatic view, Magnani would easily observe that we can somehow have internal representations of a certain form and, even without trying, we recognize that we cannot insert a round stick into a smaller, triangular hole.

Magnani's perspective on the relationship between mind and world emphasizes the role of our environment for important cognitive functions and actions. A possible description of the constitution of social facts is implicit in this perspective, but we can describe it from the analysis of artifacts in habits and rituals. This analysis is fundamental to my point of view on the processes that found social ontology. We are referring to a perspective that shows the important relationship between disembodiment and distribution of the mind to grasp the role of habits, rituals and symbols in human cognition. Magnani proposes an "eco-cognitive" perspective that is based on the notion of "manipulative abduction" in which the relationship between internal and external representations underlines the role of the environment in human cognition (Magnani, 2009). Extending, disembodying and distributing the mind means that the human mind can delegate to external representations important cognitive functions already present in our environment (Magnani, 2018, p. 147):

> The external representations can be merely mimetic (when they represent something already internally present in the mind, for example a triangle depicted in a blackboard), but also poetic, when externally represented aspects are manipulated with the aim of discovering something new by exploiting the specific constraints of the materiality of the external "mediator" chosen. Also, all fields of scientific knowledge are characterized by these kind of processes, as in the case of logic and mathematics (…) Cognitive action embedded in manipulative abduction generates information not yet available (for example in science the exploitation of external tools, devices, and instruments is widespread) with the aim of selecting and generating interesting hypotheses.

Magnani refers to Jung's psychoanalytic view to highlight a plausible relationship between rituals, habits and cognitive artifacts, including the role of tacit inferences

(Jung, 1968a). The human psyche is "differentiated" in Jung's terms in that it is connected to the interrelationship of internal and external representations. For example, ideas about religious gods or mysteries are externalized, so that they can be available to cognition, "hybridizing" them with suitable, the so-called "social objects" such as statues, icons, etc. External representations acquire important functions in human life because they "re-represent" the religious ideas carried by human life. We observe that human life is precious and is perceived as precarious in a singular way. External representations work to promote a "consolidation of consciousness" necessary to deal with the "dangers of the unconscious" always embedded in human actions and practices. The external representations stored and disembodied in the environment recall the symbols of the archetypal ideas of the well-known Jungian "collective unconscious" clearly visible in religious rituals. This is a particular view of how we apply functions to objects or people, namely state functions, and give rise to shared practices and social facts.

Another important point for our discussion is closely related to the problem of the complexity of the psyche; Jung gives more importance to "introspection" rather than "projection" to emphasize that the psyche has acquired its complexity from a series of introjections. Recalling the example of an icon of God, our neural networks work to make the external stimulation from the environment, which gave rise to that new meaning, completely superfluous. Magnani refers also to Charles Sanders Peirce's notion of "fixed belief" to clearly represent the meaning of what he calls "a fixed internal record of an external structure", which is always present even when the external structure of related origin is not present.

The manipulation of environment in human practices gives rise to new habits and rituals and represents the dissemination in the environment of what Jung calls "psychic energy", a kind of "semiotic" process like the Freudian libido (Jung, 1968b). The new activity is a "symbol" well exemplified by Magnani with the example of the Wachandi agriculture ritual, where we have the image of the "Mother Earth", the hole that represents the fertility of the land (Magnani, 2018, p. 151):

> (...) new artifactual process establishes new meanings, new possibilities of making knowledge and reasoning, with respect to the members of the collectivity that can pay attention to the ritual itself with its objects, motor aspects (dance), and the various kinds of organic and non-organic physical structures, that play significant roles. When interested humans will be able to internalize the new meanings their neural networks will be appropriately conditioned in such a way that Jung can say they will present an intrinsic "disposable energy" and we will not need further exposition to the external ritual to act as they act.

Symbols are "memory mediators" as they maximize abducibility so that people can recognize or infer by hypotheses regarding the original mental events. Magnani shows an important trait of abducibility, namely its "eco-cognitive-sensitivity". Abducibility presents important aspects:

1. *Optimization of situatedness* entails "exosomatization (or externalization of human bodily activities) that favors cognitive abductive processes of people who invented and who participate to a social practice for example a ritual.

2. *Maximization of changeability* regarding the data of the eco-cognitive situation, to favor new information and available knowledge.
3. The *flux of information* ought to be monitored as it gives rise to cyclically generated information during various stages.
4. *Multimodality* underlying the fact that abductive inferences are favored by different cognitive devices (propositions, diagrams, visualizations, simulations, artifacts, etc.).

Along the line of Jung's psychoanalytic approach, a further important aspect of Magnani's argument needs to be emphasized: the fact that symbols also work on an unconscious level. For example, some collective agencies could build and exploit suitable artifacts to encourage unconscious reactions in agents so that tacit habits are produced efficiently and unconsciously fixed. They are capable of acting tacitly in different pragmatic contexts. We can observe how this process works in everyday life where individuals are exposed to mass media or movements and political and social organizations that exploit a multimodal variety of "symbols" available to agents. We can conclude our observation on Magnani's perspective on habits and rituals by emphasizing the role of "manipulative abduction" for human knowledge and practices. We recognize the advantage of abductive processes in broadening knowledge. In the case of simple "guessing" on a state of affairs as in the case of the Wachandi ritual of agriculture mentioned by Magnani, we have habits based on ignorance. Otherwise, in the case of well-known and consolidated habits, already available and acquired as common rules of action, we can speak of knowledge-based habits. Suitable artifacts maximize abducibility as they reflect natural human activity that is in a sense "transposed" into a certain cultural context. The human being belonging to appropriate collectives can immediately collect and use new pieces of knowledge; this thesis reveals the very nature of "rituals", that is to say the fact that they become stable cultural references, (unequivocally available and shareable).

Magnani's research is mainly focused on manipulative abduction and its epistemological role in the philosophy of science. It sheds light on the relevance of artifacts in customs and rituals that play a fundamental role in the constitution of social facts in social ontology. We need a vision of habitual behavior that also shows its important function in contemporary societies to address the problems we encounter in our time (in particular climate change and the latest developments in AI). In this context, Magnani argues that the very discoverability in science is closely related to the sustainability of human creativity in an "eco-cognitive" perspective. From a plausible point of view that recalls the questions raised by Pope Francis in his encyclical *Laudato Si'* (Pope Francesco, 2015). Magnani claims the urgency of an "integral ecology" through the emphasis on the concept of abduction. It argues that an ecology of human creativity should take precedence over other needs, namely that the first ecological duty is to protect human creativity and support discovery. Improving discoverability will protect human creativity, and it is precisely human creativity that is an innovative form of abductive cognition that can promote other types of ecology (Magnani, 2022).

The role of habits and rituals in the eco-cognitive perspective suggests considering their constitutive function for social reality. Habitual behavior is essential to organize our activities in individual and social contexts. We focus on habits because they reduce the complexity of daily life and also on their corresponding activity in social life, where we take part in informal and institutionalized joint practices. Human beings cooperate to create and participate in social practices because they need to organize their lives together with other people to create common spaces that have different functions and meanings depending on the corresponding practice (for example, we all pay the ticket to take a train and many of us participate in religious rituals or similar activities).

Contemporary studies on habits in the field of neuroscience and psychology show different dimensions of habitual behavior. There is a shared tendency to go beyond the classic view of habits as simple routines to highlight their connection with goals. The integration between routine and target-directed behavior is at the center of an important study in social psychology. Wendy Wood notes that successful long-term self-regulation involves habitually engaging in actions that correspond to evaluated long-term goals; people exercise self-control to inhibit unwanted habits and repeat actions directed to goals in stable contexts in order to form the desired habits (Wood, 2016).

Ann Graybiel proposes a fruitful relationship between habits and goals because goals are explicitly present during the evolution and selection of action and the more they blur the more an action is repeated. A habit completely disengaged from a target becomes a stimulus-response pair for a non-human animal or a pathological trait for humans. Habits can be "neutral", "good" or "bad" where good habits seem to be those chosen to guide our behavior and bad habits are those that powerfully take control of our behavior. This categorization makes it possible to include goals as drivers of habits. Graybiel also argues that habits play an important role in social life; in this case they are "shaped" as mannerisms and rituals (Graybiel, 2008). Bernacer and Murillo highlight important results from an in-depth study of the Aristotelian analysis of habits in Nicomachean ethics. An acquired habit is an acquired disposition to perform certain types of actions; this disposition, usually acquired through the repetition of one or more actions, makes the exception of these actions faster, more spontaneous and autonomous from continuous supervision, which generally leads to better performance. If the habit increases the cognitive control of actions, a habit can be defined as learning; on the contrary, if it increases their rigidity, it is a habit as a routine (Bernacer & Murillo, 2014).

Routines and target-directed behavior characterize habits in both individual and social behavior. We create our habits by satisfying our basic needs and desires. But we are social beings and we must also organize our activities to participate in different social practices. Rituals have the important function of creating social spaces in which individuals can share emotions, experiences, values, norms and knowledge (Giovagnoli, 2020, 2023).

In her recent book Molly Farneth starts from the Aristotelian vision presented by Javier Bernacer and Jose Iniacio Murillo to highlight the ethical dimension of habits and rituals. Farneth refers rather to Saba Mahmood's description of women's participation in the mosque movement in Cairo, Egypt, which is an application of Aristotle's pattern of habit and virtue. Women teach lessons, conduct prayer meetings, engage in daily prayer practices and choose to wear the hijab. Their practices not only express their values; they shape and transform the women who undertake them into subjects with particular habits and dispositions. Therefore, they are the result of an intentional practice at the service of ethical training. The self-conscious conception of the habit that is taken in Islamic philosophy and influences women's understanding of what piety is and how a person becomes pious. The function of sharing experiences is fulfilled when there is a social space created by cooperation to achieve certain objectives (Farneth, 2023).

To conclude, we observe that rituals characterize human daily life (but also animals have their rituals) not only in the religious sphere. Rituals are characterized by two special features: "collective intentionality" that expresses our social nature and the "collective imposition" of a social status to an object or person. We use ordinary language to represent the state of affairs and norms, namely to understand what the conditions of satisfaction of different speech acts are (assertions, commands, promises etc.). Beyond the classical dimensions of syntax, compositionality and generativity, there is a fundamental dimension that generates public norms i.e. "deontology", which is characterized by the speech act of "declaration" (Searle, 2009). For example, if we say "This is my house" or "This is my coach", we not only represent a state of affairs, but we create a deontology that manifests itself in rights, obligations and duties as well as in the acceptance of the corresponding speech acts from the part of the interlocutors. This is the process by which a public deontology is created, namely public reasons for acting that are desire independent. Language not only describes something but it creates and "partially" constitutes what it at the same time describes and creates. Representations that are partially constitutive of institutional reality, namely the reality of government, private property, money, universities and cocktail parties, are essentially linguistic. What must be clarified is the sense of this "partially" constitutes, because language works on the basis of a prelinguistic dimension which embeds background capacities such as ability to cooperate, to act as a "We". Moreover, this very capacity is fundamental to share and constitute rituals.

We have highlighted the fundamental process of assigning functions to objects or to some non-physical entities, which is a form of symbolization aimed at creating institutional reality. This process is the basis of the institutionalization of rituals and works in every community, even if social practices in general are culturally characterized. As already observed, with the status function there are two basic notions which occur in the explanation of successful functioning and stability of institutional facts. The first is "cooperation" as a "strong" form of collective intentionality and the second is "collective recognition" as a "weak form" of it. We can understand these two forms of intentionality as corresponding to the notion of "flexibility", which implies voluntary control over our actions, and to the notion of "rigidity", which characterizes the mere rules following in the sense of routine behavior.

We can describe a very famous example of ritual (Searle often refers to), namely "marriage". Firstly, we must be moved to act in a certain way. Collective intentionality works when we want to do something together (we have a collective intention) so we can cooperate to achieve our common goal. Both collective recognition and cooperation are crucial for rituals, in our case marriage.

Now we can observe a social transition from one status to another that is performed through an institutionalized ritual:

1. We have "collective recognition", which means that the couple simply accepts the institution of marriage prior to actually getting married.
2. But, the actual marriage ceremony is an example of active cooperation, in which the couple enters in a new social situation acquiring new social statuses consequently.
3. This fact obtains by the performance of the speech act of promise.
4. The social context requires also the speech act of declaration from the part of the institutional figure who has the suitable deontic powers to celebrate the rite and to ascribe the new status to the couple.

5.4 Conclusion

We presented a brief introduction to the notion of "social fact" that starts from sociological research and is at the heart of the relevant philosophical theories in social ontology. In particular, we emphasized the role of assigning functions to objects and persons in Searle's philosophy of society. Collective intentionality is fundamental to isolate the motor of social processes and the complex structure of mental states and speech acts relevant to the construction of social and institutional facts. The eco-cognitive perspective introduced by Magnani could be considered as a plausible foundation for human knowledge in general and his account of the abduction opens up a very interesting perspective on how we assign functions to entities in our environment that become artifacts as mediators of memory. We form habits and rituals that show important epistemological and normative consequences for the ecology and creativity of human knowledge. In conclusion, we can propose a complementarity of the eco-cognitive perspective based on the extended, disembodied and distributed mind and the functionalist perspective of Searle based on the notion of status function that we use to create social and institutional facts. In addition, the dimensions of habitual behavior can provide a plausible basis for understanding collective intentionality and collective recognition that are fundamental in the creation and maintenance of social reality.

Acknowledgements I would like to thank Selene Arfini for the invitation to celebrate Lorenzo Magnani's Festschrift. I am warmly grateful to Lorenzo Magnani for our continuous and lively discussions on the very important issues he introduced in his brilliant works and for the unique possibility to publish some of my works in the Sapere Series he is directing at Springer.

References

Bernacer, J., & Murillo, J. I. (2014). The Aristotelian conception of habit and its contribution to human neuroscience. *Frontiers in Human Neuroscience, 3*(8), 883.

Farneth, M. (2023). *The politics of ritual*. Princeton University Press.

Giovagnoli, R. (2020). From habits to we-intentionality: Rituals as social habits. In R. Giovagnoli & R. Lowe (Eds.), *The logic of social practices* (pp. 185–199). Springer.

Giovagnoli, R. (2023). Habitual behavior: Reduction of complexity of human daily life. In R. Giovagnoli & R. Lowe (Eds.), *The logic of social practices II*. Springer.

Graybiel, A. (2008). Habits, rituals and the evaluative brain. *The Annual Review of Neuroscience, 31*, 359–387.

Jankovic, M., & Ludwig, K. (Eds.). (2018). *Introduction, The Routledge Handbook of Collective Intentionality*. Routledge.

Jung, C. G. (1968a). *Archetypes and the collective unconscious, in The Collected Works of C.G. Jung* (pp. 3–41). Princeton University Press.

Jung, C. G. (1968b). *On Psychic Energy, in The Collected Works of C.G. Jung*, pp. 355–384.

Magnani, L. (2009). *Abductive cognition, the epistemological and eco-cognitive dimensions of hypothetical reasoning*. Springer.

Magnani, L. (2018). *Ritual artifacts as symbolic habits, open information science* (p. 147). De Gruyter.

Magnani, L. (2022). *Discoverability. The urgent need of an ecology of human creativity*. Springer, Sapere.

Pope Francesco, Laudato Si', Lettera Enciclica 24 Maggio 2015.

Searle, J. (1983). *Intentionality. An essay on the philosophy of mind*. Cambridge University Press.

Searle, J. (1995). *The construction of social reality* (Vol. IX). The Free Press.

Searle, J. (2009). What is language. Some preliminary remarks. In R. Giovagnoli, *Prelinguistic practices, social ontology and semantics*, Special Issue Etica & Politica/Ethics & Politics.

Searle, J. (2010). *Making the social world*. Oxford University Press.

Wood, W. (2016). The role of habits in self-control. In K. Vohs & R. Baumeister (Eds.), *Handbook of self-regulation* (p. 104). Guilford.

Chapter 6
Problem-Solving in Mathematics with Theoric Transduction

Michael H. G. Hoffmann

Abstract Charles Peirce insisted that creativity is essential both in deductive and in non-deductive reasoning. For the latter, he developed the well-known inference form of abduction. Less studied with regard to creativity in deductive reasoning is in particular one concept: what he called late in his life a "theoric step" or a "theoric transformation" which consists in changing the perspective on a problem. Whereas abduction focuses on explanatory hypotheses for phenomena that call for an explanation, problems call for a solution. Reacting to Peirce's "wish [that] a historical study were made of all the remarkable theoric steps and noticeable classes of theoric steps," the present contribution provides three case studies from the history of mathematics. With these cases I will (1) show how certain problems in mathematics can be solved by finding a theoretical model by means of which the solution can be deduced; (2) conceptualize the corresponding process of reasoning as "theoric transduction"; and (3) provide a proposal to capture the inferential structure of theoric transduction and a preliminary typology of its forms. All this will enrich our understanding of human reasoning and creativity and might be useful for the design of artificial intelligence (AI).

6.1 Introduction

Within the extensive body of research that Lorenzo Magnani dedicated to scientific creativity, what stands out are his contributions to abduction and model-based reasoning, and his development of an "eco-cognitive" approach to the relationships and interactions between cognitive agents, their environment, and mediating cognitive tools (e.g. Magnani, 2009, 2017). The present contribution adds to this work a novel

M. H. G. Hoffmann (✉)
Jimmy and Rosalynn Carter School of Public Policy, Georgia Institute of Technology, Atlanta, GA, USA
e-mail: michael.hoffmann@pubpolicy.gatech.edu

S. Arfini (ed.), *Scientific Cognition, Semiotics, and Computational Agents: Essays in Honor of Lorenzo Magnani - Volume 2*, Synthese Library 506,
https://doi.org/10.1007/978-3-031-96688-0_6

perspective by focusing on problems. Whereas abduction focuses on explanatory hypotheses for phenomena that call for an explanation, problems call for a solution.

A problem is usually understood as the experience of a situation in which one

> has a goal but does not know how this goal is to be reached. Whenever one cannot go from the given situation to the desired situation simply by action, then there has to be recourse to thinking. (By action we here understand the performance of obvious operations.) Such thinking has the task of devising some action which may mediate between the existing and the desired situations. (Duncker, 1945, p. 1; see also Newell & Simon, 1972, p. 72).

Such a problem could be called an epistemic problem because it is about not knowing. (Other types of problems such as wicked problems are not of an epistemic nature; instead, they are socially problematic). Focusing on debates in philosophy of science, Thomas Nickles provided another answer to the question of what an epistemic problem is. His "constraint-inclusion model" of problem is based on the uncontroversial assumption "that problems do not arise apart from goals and demands" and his own consideration that the definition of a particular problem is supported by knowing all the constraints on solutions that rule out "*some* conceivable solution as inadmissible." Based on this, his model of problem "consists of *all* the conditions or *constraints* on the solution plus the demand that the solution (an object satisfying the constraints) be found" (Nickles, 1981, p. 109; his italics). It should be noted that this account of problem does not provide a general definition of what a problem is; instead it allows us to individuate "one problem from another" by comparing their sets of constraints (p. 111). But a general definition of problem can be derived from these considerations: A problem consists of a demand for a solution and a set of constraints that this solution needs to satisfy.

Various schemata have been developed to describe how we deal or can/should deal with epistemic problems. Karl Popper proposed one that focuses on "shifting" from an initial problem to an improved one based on "a severe critical examination of our conjecture" regarding "tentative solutions"[1]:

initial problem → tentative solutions → error elimation → new problem

While for Popper the focus is not on "solving" a problem, this is the case for the theory developed by Allen Newell and Herbert Simon in *Human Problem Solving*. Leaving aside the processes of problem solving that are at the center of their theory, it seems that its most important components and how they are related can be represented as in Fig. 6.1.

Newell and Simon explain problem-solving behavior as adaptation to what is "demanded by the task environment" (pp. 79–80). In their three case studies which form the core of their book—about an algebra-like puzzle, chess, and a problem from symbolic logic—they determine themselves in an introductory chapter for each study what is demanded in a situation by performing a task analysis which is then followed by chapters on problem-solving behavior. This means that their approach to problem solving is limited to cases in which expert solutions to

[1] Popper (1972, pp. 164 and 243); see also Gillies (2009).

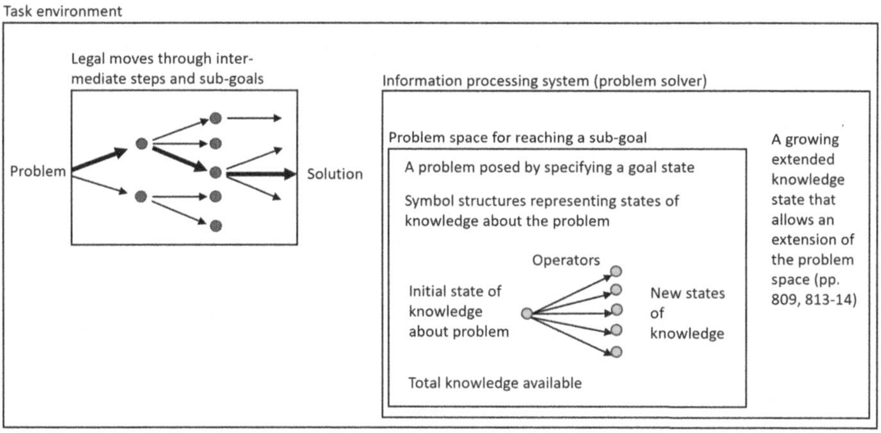

Fig. 6.1 According to the geographical situation depicted in Fig. 6.8, each of the four land masses distinguished there has a particular number of bridges connected to it. The number of bridges, again, determines how often a letter referring to a specific land mass occurs in the sequence of letters describing a path across the bridges

problems are already known. However, the model can be generalized by just focusing on what is "inside" the information processing system (Fig. 6.1). The "legal moves"—whose legality is determined by socially shared standards—and other things provided by the task environment would then be represented as parts of the problem solvers "total knowledge available." Crucially, if the knowledge available to a problem solver is not sufficient to solve the problem, then problem spaces "can be changed and modified during the course of solving" (p. 809). Newell and Simon assume that there are "extended knowledge states"—which are not immediately available within the problem space as part of the "total knowledge"—that are part of an external memory such as what has been written on a piece of paper or in a book (pp. 813–14).

However, their empirical studies show that problem spaces are usually "closed" in the sense that "the problem solver usually evokes an extremely limited part of his internal store, and attends to an extremely limited part of his external environment." The authors hypothesize

> that the kinds of information produced by the problem solving processes—namely, the knowledge states—are all of a piece with the information initially evoked, and do not have associations, via the productions, with other materials in memory to evoke new information that would continually enlarge the problem space. If only algebraic operations are performed on algebraic expressions, then only new algebraic expressions are produced, and these evoke, in turn, only algebraic operations. (Newell & Simon, 1972, p. 819)

But even in the time-constrained experiments that Newell and Simon performed they found examples of subjects "generating ideas that go outside the standard problem space" (pp. 819–20). This is what the present study is about. I am mainly interested in those epistemic problems whose solution requires going beyond the limits of a closed problem space and finding or creating theoretical tools that are initially not related to the problem.

The significance of looking outside of the problem space for the solution to a problem becomes clear when we compare the approach pursued here with what Carlo Cellucci defined as the "analytical method." "The analytic method is the method according to which, to solve a problem, one looks for some hypothesis that is a sufficient condition for solving it." (Cellucci, 2013, p. 55). Looking for something that is sufficient to solve the problem is—in contrast to Popper—also at the core of the problem-solving schema discussed here. But for Cellucci the "hypothesis is obtained from the problem, and possibly other data already available, by some non-deductive rule." This is not sufficient for the kind of problems I am interested in. Nothing is "already available" or can be obtained from an analysis of the problem alone. My approach is more closely related to the challenge of finding the right perspective on a problem and the role of representations in problem solving that Emily Grosholz highlighted (Grosholz, 2007), as well as to her claim "that some of the most important ampliative reasoning—reasoning that extends knowledge—in mathematics and the sciences takes place when heterogeneous discourses are brought into novel, and rational relation"; a claim that she substantiates in her book *Starry Reckoning* with many examples (Grosholz, 2016; see also Nersessian, 2022).

The kind of epistemic problems I am interested in and the problem solving schema on which my approach is based can be described with a simple example. Assume you want to cut a wooden block into 27 equal cubes. You will need six cuts to achieve the result that is depicted in Fig. 6.2. The question is: Is there a way to do it with fewer than six cuts? There might be an alternative strategy. For example, after a first cut you could take the slice that you created, lay it on top of the

Fig. 6.2 Cutting a wooden block. Created by Aria, an AI assistant provided in the Opera browser, based on OpenAI's GPT-4o, Gemini 1.5, and Imagen-3 models

remaining piece, and cut again through the combined pieces. Is there a sequence of cutting-moving-cutting that leads to 27 cubes with fewer than six cuts?[2]

If you did not come up with the correct answer right away, then you experienced what it means to have a problem as described by Duncker: The experience of a situation in which you cannot apply a readily available *habit* to achieve a certain goal.

For our riddle, there is a simple "move" to get from the problem state to the goal state, the solution. You only have to consider that there is always one cube in the middle of the wooden block. Since this cube has six sides, it is impossible to cut the block with fewer than six cuts. This example provides insights into three remarkable features of solving the sort of problems represented by this case:

1. To solve the problem, you have to step back from the given problem formulation and look at the problem from a different point of view. Charles Peirce conceptualised this move as performing a "theoric step." This step consists "in the transformation of the problem,—or its statement,—due to viewing it from another point of view" (Peirce MS 318, CSP 68 = ISP 225). A theoric step means *changing the perspective* (Magnani, 2009, pp. 181–83; Hoffmann, 2010; Pietarinen & Bellucci, 2014).
2. In order to think about the problem, you need to represent it. This "representation of the problem guides the search for a possible solution" (Bassok & Novick, 2012, p. 414). But this guidance does not necessarily support your efforts. The guidance provided with the idea of moving pieces before cutting, for example, is a distraction. It tricks you into believing that a specific rearrangement of the pieces after each cut might reduce the number of cuts needed, and it invites you to test this empirically in thought experiments. The solution, however, requires getting rid of this guidance and focusing instead on the cube in the middle. The solution—that it is impossible to get it done with fewer than six cuts—can be derived analytically and with logical necessity from the concept of a cube without any empirical testing. Thus, the challenge is not just to represent the problem, but to *find a representation of the problem that is conducive to its solution.*
3. The crucial step for solving the problem is finding the right perspective that connects the problem state with the goal state. Since this mediating element is not an explicit component of the problem presentation, you need to get it from some other place. I call the required process of reasoning that takes something from outside of a given problem formulation to solve a problem a *theoric transduction.*

The novel term "theoric transduction" is formed in alignment with Peirce's distinction of three types of inference—deduction, induction, and abduction—that he presented as fundamental for scientific reasoning. I will argue that theoric transduction should be added—not as a fourth form of inference, but, following Peirce's lead, as a sub-type of deductive reasoning. I call this a *theoric* transduction to make clear that there is no relation to "transduction" or "transductive inference," a term that is

[2] The idea of this example has been developed by Jacobs (1970, pp. 23–24).

widely used in statistics, computer science, and many other sciences. It has been introduced, it seems, by Vladimir N. Vapnik who writes:

> *transductive inference* … is different from inductive inference. In contrast to the inductive inference where one uses given empirical data to find the approximation of a functional dependency (the inductive step) and then uses the obtained approximation to evaluate the values of a function at the points of interest (the deductive step), we will try to estimate the values of a function at the points of interest in one step" (Vapnik, 1998, p. 339).

Vapnik's transduction is closely related to induction insofar as it allows to determine the unknown values of a function without first determining this function by means of induction. Theoric transduction, by contrast, is a form of deduction. "Theoric" has been introduced by Peirce who formed it based on the Greek "*theoria*" which originally meant "vision." He translates this term as "the power of looking at facts from a novel point of view" (Peirce MS 318, CSP 50 = ISP 42).

But why shouldn't we just use another term here that is well-known in research on problem-solving: analogical transfer? For Keith Holyoak, analogical transfer is the process of using knowledge about a particular "pattern of relationships" among the constituent elements of a "source" situation "to generate inferences about the target" (Holyoak, 2012, p. 234). As an example, he points at "the earliest major scientific analogy, dating from the era of imperial Rome," when familiarity with water waves was used for "a deeper understanding of sound" (p. 234; see also Holyoak & Thagard, 1995, pp. 10–12).

However, for Holyoak analogy "is an inductive process, and hence analogical inferences are inevitably uncertain. The wave analogy for sound proved successful; an alternative 'particle' analogy did not" (Holyoak, 2012, p. 235). This is not what we observe when solving the problem of cutting the wooden block. Once we consider the cube in the middle of the block, the solution to the problem is not uncertain; it is necessarily true. Even though much can be learned from research about analogical reasoning and analogical transfer—especially regarding strategies for, and cognitive processes involved in the creative part of problem-solving—it should be important to keep things conceptually apart that can be clearly distinguished. Since problem solutions that can be proved with deductive necessity differ from those that can be refuted by additional evidence, we should use different terms to refer to each type of problem solution.

Whereas Holyoak and many others treat analogical reason as a form of *induction*, research that builds on Peirce's distinction of the three basic forms of scientific reasoning often connects the creativity involved in cases like the wooden block with *abduction*.[3] For Peirce, abduction "is the process of forming an explanatory hypothesis. It is the only logical operation which introduces any new idea; for induction does nothing but determine a value, and deduction merely evolves the necessary consequences of a pure hypothesis" (Peirce CP 5.171; 1903). Peirce determines the inferential form of abduction as follows: "The surprising fact, C, is observed; But if A were true, C would be a matter of course. Hence, there is reason to suspect that A

[3] Magnani (2009, ch. 2); Pedemonte and Reid (2011), Pietarinen and Bellucci (2014), Woods (2023).

is true" (Peirce CP 5.189). These quotes make clear, first, that for Peirce abduction focuses on phenomena that call for an explanation, not on problems that should be solved; second, the explanatory hypothesis in the conclusion remains a conjecture: we only have "reason to suspect" that it is true, no certainty. This is not the case with the theoric step that leads to the solution of the cube problem. As soon as we realize that there is this cube in the middle whose six sides need to be cut, the solution of the problem is necessarily true; it is impossible to do it with fewer than six cuts.

I know only of one remark by Peirce in which he connects creative problem-solving with abduction. In a manuscript he writes that "theôric reasoning … is very plainly allied to retroduction [earlier called abduction], from which it only differs as far as I now see in being indisputable."[4] But this is part of a discussion on various kinds of *deductive* reasoning. This makes sense because the result of theoric reasoning is "indisputable" whereas the result of abductive reasoning is just a conjecture. As I showed elsewhere (Hoffmann, 2010), Peirce developed the term "theoric" only late in life. The main motivation was, it seems, to conceptualize the creative step of developing the "plan" for a mathematical proof (Peirce CP 4.613–14; 1908). Once a mathematician succeeds in finding an appropriate proof idea, the theorem in question can be proved with logical necessity.

It is, of course, possible to extend the Peircean notion of abduction. Following Gabbay and Woods (2005), Lorenzo Magnani characterizes abductive reasoning as "a response to an ignorance-problem: one has an ignorance-problem when one has a cognitive target that cannot be attained on the basis of what one currently knows" (Magnani, 2009, p. 65). The notion of an ignorance problem is broad enough to cover the experience of surprising facts, of problems for which one doesn't know a solution, and of other forms of creative reasoning. But why should we replace Peirce's classification which has already been widely adopted by another one? Why should we subsume all forms of creative thinking under abduction after Peirce suggested that some of them should be treated as forms of deduction?

Following Peirce's lead, this contribution aims at clarifying his notion of "theoric reasoning"—which he presents as a form of deductive reasoning—with the introduction of the inference form theoric transduction. Whereas Peirce claimed in the quote from 1903 above that abduction is "the only logical operation which introduces any new idea," I will show that his considerations on theoric steps that he performed a few years later should lead to the conceptualization of another such operation: theoric transduction. But what exactly is theoric transduction? How does it work? Peirce concluded his considerations about the role of theoric steps in creating the plan for a mathematical proof as follows:

> I wish a historical study were made of all the remarkable theoric steps and noticeable classes of theoric steps. I do not mean a mere narrative, but a critical examination of just what and of what mode the logical efficacy of the different steps has been. Then, upon this work as a foundation, should be erected a logical classification of theoric steps; and this should be crowned with a new methodeutic of necessary reasoning. (Peirce CP 4.615; 1908)

---

[4] Peirce MS 754 (1907), ISP 8; his highlighting; see also ISP 3.

A first step in this direction is the goal of the following considerations. I am going to analyze three cases of theoric reasoning in the history of mathematics to develop a better understanding of theoric transduction.[5] The goal is a clear definition of theoric transduction, a determination of its inferential structure, and a preliminary typology of different forms of theoric transduction. What will not be discussed are possible rules or heuristics for solving problems (e.g. Ippoliti, 2018), the question of whether mathematics should be described as an axiomatic system of theorems or as the activity of problem-solving,[6] or the idea that scientific inquiry in general should be understood as problem-solving.[7]

6.2 First Case: Gauss's Summation of an Arithmetic Series

There is a wonderful, often-told story in which the later famous mathematician Carl Friedrich Gauss, as a very young boy, puts his teacher Büttner to shame by teaching him a lesson about mathematical creativity. In what seems to be the oldest report of the event, Wolfgang Sartorius, a professor of mineralogy and geology at the University of Göttingen, where Gauss spent his entire academic career, writes in a memorial volume published in 1856, just a year after Gauss's death, that Gauss reached the arithmetic class at the age of nine.

> Here occurred an incident which he often related in old age with amusement and relish. In this class the pupil who first finished his example in arithmetic was to place his slate in the middle of a large table. On top of this the second placed his slate and so on. The young Gauss had just entered the class when Büttner gave out for a problem [the summing of an arithmetic series]. The problem was barely stated before Gauss threw his slate on the table with the words (in the low Braunschweig dialect): "There it lies." … At the end of the hour the slates were turned bottom up. That of the young Gauss with one solitary figure lay on top. When Büttner read out the answer, to the surprise of all present that of young Gauss was found to be correct, whereas many of the others were wrong. (Hayes, 2006, p. 201)

This translation of the German original was provided by Helen Worthington Gauss, a great-grand daughter of the mathematician.[8] The English translation is quite liberal which leads, as Brian Hayes points out in his analysis of all the material available about the story, to a problem with the phrase in brackets. Here, Worthington Gauss writes that "Büttner gave out for a problem the adding of a series of numbers from 1 to 100," where Sartorius just talks about "die Summation einer arithme-

[5] It should be noted that Peirce himself pointed repeatedly at a fourth case: the proof of Desargues's theorem which he called the "ten point theorem." I provided an analysis of this case in Hoffmann (2010, pp. 579–584).

[6] See for example Hoffmann (2005, pp. 85–122), Cellucci (2017, pp. 273–292).

[7] See for example Lakatos (2015 <1976>), Laudan (1977, 1981), Simon (1989), Cellucci (2013, 2018).

[8] Hayes published this translation and Sartorius's original also at http://bit-player.org/wp-content/extras/gaussfiles/gauss-snippets.html. There he collected 145 snippets of the story, the last one in 2018 (accessed November 2024).

tischen Reihe—the summing of an arithmetic series," as Hayes corrects the translation. This means that the oldest version of the story neither mentions which numbers the students were asked to add nor the specific method that Gauss used to get so quickly to the correct answer.

Most later renderings of the story, however, do indeed assume that the task was to add all numbers from 1 to 100. And with regard to the method Gauss might have used, this is usually explained, as Hayes writes, "in terms of adding pairs of elements from opposite ends of the sequence, so that all the pairs have the same sum" (Hayes, 2006, p. 204). One way of doing this is by visualizing the sequence of numbers in a spatial structure that is similar to the one depicted in Fig. 6.3.

As Hayes shows, the research literature provides a large variety of methods to solve the problem more efficiently than adding one number after the other. Four main groups of methods can be distinguished: (1) those that use a particular spatial presentation of the sequence of numbers as in Fig. 6.3; (2) the method of multiplying the average of all elements of the sequence by the number of these elements which requires the insight—to get to the average without first adding all the numbers—that in an arithmetic progression the average over the entire series is equal to the average of the first and last elements; (3) a method that is based on an arrangement of number sequences in the form of a triangle (more on this in a second); and finally (4) there are all sorts of pattern-based shortcuts that suggest themselves when you actually do the summing up of the numbers one by one (for examples see Hayes, 2006, p. 204).

Among these four types of methods, the one with the "triangle" is of particular interest because we can assume that Gauss was familiar with its basic idea from classes in Greek or Latin if these covered the Pythagorean doctrine of the so-called *tetractys*. As Leonid Zhmud writes:

> The sum of the numbers in a tetractys is equal to ten, and the Pythagoreans supposedly considered this number "perfect". One of the Pythagorean "symbols" identifies the tetractys with the most important fount of wisdom, the oracle of Delphi: "What is the oracle at Delphi? The tetractys, which is the harmony in which the Sirens sing". Finally, the tetractys figures in the famous Pythagorean oath, which is broadly represented, with slight variations in form, in the literature of the Imperial era. The Pythagoreans swore not by the gods but by their Master, and regarded his discovery of the tetractys as his principal achievement: "No, I swear by him who gave the tetractys to our head, which has the source and root of everlasting nature". (Zhmud, 2012, p. 301)

The tetractys represents the first four numbers in the spatial form of a triangle like the one in Fig. 6.4. This way the harmonic relationships of the fourth (4:3), the fifth

$$
\begin{array}{ccccccccccccccc}
 & & 1 & & 2 & & 3 & & 4 & & 5 & & & & \\
 & + & 10 & & 9 & & 8 & & 7 & & 6 & & & & \\
\hline
S_{10} & = & 11 & + & 11 & + & 11 & + & 11 & + & 11 & = & 5\times 11 & = & 55
\end{array}
$$

Fig. 6.3 Adding the first ten integers by "folding" the sequence in half

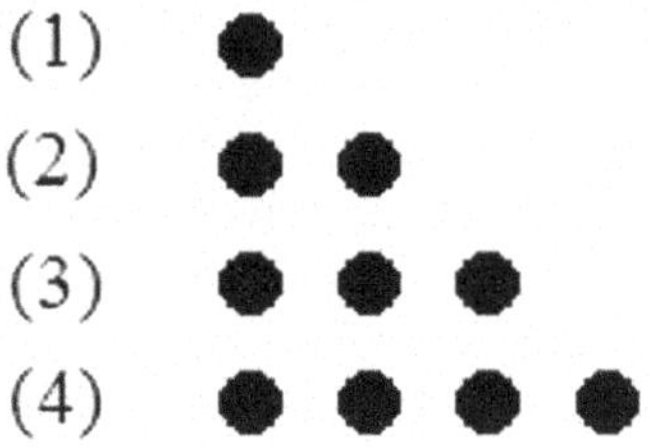

Fig. 6.4 The Pythagorean tetractys

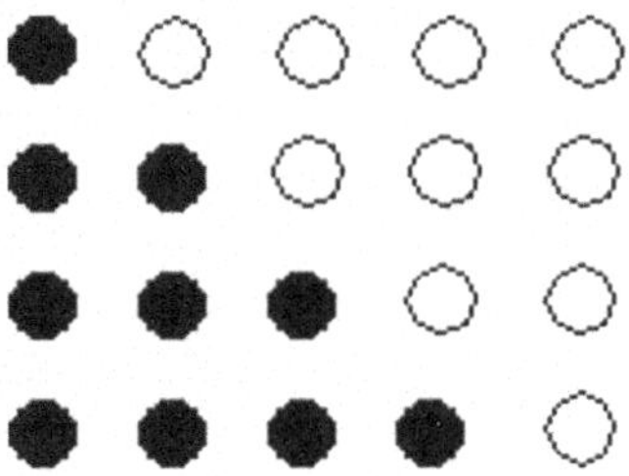

Fig. 6.5 Doubling the tetractys

(3:2) and the octave (2:1) can be represented. As Sextus Empiricus continues, after quoting the oath mentioned by Zhmud:

> By "the man who handed down [who gave the tetractys to our head]" they mean Pythagoras (for they made him a god), and by the "tetractys" they mean a certain number. Since it is put together out of the first four numbers it produces the most perfect number, ten; for one and two and three and four make ten. And this number is the first tetractys and is called a "spring [source and root] of everlasting nature" in so far as the entire universe, according to them, is administered in accordance with harmony. And harmony is a system of three musical intervals, the fourth, the fifth, and the octave; and the proportions of these three intervals are found in the four numbers mentioned before – one, two, three, and four. (Sextus Empiricus, 2005, p. 20 [= *adv. math.* VII, 94-5]; insertions in brackets provide the terms that Zhmud used in his translation)

We can imagine that the young Gauss experimented with the tetractys already in his classical studies to figure out whether it can be used not only to *visualize* the sum of the first four numbers, but also to *determine* the sum of any arithmetic series. A simple way to do this is to double the tetractys as in Fig. 6.5 and then to calculate the sum with the general formula $S_n = 1/2\ (n^2 + n)$ in which "n" stands for the largest number in the series. The $n^2 + n$ part of this formula can be "seen" in Fig. 6.5 when we realize that all the numbers above the four black circles at the bottom form a square (4x4) to which just the four white circles on the right need to be added to get to the doubled tetractys. Along these lines it is easy to determine the sum of the first hundred numbers: $S_{100} = 1/2(100^2 + 100) = 5{,}050$.

The solution to the problem is 5050. It can be calculated by various methods. Just adding one number after the other is one way to do it. Of course, this is very cumbersome and error prone. So, the challenge is, and this leads to a modification of the problem formulation, to figure out whether there is a more efficient method to get to the sum. Whatever method Gauss used, we can now refine the three required features of the problem-solving process that we distinguished in the beginning with reference to the task of cutting the wooden block: *Changing the perspective* in a theoric step; *finding a representation of the problem that is conducive to its solution*; and taking something from outside of the given problem formulation to solve the problem in a *theoric transduction*.

To solve the problem, Gauss needed to step back from the given problem formulation and find a new perspective or an alternative representation of the problem as they are available in the folding strategy presented in Fig. 6.3, in the idea of multiplying the number of all numbers with the average of the first and last number, and in the method of doubling the tetractys. Each of these representations of the problem offers a strategy that can be realized step-by-step to calculate the solution. It is this strategy that is transduced from each of these representations to the problem in question. However, since each of these arithmetic strategies can be represented in an algebraic formula, the theoric transduction can also refer to a particular formula that is taken from *algebra* as a particular system of representation to solve an *arithmetic* problem:

- $S_n = n/2(n + 1)$ for the folding strategy
- $S_n = n(n + 1)/2$ for the method that uses the average of all numbers, and
- $S_n=1/2\ (n^2 + n)$ for the tetractys-based method

With any of these formulas or strategies the problem can be solved quickly, efficiently, and correctly. The object of the theoric transduction—what is "taken" from *A* to *B*—is either a strategy that can be formulated by the means available in the same representational system as the one in which the problem is formulated or a formula or algorithm from a different system of representation. If we believe that Gauss was able solve the problem quickly, then we have to assume that he was already familiar with one of these. The theoric step that enabled his transduction was possible because the searched-for idea was already part of the "total knowledge available" to him in this situation, as Newell and Simon put it (see Fig. 6.1). So, in this case there was no need to change the "problem space." This will be different in the following two cases of problem-solving. In both it was necessary to enlarge the problem space. In the first one this was done by using knowledge that had been developed in a different area of research, and in the second one by creating something entirely new.

6.3 Second Case: Proving Incommensurability

The discovery that certain lengths are "incommensurable"—meaning: they cannot be measured with a common unit—was of tremendous importance for the evolution of mathematics. It led eventually to the introduction of irrational numbers and, thus, to an extension of the system of numbers.

The first known proof of incommensurability has been preserved in what is now an appendix to the tenth book of Euclid's *Elements*. The proof is easier to understand if we simplify a few formulations, use modern terminology instead of language that nobody uses anymore, and explicate a few points that are presupposed.[9]

> Let *ABCD* be a square and *AC* its diameter. I say that *AC* will be incommensurable with *AB* in length. For let us assume that it is commensurable. I say that it will follow that the same number is at the same time even and odd. It is clear that the square on *AC* is double the square on *AB* [because, according to the Pythagorean theorem, $a^2 + b^2 = c^2$ so that, if the sides a and b are equal as in the square, the square on the diagonal c equals double the square of the side.
>
> If we assume that] *AC* is commensurable with *AB*, (then there are two integers, say m and n, which are mutually prime, meaning: the only positive integer that is a divisor of both of them is 1. From $AC/AB = m/n$ follows that $\frac{AC^2}{AB^2} = m^2 / n^2$. However, since $AC^2 = 2AB^2$ based on the Pythagorean theorem, it follows that $m^2 = 2n^2$. From this it follows that m^2 must be even and with it m, as implied by Euclid, 1956 IX Prop. 29. But, if m is even, n must be odd, according to our initial supposition that they are mutually prime. Assuming that $m = 2k$ because it is even, we can infer that $2n^2 = 4k^2$ or $n^2 = 2k^2$; and from this it can be shown by a repetition of the reasoning used above that n must be even.) But it has also been demonstrated that n must be an odd number, which is impossible. It follows, therefore, that *AC* cannot be commensurable with *AB*, which was to be demonstrated.

As a so-called *reductio ad absurdum* or *reductio ad impossibile*, this proof laid the foundation for a whole class of mathematical proofs. The fact that incommensurability is deduced from an "absurdity" is made clear right in the beginning when the author writes that from assuming a counterfactual—that side and diagonal are commensurable—it would follow "that the same number is at the same time even and odd," which is obviously absurd. Since commensurable and incommensurable are mutually exclusive attributes, demonstrating that commensurability leads to an impossibility is equivalent to proving incommensurability.

This arithmetic proof represents a major achievement in the history of mathematics. But how has it been created? Since we are interested in problem-solving, it is important to note that there are also geometrical proofs for incommensurable lengths. In the following I will argue that Greek mathematicians began, at some point in time, to perceive these geometrical proofs as problematic, and that it was *this problem* that motivated the search for a better demonstration.

For this we have to dig deeper into a history that is, unfortunately, controversial. The problem is, first, that there is not enough clear evidence of what exactly happened when and, second, that testimonies are often using conceptual frameworks that are difficult to understand, or they are so brief that they leave a lot to interpretation. Hence, the early history of mathematics depends to a large degree on speculation. Since the points that are needed for the proposed argument should be

---

[9] This annotated paraphrase uses only in the beginning and at the end Kurt von Fritz's "literal translation" of the text (von Fritz, 1945, pp. 254–255, Fn. 60; for another literal translation see Knorr, 1975, p. 23). Passages in brackets are additions that are based on interpretations by Kneale and Kneale (1988 <1962>) (p. 8), van der Waerden, 1979 (p. 398), and myself. The parts in parentheses replace von Fritz's formulations by ones that are quoted from these authors or my own.

convincingly justified to strengthen them against possible objections, they need to be justified in some detail. I start with the earliest known references to incommensurability.

The arithmetic proof quoted above must have been developed before Euclid because Aristotle already mentions it as the proof "that the diagonal of a square is incommensurable with the sides by showing that if it is assumed to be commensurable, odd become equal to even numbers" (Aristotle *An. Pr.* 41a26–28). Plato, again, writes in old age—using the voice of the "Athenian" in his *Laws*—about the "shame" that he experienced after learning that there are lengths that cannot be measured against lengths, areas not against areas, and volumes not against volumes (Plato *Leg.* VII 819e–820e). This, however, is surprising because in his earlier dialogue *Theaetetus* the young mathematician of this name mentions that his teacher Theodorus demonstrated with the aid of diagrams that the "*dýnamis*," that is the "square identified with its side," "of three square feet and the *dýnamis* of five square feet are not commensurable in length with the *dýnamis* of one square foot."[10] Theaetetus, whose death is alluded to in the beginning of the dialogue, probably died in 369 BC so that it is generally assumed that Plato wrote the book shortly after that. The *Laws*, by contrast must have been written much closer to Plato's death around 347 BC (Cooper, 1997, pp. xiii-xiv). Since Plato places the dialogue that the *Theaetetus* envisions shortly before the death of Socrates in 399 BC, it is clear that a "diagrammatic" or geometrical proof of the incommensurability of the side of a two square feet square, which is not even mentioned in the dialogue, must have been well-known around 400 BC (Knorr, 2004 <1981>, p. 89). So, it is surprising to see that Plato confesses in the *Laws*, in the voice of the "Athenian" who leads the dialogue, that he himself "heard only quite lately (*οψε*) about these things and was stunned about our misconceptions" (819d6) while he indicates in the *Theaetetus* that incommensurability was proven already about 50 years earlier.

The surprise, however, disappears when we take into account that Plato presents in the *Laws*, just before the passage mentioned above, the following as an admirable part of Egyptian education: "After this, [the teachers] remove in measurements everything that has length, area, and volume, about which in all human beings a ridiculous and shameful ignorance is given by nature" (Plato *Leg.* VII 819c7-d3). On the one hand, this remark refers to the obviously ancient problem in teaching mathematics that, even though it is a good idea to start with apples and cakes that need to be distributed and partitioned (819b, c), at some point mathematics need to be taught as a more abstract endeavor. But it is exactly this formulation that forms the bridge to the Athenian's talk about incommensurability as an example of "shameful ignorance." If we assume that his recent astonishment "about these

[10] Plato *Theaet.*, 147d, with keeping the Greek *dýnamis* instead of "power." The translation of *dýnamis* as the "square identified with its side" has been suggested by Høyrup (1990) who showed that this interpretation of the term can be found also in Babylonian mathematical thinking. Even though this framing of mathematical relations is very different from what we are used today, it seems to be the only one that allows a consistent interpretation of the mathematical term *dýnamis* across various passages in the *Theaetetus* and beyond.

things" refers to a proof of incommensurability that "got rid of" working with lengths, areas, and volumes (819d3)—that is, to the arithmetic proof—then there is no tension between references to events that happened half a century apart. While the *Theaetetus* refers explicitly to proofs by means of diagrams (147d3), the *Laws* might refer to a purely arithmetic proof.[11]

This assumption can be supported when we interpret Plato's excitement about getting rid of "length, area, and volume" in measurements in the context of Aristotle's observation that arithmetic is "more exact" than geometry because those scientific disciplines "which are based on fewer principles are more exact than those which include additional principles."[12] The arithmetic proof that we find in Euclid is indeed perfectly exact. Now, let's have a look at a geometric proof.

Kurt von Fritz suggested in a classical paper that the first proof of incommensurability was performed on a regular pentagon, not a square. There is no historical evidence for any particular and completely formulated geometric proof in pre-Euclidean mathematics. What speaks for this proof is the fact that the discovery of incommensurability has been ascribed to Pythagorean mathematicians early on,[13] and that it is well known that the Pythagoreans used the pentagram as a symbol of mutual recognition in their secretive community (Burkert, 1972 <1962>, pp. 176, 452, 468). The pentagram is a regular pentagon with its sides prolonged to the point of intersection.

The proof requires the method of "*anthyphairesis*" or reciprocal subtraction.[14] The goal of this method is to determine the greatest common measure of two

[11] As John M. Risk argues, Aristotle's *Prior Analytics,* in which the arithmetic proof is first mentioned, can be dated at about 341/40 BC (Rist, 1989, p. 83), about seven years after Plato's death. It is, thus, possible that this proof has been discovered around the time Plato wrote the *Laws*.

[12] Aristotle *Met.* A2, 982a26–28. See also the remark by the Pythagorean mathematician Archytas: "λογιστικά [should be translated as "arithmetic"; see Huffman (2020), Sect. 3.2] seems to be far superior indeed to the other arts in regard to wisdom, in particular compared to geometry since it can treat what it wants to treat more clearly. … where geometry fails, λογιστικά realize proofs, as well as proofs of the forms, if there is a real treatment of the forms" (Diels & Kranz, 1989, DK 47B4). Note that Archytas uses Plato's term εἶδος here, "form" or "kind," not "shape," as Huffman translates.

[13] See von Fritz (1945, p. 245). Testimonies include references by Pappus, Proclus, Iamblichus, and Athenagoras, and a scholia on Euclid (Knorr, 1975, pp. 20–21; Burkert, 1972 <1962>, pp. 461–62). Even though he doubts that Pythagoras himself could have been involved, Knorr presupposes in his historic reconstruction a Pythagorean proof (pp. 24–26). Critical about all these testimonies is Fowler, 1999 <1987>, pp. 356–69. But his main interest in the matter—as it was already the case for Burkert, 1972 <1962>, pp. 454–65—is to dispel the myth that the discovery of incommensurability led to serious doubts about the "foundation of mathematics" among Pythagoreans. Like Burkert but unlike Knorr, he is not interested in the question of how the discovery was made. For this question, I think, we should take whatever testimony we can get.

[14] Szabó (1978) assumes that the method originated in the Pythagorean theory of music (pp. 136, 199–201). von Fritz (1945) calls it an "old time-honored method" (p. 259). Fowler, (1979) finds evidence in Pappus that Theaetetus used anthyphairesis for "the development of a procedure for defining ratio" (p. 832). Knorr (1975, pp. 29–36) doubts that the first geometric proof of incommensurability used the method even though he argues that the method was known toward the end of the fifth century. But his two arguments that anthyphairesis was not used to proof incommensu-

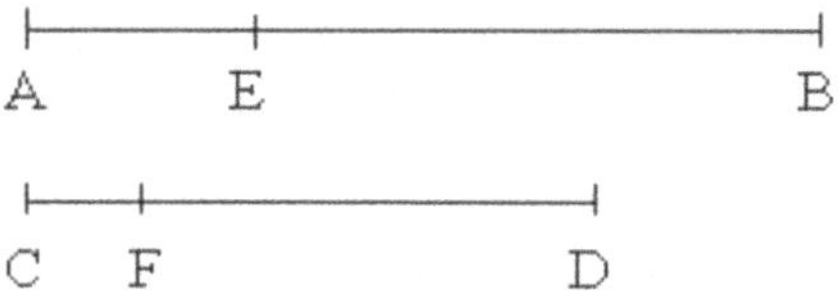

Fig. 6.6 Determining the greatest common measure of AB and CD by reciprocal subtraction

lengths. Looking at Fig. 6.6, this common unit can be determined by subtracting the shorter line *CD* from the longer *AB* so that *AE* remains. *AE*, in turn, fits twice into *CD* with *CF* as the remainder. In so far as *CF* measures *AE* twice without leaving any remainder, *CF* is the greatest common measure of the two lines.

The problem of incommensurability appears when we try to determine the ratio of side and diagonal in a regular pentagon with this method. In order to determine the ratio of *DE* to *DA* in Fig. 6.7—which is copied here from von Fritz, 1945—we start by subtracting the former from the latter. Since *DE* is equal to *DC'*, the remainder of the first subtraction is *C'A*. However, *C'A* is equal to *C'E'* so that with the first difference between the side and the diagonal of the original pentagon we have obtained the *diagonal* of the next smaller one which appears "upside-down" in the middle of the original pentagon.

The next step in the process of reciprocal subtraction is then to subtract *C'A* from *DE*. Since *DE* is equal to *EA* which again is equal to *ED'*, and *C'A* is equal to *C'E*, we get *C'D′* as the remainder of the second subtraction. Since *C'D′* is the *side* of the next smaller "upside-down" pentagon in Fig. 6.7, it becomes clear that the process of reciprocal subtraction, by which we try to determine the ratio of side and diagonal in the original pentagon, leads after two steps to another regular pentagon in which we have to subtract again the side from its diagonal. One "sees at once," as von Fritz writes, that this method to determine a common measure leads from one regular pentagon to the next one, "and so on in an infinite process" (von Fritz, 1945, p. 257). The method reveals the recursive structure of a pentagon within a pentagon within a pentagon, and so on. But if the process of reciprocal subtraction can be continued without end, then this means that there is no common measure of the diagonal and the side of the regular pentagon. Both lengths are "incommensurable."

Even though we don't know for certain by which geometric process the incommensurability of two lengths was first discovered, two things seem to be certain: first, that a geometric proof was known when Plato wrote his *Theaetetus* and, secondly, that the geometric approach was realized prior to the arithmetic one. As von Fritz argues with regard to the second point, the fact that the arithmetic proof, as

rability are not convincing. The first one says that the corresponding proof is more "elaborate" than the "simple" one that he suggests, and the second one that there is no evidence that anthyphairesis has been used in this context. But no evidence, obviously, cannot be evidence for the contrary (this applies also to Burkert, 1972 <1962>, p. 459 who makes the same argument). Anyway, the argument that I am developing here does not depend on a particular geometric procedure to prove incommensurability. It only requires that a geometric proof preceded the arithmetic one. For this assumption Knorr develops a convincing argument (pp. 24–28).

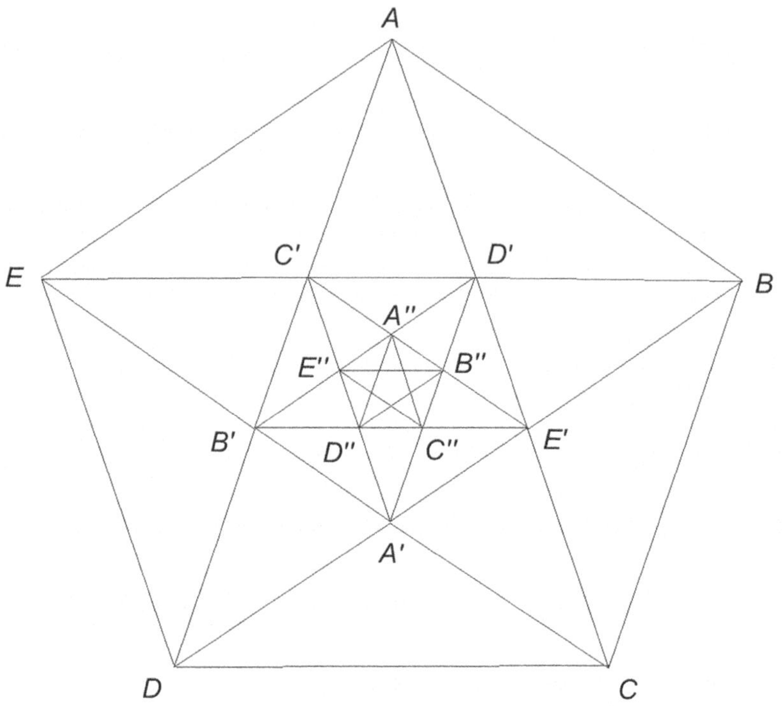

Fig. 6.7 Determining the ratio of side and diagonal in a regular pentagon with the method of reciprocal subtraction. Source: von Fritz (1945, p. 257). Reprinted with permission from the Annals of Mathematics, Department of Mathematics, Princeton University

presented above, uses the terms commensurable and incommensurable as the starting point "seems to presuppose that incommensurability was already known when the demonstration was elaborated" (p. 256).

More interesting in our context is that von Fritz provides additionally an argument that is based on an assumption about human cognition. Referring to the arithmetic proof, he points out that

> it is difficult to believe that the early Greek mathematicians should have discovered the incommensurability of the diameter of a square with its side by a process of reasoning which was obviously so laborious for them if they had no previous suspicion that any such thing as incommensurability existed at all. If, on the other hand, they had already discovered the fact in a simpler way, it is perfectly in keeping with what we know of their methods to assume that they at once made every effort to find out whether there were other cases of incommensurability. (von Fritz, 1945, p. 256)

Without being convinced that there are indeed incommensurable lengths, there would not be a motive to prove the impossibility of the counterfactual as it happens in the arithmetic proof. But why did ancient mathematicians feel the necessity to create another proof besides a geometrical one? The answer to this question leads us back to Aristotle's observation that arithmetic is "more exact" than geometry.

There are two problems with the geometric proof presented above. The first one is that it presupposes that lines such as *DE* and *DC'*, *C'A* and *C'E'*, and *DE* and *EA* are of equal length. As von Fritz shows, there is evidence that already Thales formulated the theorem

> that in an isosceles triangle the base angles are equal. ... Aristotle refers to an archaic proof of this proposition. He does not quote all the steps of this proof, but what he quotes shows that "mixed angles," i.e., angles formed by a straight line and the circumference of a circle, were used in the demonstration, and that in all likelihood the proof was based on a rather primitive method of superimposition. It is clear that with this latter method the converse of the proposition could be proved without difficulty. It follows that the equality of *AE* with *AB'* and of *B'D* with *B'E'* could be derived from the equality of [the angle] *AEB'* with [the angle] *AB'E* and of [the angle] *B'DE'* with [the angle] *B'E'D*, if these angles could be proved to be equal respectively. (von Fritz, 1945, p. 258)

As he continues, the evidence for this latter proof "is somewhat less definite," but he is able to show that the equality of these angles "can be very easily shown" by means of the proof that the sum of the internal angles in any triangle is equal to two right angles, which is attributed to early Pythagoreans (p. 258). However, the problem remains that "archaic" proofs such as the "primitive method of superimposition" do not have the power of logical necessity; they remain visual or empirical proofs.

The same "defect" needs to be acknowledged with regard to what von Fritz describes as "seeing" that the series of congruent pentagons in Fig. 6.7 is infinite. As Árpád Szabó points out with regard to another geometrical proof that uses the method of reciprocal subtraction to determine a common measure of side and diagonal in a square, "the fact that the process can be continued to infinity is not proved rigorously" (Szabó, 1978, p. 211). However, it is important to note that what counts as a "rigorous" proof is not fixed. We know that proof standards evolved substantially since Thales used visual methods (Mittelstraß, 1965; Szabó, 1978, p. 197). As von Fritz emphasizes with regard to the geometrical proof in Fig. 6.7, the question is not whether the creator of this proof "could give a demonstration which in all its steps would have satisfied Euclid or Hilbert, but whether he was able to find a proof which at the level which mathematical theory had reached in his time was considered absolutely convincing, and as to this there can be no doubt" (p. 259).

What we can learn from all this is that Aristotle's remark that arithmetic is "more exact" than geometry is evidence for an important change in what counts as satisfactory proof. At some point visual evidence was no longer considered sufficient. Based on a change in proof standards we get into a situation that must have been experienced as a *problem* at the time: Since the truth of the theorem that there are incommensurable lengths seems to be evident, how can it be proven so that the most recent proof standards are met?

The arithmetic proof quoted above is so impressive because the gap between this *reductio ad absurdum* argument and the geometric proof depicted in Fig. 6.7 is so large. However, it is possible to bridge this gap if we assume that the mathematicians who created the arithmetic proof were familiar with the proof strategy of a *reductio ad absurdum*. This strategy was certainly known at the time in the form that Plato ascribes to Zeno of Elea in his dialogue *Parmenides*. In this dialogue, a still very young Socrates (Plato *Parm.* 127c) determines in a conversation with Zeno that the main goal of Zeno's writing was to defend the core thesis of his teacher Parmenides (128c) that "the all is one" (128a8-b1). The strategy that he used for this defense was to assume the opposite—that "the beings are many"—and then to demonstrate that these beings, under this assumption, "must be similar and not similar at the same time, but that is impossible" (127e1–3). The general proof strategy of Zeno—proving a thesis by showing that it's opposite leads to "absurdity by deducing from [it] contradictory consequences" (Kirk & Raven, 1964 <1957>, p. 287)—is the same as in the arithmetic proof quoted above, even though the details are different.[15]

It is clear that the geometrical and the arithmetic proof follow two completely different lines of valid reasoning to get to the same conclusion: certain lengths are incommensurable. But the standard of validity that each proof realizes is fundamentally different. The geometrical proof assumes that visual evidence, intuition, and empirical reasoning are sufficient. The historic development reconstructed above shows that based on a shift from empirical to logical standards the older geometrical proof was perceived as problematic at some point. It no longer satisfied the new proof standard. My thesis is that the solution of this problem was possible by *theoric transduction*. Mathematicians took the idea of the *reductio ad absurdum* proof strategy as it had been developed by Zeno and used it for developing the arithmetic proof of incommensurability. They transduced a proof strategy from eleatic logic to arithmetisized geometry.

6.4 Third Case: Euler and the Seven Bridges of Königsberg

In 1735, Leonhard Euler gave a presentation to the members of the Petersburg Academy in which he demonstrated that a popular puzzle about the seven bridges of the Prussian city of Königsberg (now Kaliningrad) does not have a solution. In the paper, which was eventually published in the proceedings of the Petersburg Academy (Euler, 1741), Euler presented the puzzle with the map that is depicted in Fig. 6.8. The question was: is it possible to devise a path so that one could cross each of the seven bridges exactly once and return home?

The title of Euler's paper, "The solution of a problem relating to the geometry of position" (an English translation is available in Euler, 1953 <1741>), makes it clear

[15] Using the same strategy, Zeno argued "that the same things appear to his listeners to be both like and unlike, both one and many, both at rest and in motion" (Plato *Phdr.* 261d6–8). Especially his four paradoxes about motion are well known. Szabó (1978) showed convincingly, I think, that the mathematical *reductio ad absurdum* originated in Eleatic philosophy (e.g., pp. 216–220; 248–250) in contrast to those who argue—as most recently Panchenko, 2017—for a reversed chronology, that is, that Zeno found the idea of the *reduction* in Hippasus who allegedly developed the arithmetic proof of incommensurability. Whoever argues along these lines would need to refute my argument about the chronology that is based on Plato's writings about incommensurability.

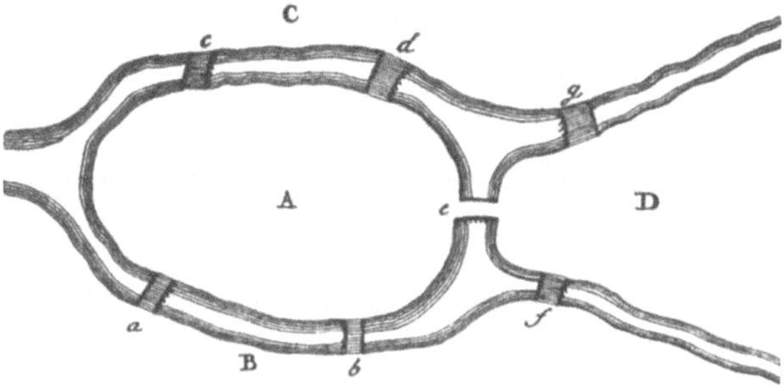

Fig. 6.8 The seven bridges (**a**–**f**) of Königsberg. "A" refers to the island Kneiphof around which the Pregel River branches into two parts (Source: Euler, 1741, his Fig. 1)

that its focus is on a new branch of geometry which is known today as graph theory. It is generally acknowledged that in this paper Euler laid the foundations of graph theory (Alexanderson, 2006). Euler himself starts the paper as follows:

> The branch of geometry that deals with magnitudes has been zealously studied throughout the past, but there is another branch that has been almost unknown up to now; Leibnitz [sic] spoke of it first, calling it the "geometry of position" (geometria situs). This branch of geometry deals with relations dependent on position alone, and investigates the properties of position; it does not take magnitudes into consideration, nor does it involve calculation with quantities. But as yet no satisfactory definition has been given of the problems that belong to this geometry of position or of the method to be used in solving them. (Euler, 1953 <1741>, p. 66)

Euler took the seven-bridges problem as a paradigmatic example for this new "branch" of geometry since it has nothing to do with either measurement or calculation. "In this paper," he writes, "I shall give an account of the method that I discovered for solving this type of problem which may serve as an example of the geometry of position." In contrast to the two other cases discussed so far, Euler provides an explicit reflection on the reasoning process that led him to his discovery:

> The particular problem of the seven bridges of Koenigsberg could be solved by carefully tabulating all possible paths, thereby ascertaining by inspection which of them, if any, met the requirement. This method of solution, however, is too tedious and too difficult because of the large number of possible combinations, and in other problems where many more bridges are involved it could not be used at all.... Hence I discarded it and searched for another more restricted in its scope; namely, a method which would show only whether a journey satisfying the prescribed condition could in the first instance be discovered; such an approach, I believed, would be simpler. (Euler, 1953 <1741>, p. 67)

Euler started his search for such a method by simplifying the representation of the problem. Abstracting from the topography depicted in Fig. 6.8, he suggested reducing the land masses to points that are signified by capital letters and to represent the bridges not as such, but only as connections between these points. "ABDC" represents, thus, the walk from A to B, crossing a bridge, to D, and then to C, again by crossing bridges. From this it becomes clear that the number of points is always one greater than the number of bridges, and that the solution to the problem, if there is one, can be represented as a sequence of eight letters. This way, the problem is "now reduced to whether from the four letters A, B, C and D a series of eight letters can be formed in which all the combinations just mentioned [the number of pairs AB or

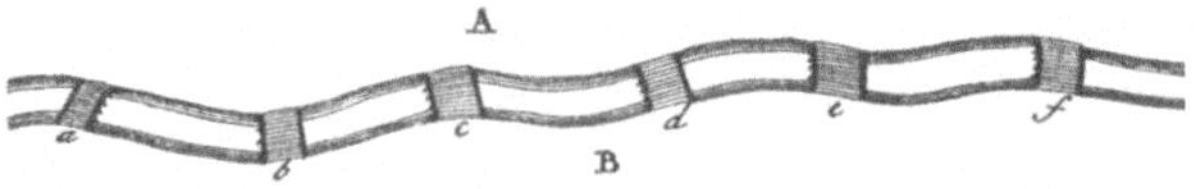

Fig. 6.9 Just two land masses A and B with bridges (Source: Euler, 1741, his Fig. 2)

BA in this sequence, for example, depends on the number of bridges] occur the required number of times" (p. 67). Before that, however, it would be useful to determine whether it is possible at all to find such an arrangement. So, the task that Euler puts in front of him is to find "a rule that would determine without difficulty" whether "the required arrangement of letters is feasible" for "this and all similar questions" (p. 67).

To find such a rule, he starts by reducing the problem's complexity by focusing on just two land masses as depicted in Fig. 6.9. Assume that there is just one bridge that may be crossed only once. It does not matter where you start, in the series of letters that describe the situation, "the letter A will appear exactly once" (p. 68). Similarly, if there are three bridges, then the letter A will occur twice, and if five bridges, then three times. Using this consideration, Euler goes back to the situation in Königsberg and represents the number of bridges that are connected with each land mass according to Fig. 6.8 and the corresponding occurrences of the letter representing this land mass in the final sequence in a table that I present here in a slightly modified form (Table 6.1).

At this point, the impossibility of finding a path on which each bridge is crossed only once becomes evident. We mentioned above that seven bridges require a sequence of eight letters, but if you add up all the occurrences of each letter in the table, then you get nine letters. "Thus it is apparent that a crossing of the seven bridges of Koenigsberg in the manner required cannot be affected," concludes Euler (p. 68). This particular proof can be generalized in the following rule:

> Using this method we are always able, whenever the number of bridges leading to a particular region is odd, to determine whether it is possible in a journey to cross each bridge exactly once. Such a route exists if the number of bridges plus one is equal to the sum of the numbers which indicate how often each individual letter must occur. (Euler, 1953 <1741>, p. 68)

With further considerations, Euler determines more rules such as about situations when the number of bridges is even. At the end, he is able not only to determine for all possible numbers of areas and bridges whether a path is possible, but also where to start the journey if there is such a path.

What kind of theoric transduction do we encounter in this case? In the case of Gauss, the algebraic formulas or strategies that he could have used were already given in algebra or in other fields of study like the classics regarding the tetractys. In the case of proving incommensurability, what got transduced from eleatic logic is the *reductio ad absurdum* strategy. The case of Euler is different because he had to *create* the perspective on the problem and a useful interpretation and representation of the problem situation that made it possible to prove the solution.

Table 6.1 The corresponding occurrences of the letter representing the land mass in the final sequence

Land mass	Bridges	Occurrences of letter
A	5	3
B	3	2
C	3	2
D	3	2

Two observations were crucial for Euler's discovery: (1) that the number of areas touched is always one more than the number of bridges, and (2) that the number of occurrences of each letter representing an area can be derived from the number of bridges that are connected to this area. This means, nothing is taken from another field of study or from a different system of representation—as when Gauss used algebra to solve a problem in arithmetic or the Greek mathematicians used eleatic logic to prove incommensurability in arithmetisized geometry. Instead, what is used comes from a deeper analysis of the problem situation. This is similar to the case of cutting the wooden block: The essential theoric step is looking at the problem from a new point of view and to create an *interpretation* of the problem situation that allows us to deduce the solution. I would still call this process of reasoning a theoric "transduction" because this interpretation was not explicitly or directly given in the problem formulation. The task description for cutting the block did not mention anything about the inner cube, and the formulation of the seven-bridges problem did not mention anything about the number of areas touched on the journey or the number of occurrences of each letter representing an area in the final sequence of areas touched. Euler transduced from a particular *interpretation* of the problem situation—which required a deeper analysis of particular components of the problem—a description of this problem situation (summarized in the two points above) from which the solution can be deduced.

6.5 Conclusion: Defining Theoric Transduction and a Typology

The example of cutting a wooden block into 27 equal cubes that I presented in the beginning of this contribution led to the distinction of three activities that seem to be required to solve this kind of problem:

1. The *theoric step*: We need to step back from a given problem formulation and look at the problem from a different point of view.
2. Finding a *representation* of the problem that is conducive to its solution.
3. *Theoric transduction*: Taking from outside of a given problem formulation something that allows us to solve the problem.

My analysis of the three cases from the history of mathematics was motivated by the question of what exactly theoric transduction is and Peirce's request for a historical study "of all the remarkable theoric steps and noticeable classes of theoric steps"

(Peirce CP 4.615). As becomes clear after the case studies, the notion of theoric transduction is more important than that of theoric steps and finding a suitable representation. Both these activities should be conceptualized as *preconditions* for theoric transductions. A shift of perspective cannot be sufficient because it does not include the identification of a means to solve the problem as theoric transduction does; nor is finding a representation sufficient because being conducive to a solution does not entail that such a representation implies a solution. For this reason, we should focus on a definition of *theoric transduction* and a classification of its various forms.

With regard to the latter, the case studies suggest two different distinctions: one according to the object that gets transduced, and another one regarding the question of whether the transduced object is historically new or already known. Both distinctions can be combined in the following Table 6.2 that represents various forms of theoric transduction.

This table can summarize only the beginning of work towards a typology. While the distinctions of the novelty dimension seem to be complete, there might be more objects that can be transduced.

This leaves the question of how we should *define* theoric transduction. Starting with the brief analysis of the wooden block example in the beginning, I used only a preliminary definition of theoric transduction as a "process of reasoning that takes something from outside of a given problem formulation to solve a problem." This definition should be refined with regard to the "something" that is taken, and it can be refined by replacing "process of reasoning" with "inference" as Peirce did when he not only conceptualized deduction and induction as inferences, but also abduction.

Let's start with what gets transduced, the object of a theoric transduction. The first column in Table 6.2 presents what we found in the cases discussed. To signify all these things with just one concept, I suggest "theoretical model," understood in the sense determined by Gerhard Schurz in his analysis of various types of abductive reasoning. Schurz defined the task of "theoretical model abduction" as "*finding theoretical (initial and boundary) conditions* which describe the causes of the

Table 6.2 Nine types of theoric transduction with regard to (1) the content of what gets transduced and (2) its novelty

		Known from other contexts	
What gets transduced is	Historically new	Culturally	For an individual
Formula, algorithm			Gauss
Proof strategy	Desargues's theorem[a]	Incommensurability[b]	Gauss
Interpretation of the problem	Euler		Wooden block

If the transduced object is not historically new, then it might be known in a culture but not by a particular individual. The latter distinction is important for distinguishing individual from cultural learning of a particular theoric transduction. For all types, what gets transduced is never part of the explicit problem formulation

[a]This case has been analyzed in Hoffmann (2010, pp. 579–584)

[b]The proof strategy was "culturally known" (from Zeno) but was "historically new" for solving the problem of proofing incommensurability

phenomenon" in the language of a scientific theory that is "*already established.*" For Schurz, the "semantic content" of the factual or lawlike statements that express these theoretical conditions is called "a theoretical model for a particular kind of phenomenon within an already given theory" (Schurz, 2008, p. 213; his italics). As an example of a theoretical model that provides an explanatory hypothesis he points to Archimedes' explanation of the phenomenon that certain substances like stones or metals sink in water while others like wood or ice swim.

> Archimedes' ingenious abductive conjecture was that the amount of water which is supplanted by the swimming or sinking body tends to lift the body upwards, with a force … which equals the weight of the supplanted water (…). If this force is greater than the weight of the body (…) the body will swim, otherwise it will sink. Since the volume of supplanted water equals the volume of the part of the body which is under water, and since the weight is proportional to the mass of a body, it follows that the body will sink exactly if its density (mass per volume) is greater than the density of water. (Schurz, 2008, p. 213)

This model describes exactly what will happen in dependence of the densities of the water and the object. This is a *theoretical* model because a theory is required to determine the relations of the components of the model. In the same sense, the idea of the six-sided cube in the middle of the wooden block is a theoretical model because it is determined by our "theory" of cubes; the strategies that Gauss might have used—as they are represented in Figs. 6.3 and 6.5 and the description of using averages—or the corresponding three formulas are theoretical models because the relations among the components of the models are determined by arithmetic or algebra; the proof strategy of *reductio ad absurdum* used in the arithmetic proof of incommensurability is a theoretical model because its sequence of steps—starting with assuming the opposite of the theorem to be proved and then demonstrating that it implies a logical contradiction—is determined by logic; and Euler's discoveries about the relationships between land masses and connecting bridges represented theoretical models of a theory whose scope did not reach further—at this point in time—than the models themselves.

Using this notion of theoretical model as developed by Schurz with regard to abduction we can define theoric transduction now as *a process of reasoning—or as an inference—that consists in finding a theoretical model that can solve a given problem and using it to solve the problem, whereby the model is not already part of the formulation of the problem*. It should be noted that the determination of whether a model is part of the problem formulation or not is not necessarily a question of the language used in the description of model and problem, and it does not require a complete description of the problem. For the incommensurability problem, for example, I never provided a concrete formulation of this problem. It was sufficient to mention that a relevant group of experts—the mathematicians in Athens a few years before Plato's death or just one of them—felt that the geometric proof did not meet the most recent standard for a mathematical proof. The problem, though, could be described in many different ways. It is questionable, for example, whether scholars at the time had the notion of "proof standard" in their conceptual toolbox. Maybe mathematicians would just have considered some of the problems of the geometrical proof that I discussed above. However, what is clear in this example is that the relevant proof strategy—the *reductio ad absurdum*—was in no way indicated in any of the many problem formulations that we can imagine.

Similarly, in the problem formulation of the wooden block example ("Is there a way to cut a cube into 27 cubes with fewer than six cuts?") there is no hint at the cube in the middle. And the same separation between problem formulation and theoretic model can be seen in the examples of Gauss ("How can I solve the task of adding the numbers from 1 to 100 in a less cumbersome way than just adding one by one?") and Euler ("Is it possible to devise a path so that one could cross each of the seven bridges of Königsberg exactly once and return home?"). In the last three cases the truth of the statement that the relevant theoretical model has not been part of the problem formulation can be determined by checking the elements to which the language of problem and model formulation refer—which requires that we need an agreement on what exactly the problem is and how it should be formulated. An argument for the claim, by contrast, for "not being part of the problem formulation" in the case of incommensurability would need to show that it is very unlikely that any problem formulation would ever include a reference to Zeno's proof strategy. So, depending on whether we have an agreed upon problem formulation available or not, we need a different argumentation strategy. It should be noted that there might be cases where the separation of problem formulation and theoretical model is far less clear than in the examples discussed here.

If we conceptualize theoric transduction as an inference, what is its structure? Since it involves a creative step just like abduction, we can structure the inferential form of a theoric transduction in analogy to the structure of abduction as Peirce suggested it in his *Lectures on Pragmatism*—see Table 6.3.

With this our task is completed. The three case studies from the history of mathematics illustrate how theoric transduction works in mathematical discoveries. I used these case studies to develop a definition of theoric transduction, to determine its inferential structure, and to distinguish various forms. Moreover, the case studies show that although theoric transduction requires creativity just as abduction does, it makes sense to conceptualize it as a sub-type of deduction based on the fact that the outcome is a solution to a problem that is necessarily true. I mentioned in the abstract of this chapter that an enriched understanding of human reasoning and creativity

Table 6.3 Theoric transduction structured as an inference in a form similar to the one suggested by Peirce for abduction

	Abduction (Peirce CP 5.189)	Theoric transduction
Premise 1	The surprising fact, C, is observed	A problem P is formulated
Premise 2	But if A were true, C would be a matter of course	If a theoretical model TM were found from which a solution of P could be deduced, then P would be solved
Premise 3		TM is found
Conclusion	There is reason to suspect that A is true	P is solved

might be useful for the creation of AI. The challenges would be huge. They range from how to deal with a plurality of different problem formulations for the same problem—which includes the challenge of how to draw the boundary between "the" problem and its context—to searching for theoretical models and testing whether these models are able to solve "the" problem. But even if it seems unlikely that any AI will ever be able to perform a complete theoric transduction for a historically new problem solution, there would already be much to gain if at least parts of the process could be realized by AI—because this can augment human reasoning. It would be helpful, for example, if large language models were able to identify all proof strategies that have ever been developed in whatever discipline or field of study. To do any of this, more needs to be done for a better understanding of reasoning.

Acknowledgements I want to thank Clint Johnson, J.P. Nelson, and two anonymous reviewers for their valuable comments. Special thanks to Michael Otte, Mircea Radu, and Johannes Lenhard for what they taught me about mathematical reasoning, and to Lorenzo Magnani for his many wonderful conferences and opportunities to learn.

References

Alexanderson, G. L. (2006). About the cover: Euler and Königsberg's Bridges: A historical view. *Bulletin of the American Mathematical Society, 43*(04), 567–574. https://doi.org/10.1090/s0273-0979-06-01130-x

Aristotle. *Met. The Metaphysics. In: Aristotle in 23 Volumes, Vols. 17, 18, translated by Hugh Tredennick,* 1989 <1933>. The Loeb classical library, vol. 271, 287. Cambridge (Mass.); London: Harvard University Press.

Aristotle. *An. Pr.* Analytica Priora. Edited and translated by Hugh Tredennick. In *Aristotle. The categories. On interpretation. Prior analytics*, 24a10–70b. The Loeb classical library 325. Cambridge, MA, London (1996 <1938>): Harvard University Press; Heinemann.

Bassok, M., & Novick, L. R. (2012). Problem solving. In K. J. Holyoak & R. G. Morriso (Eds.), *The Oxford handbook of thinking and reasoning* (pp. 413–432). Oxford University Press.

Burkert, W. (1972 <1962>). *Lore and science in ancient Pythagoreanism*. Harvard University Press.

Cellucci, C. (2013). *Rethinking logic. Logic in relation to mathematics, evolution, and method.* Springer.

Cellucci, C. (2017). Rethinking knowledge: The heuristic view. In *European studies in philosophy of science; volume 4*. Springer.

Cellucci, C. (2018). Theory building as problem solving. In D. Danks & E. Ippoliti (Eds.), *Building theories: Heuristics and hypotheses in sciences* (pp. 63–79). Springer.

Cooper, J. M. (1997). Introduction. In J. M. Cooper & D. S. Hutchinson (Eds.), *Plato: Complete Works*. Hackett Publishing Company.

Diels, H., & Kranz, W. (1989). *DK. Die Fragmente der Vorsokratiker*. Erster und Zweiter Band. 18. Aufl. Zürich 1989: Weidmann.

Duncker, K. (1945). *On problem-solving*. Trans. Lynne S. Lees. The American Psychological Association.

Euclid. (1956). *The thirteen books of the elements*. Ed. and transl. by Thomas L. Heath in 3 vols. Dover Publications

Euler, L. (1741). Solutio problematis ad geometriam situs pertinentis. *Commentarii academiae scientiarum Petropolitanae*. Retrieved from https://scholarlycommons.pacific.edu/euler-works/53/

Euler, L. (1953 <1741>). The Koenigsberg Bridges (transl. of Solutio problematis ad geometriam situs pertinentis). *Scientific American, 189*, 66–70.

Fowler, D. H. (1979). Ratio in early Greek mathematics. *Bulletin of the American Mathematical Society, 1*(6), 807–846. https://doi.org/10.1090/s0273-0979-1979-14684-6

Fowler, D. H. (1999 <1987>). *The mathematics of Plato's Academy: A new reconstruction*. 2nd ed. Clarendon Press; Oxford University Press.

von Fritz, K. (1945). The discovery of incommensurability by Hippasus of Metapontum. *Annals of Mathematics, 46*(2), 242–264.

Gabbay, D. M., & Woods, J. (2005). The reach of abduction. Insight and trial. In *A practical logic of cognitive systems* (Vol. 2). Elsevier.

Gillies, D. (2009). Problem-solving and the problem of induction. In Z. Parusniková & R. S. Cohen (Eds.), *Rethinking Popper* (pp. 103–115). Springer Netherlands.

Grosholz, E. (2007). *Representation and productive ambiguity in mathematics and the sciences*. Oxford University Press.

Grosholz, E. (2016). Starry reckoning: Reference and analysis in mathematics and cosmology. In *Studies in applied philosophy, epistemology and rational ethics, volume 30*. Springer.

Hayes, B. (2006). Gauss's day of reckoning. A famous story about the boy wonder of mathematics has taken on a life of its own. *American Scientist*. Retrieved from https://www.jstor.org/stable/pdf/27858762.pdf

Hoffmann, M. H. G. (2005). *Erkenntnisentwicklung. Ein semiotisch-pragmatischer Ansatz*. Klostermann.

Hoffmann, M. H. G. (2010). "Theoric transformations" and a new classification of abductive inferences. *Transactions of the Charles S Peirce Society, 46*(4), 570–590. https://doi.org/10.2979/trancharpeirsoc.2010.46.4.570

Holyoak, K. J. (2012). Analogy and relational reasoning. In K. J. Holyoak & R. G. Morrison (Eds.), *The Oxford handbook of thinking and reasoning* (pp. 234–259). Oxford University Press.

Holyoak, K. J., & Thagard, P. (1995). *Mental leaps. Analogy in creative thought*. MIT Press.

Høyrup, J. (1990). Dýnamis, the Babylonians, and Theaetetus 147c7–148d7. *Historia Mathematica, 17*(3), 201–222. https://doi.org/10.1016/0315-0860(90)90001-T

Huffman, C. (2020). Archytas. *The Stanford Encyclopedia of Philosophy (Winter 2020 Edition)*. Retrieved from https://plato.stanford.edu/archives/win2020/entries/archytas/

Ippoliti, E. (2018). Heuristic logic. A Kernel. In D. Danks & E. Ippoliti (Eds.), *Building theories: Heuristics and hypotheses in sciences* (pp. 191–211). Springer.

Jacobs, H. R. (1970). Mathematics. A human endeavor. A textbook for those who think they don't like the subject. In *A series of books in mathematics*. Freeman.

Kirk, G. S., & Raven, J. E. (1964 <1957>). *The presocratic philosophers. A critical history with a selection of texts*. Cambridge Univ. Press.

Kneale, W., & Kneale, M. (1988 <1962>). *The development of logic*. Clarendon Press.

Knorr, W. R. (1975). *The evolution of the Euclidean elements: A study of the theory of incommensurable magnitudes and its significance for early Greek geometry*. Synthese historical library, vol. 15. Dordrecht, Holland; Boston: D. Reidel Pub. Co.

Knorr, W. R. (2004 <1981>). On the early history of axiomatics: The interaction of mathematics and philosophy in Greek antiquity. In Christianidis, J. (Ed.), *Classics in the history of Greek mathematics* (pp. 81–109). Springer Netherlands.

Lakatos, I. (2015 <1976>). *Proofs and refutations: The logic of mathematical discovery*. Edited by John Worrall and Ellie Zahar. Cambridge philosophy classics edition. Aufl. Cambridge University Press.

Laudan, L. (1977). *Progress and its problems: Toward a theory of scientific growth*. Routledge & Paul.

Laudan, L. (1981). A problem-solving approach to scientific progress. In I. Hacking (Ed.), *Scientific revolutions* (pp. 144–155). Oxford University Press.

Magnani, L. (2009). *Abductive cognition. The epistemological and eco-cognitive dimensions of hypothetical reasoning. Cognitive Systems Monographs* (Vol. 3). Springer.

Magnani, L. (2017). *The abductive structure of scientific creativity: An essay on the ecology of cognition*. Studies in applied philosophy, epistemology and rational ethics. Springer: Imprint.
Mittelstraß, J. (1965). Die Entdeckung der Möglichkeit von Wissenschaft. *Archive for History of Exact Sciences, 2*, 410–435.
Nersessian, N. J. (2022). *Interdisciplinarity in the making: Models and methods in frontier science*. MIT Press.
Newell, A., & Simon, H. A. (1972). *Human problem solving*. Prentice-Hall.
Nickles, T. (1981). What is a problem that we may solve it? *Synthese*, 85–118.
Panchenko, D. (2017). Zeno's debt to Hippasus. *Hyperboreus, 23*(2), 208–223.
Pedemonte, B., & Reid, D. (2011). The role of abduction in proving processes. *Educational Studies in Mathematics, 76*(3), 281–303. https://doi.org/10.1007/s10649-010-9275-0
Peirce. CP. *Collected Papers of Charles Sanders Peirce (Volumes I-VI, ed. by Charles Hartshorne and Paul Weiss, 1931–1935, Volumes VII-VIII, ed. by Arthur W. Burks,* 1958*; quotations according to volume and paragraph)*. Cambridge, Mass.: Harvard UP.
Peirce. MS. *The Charles S. Peirce Papers. Manuscript Collection in the Houghton Library, Harvard University*. numbered according to Richard S. Robin, Annotated Catalogue of the Papers of Charles S. Peirce. Worcester, Mass., 1967: The University of Massachusetts Press. Available in the Peirce Microfilm edition. Pagination: "CSP" refers to Peirce's own pagination, "ISP" to that of the Institute for Studies in Pragmaticism.
Pietarinen, A.-V., & Bellucci, F. (2014). New light on Peirce's conceptions of retroduction, deduction, and scientific reasoning. *International Studies in the Philosophy of Science, 28*(4), 353–373. https://doi.org/10.1080/02698595.2014.979667
Plato. *Parm.* Parmenides. In *Platonis Opera*, ed. Ioannes Burnet, 126–166. Oxford 1900a ff.: E Typographeo Clarendoniano.
Plato. *Phdr.* Phaedrus. In *Platonis Opera*, ed. Ioannes Burnet, 227–279. Oxford 1900b ff.: E Typographeo Clarendoniano.
Plato. Leg. Leges. In *Platonis Opera*, ed. Ioannes Burnet, 624–969. Oxford 1907: E Typographeo Clarendoniano.
Plato. Theaet. Theaetetus. In *Complete Works*, eds. John M. Cooper, and D. S. Hutchinson, 157–234; quoted according to the standard Stephanus-pagination. Indianapolis; Cambridge 1997: Hackett Publishing Company.
Popper, K. R. (1972). *Objective knowledge: An evolutionary approach*. Clarendon Press.
Rist, J. M. (1989). *The mind of Aristotle: A study in philosophical growth*. Phoenix supplementary volume series, vol. 25. Toronto; London: University of Toronto Press.
Schurz, G. (2008). Patterns of abduction. *Synthese, 164*(2), 201–234. https://doi.org/10.1007/s11229-007-9223-4
Sextus Empiricus. 2005. *Against the logicians*. Trans. Richard Bett. Cambridge texts in the history of philosophy. Cambridge University Press.
Simon, H. A. (1989). The scientist as problem solver. In D. Klahr & K. Kotovsky (Eds.), *Complex information processing: The impact of Herbert A. Simon*. Psychology Press.
Szabó, Á. Á. (1978). *The beginnings of Greek mathematics*. Synthese historical library Vol. 17. D. Reidel Pub. Co.
Vapnik, V. N. (1998). *Statistical learning theory. Adaptive and learning systems for signal processing, communications, and control*. Wiley.
van der Waerden, B. L. 1979. *Die Pythagoreer: Religiöse Bruderschaft und Schule der Wissenschaft*. Die Bibliothek der Alten Welt: Reihe Forschung und Deutung. Artemis.
Woods, J. (2023). How abduction fares in mathematical space. In L. Magnani (Ed.), *Handbook of abductive cognition* (pp. 115–157). Springer.
Zhmud, L. (2012). *Pythagoras and the early Pythagoreans*. Oxford University Press.

Chapter 7
How to Solve It Afresh. Some Remarks on How to Solve the Problem of… How to Solve a Problem

Emiliano Ippoliti

Abstract The study of knowledge acquisition and scientific discovery has evolved into its own field thanks to the development of the heuristic perspective. This approach suggests that these processes can be understood through logic, rationality, and problem-solving techniques. Unlike traditional views, which conceive the justification of knowledge logic and rational, the heuristic perspective conceives problem-solving as a means to generate new knowledge in a logical and rational way.

Recent advancements in this view have improved our understanding of problem-solving, proposing new methods and even algorithms that could potentially automate these processes. This perspective builds on ideas from philosophers like Plato, Aristotle and Lakatos and has been further developed by scholars such as Popper, Simon, Cellucci, and Magnani. The paper examines the heuristic view, its fresh conceptualization of problem-solving and discovery, and its implications for fields like Big Data and machine learning.

Keywords Problem-solving · Philosophy · Scientific discovery · Heuristics · Machine-learning

7.1 The Problem of Knowledge Advancement and the Science of Problem Solving

Although there are illustrious precursors in the development of the science of problem solving and scientific discovery (e.g., Plato, Aristotle, Bacon, Mill), it is only recently that this issue has received all the attention it deserves—see for example several works of Popper, Polya, Lakatos, Magnani, Cellucci, Ippoliti, Gillies, Grosholz. The fundamental obstacle to developing this science has been removing

E. Ippoliti (✉)
Department of Philosophy, Sapienza University of Rome, Rome, Italy
e-mail: emiliano.ippoliti@uniroma1.it

S. Arfini (ed.), *Scientific Cognition, Semiotics, and Computational Agents: Essays in Honor of Lorenzo Magnani - Volume 2*, Synthese Library 506,
https://doi.org/10.1007/978-3-031-96688-0_7

the idea of the romantic genius and its limiting, black-box account of it. There are various versions of the science of problem-solving (different problem-solving approaches), which differ on several fundamental issues, and one is the possibility of automating problem-solving and scientific discovery, leading to the development of algorithms capable of doing so. This aspect is particularly relevant due to its implications for the Big Data Revolution and Machine learning.

The problem of knowledge acquisition and scientific discovery has gradually become a topic for philosophical and scientific investigation, and a science for itself. This is a merit of the developed of the heuristic view (see Lakatos, 1976, 1978; Cellucci, 2013, 2015, 2017; Magnani 2003; Magnani et al, 1999; Ippoliti, 2015, 2018a, 2018b, 2022), which maintains that is possible to account for it in terms of 'logic', 'rationality', and inference. This view contrasts the mainstream view (see e.g. Frege, 1960; Einstein, 1958), which argues that is possible to provide an account in terms of logic and rationality only for justification of knowledge. In more detail, the heuristic view has shown that we can treat the problem of knowledge acquisition and scientific discovery in terms of problem-solving and has contributed to develop a better and better science of problem solving. Thus, this view has shown that we can answer the question 'how can we produce new knowledge?' by answering the question 'how can we solve scientific problems?', and has provided rational, replicable procedures to do this (that is, *finding* new problem and *solving* existing, unsolved ones). Basically, it argues that we are justified in moving the inquiry from *How to discover it?* to *How to solve it?*

While *how to solve it?*—to use Polya's (1945) famous title—is the crucial question that we should consider, the answer to this question has evolved and improved, and a new, fresh way of 'solving it' has developed. In particular, this new way has naturalized the process of problem-solving—what I call *how to solve it afresh.*

The roots of the more developed version of the heuristic view can be traced back to the work of Lakatos in the philosophy of mathematics. Furthermore, in the broader history of science and philosophy, this view can be linked to Plato's ideas on the analytic method. Over the past few decades, the heuristic view has undergone significant evolution, with notable contributions from scholars such as Popper, Simon, Poincaré, Polya, Lakatos, Nickles, Cellucci, Gillies, and Magnani. These contributions have yielded valuable insights into both aspects of knowledge acquisition.

The heuristic view maintains that knowledge acquisition is based on problem solving and this is seen as a rational process. This paper provides a heuristic view of the Heuristic View, that is, it examines how the problem of problem-solving has been conceptualized, what are the main epistemological and ontological hypotheses, and the several methods, developed to address this issue, and assesses their plausibility. A fundamental issue is the possibility of automating problem-solving and scientific discovery, leading to the development of algorithms capable of doing so. This aspect is particularly relevant for its implications for the Big Data Revolution and machine learning.

Now, the central epistemic and methodological question of the problem of how to solve a problem is: how can we produce *new* knowledge?

To answer this question, first we must clarify what we mean by *new* knowledge. Novelty can be characterized at least in two ways—*strong* and *weak* (see e.g. Ippoliti, 2024)—and, moreover, it is a *positional* concept. This means that knowledge can be defined as 'new' in relation to something *already* known at a *certain time*. What we know is the basis for exploring the 'unknown' and therefore solving unsolved problems often requires us to attempt to *break conceptual constraints* posed by what we know (i.e. current theories). In effect, the heuristics of *constraint-breaking* is one of the most powerful heuristic procedures that scientists follow in their attempt to advance the frontier of knowledge (Nickles, 1980a, 1980b, 1981, 2013, 2015, 2016, 2018; Ippoliti, 2015, 2016).

The heuristic philosophy of science showed us that to answer the central question of how we can produce new knowledge, we need to answer the question how we can solve scientific problems and has developed procedure and inferential tools to do it. Under this respect, the first step to solve a problem is to find the relevant variables for a candidate solution for the problem at hand. As we will see, I argue that we need a theory to do this, as an inductive, pure mechanical, algorithmic procedure is not enough. In particular, an algorithm cannot solve circular problems such as the *circularity between similarity and relevance* (see Sect. 7.6.1). Thus, we can automatize only a part of this process.

The second step is to identify, or better construct, the element of the problem: initial state, final state, constraints, operators, and rational and inferential procedures to generate hypotheses to solve the problem (see Simon, 1977, 1987; Cellucci, 2013). Thus, the crucial question of the science of problem-solving (and scientific discovery) is how to produce a rational account for hypothesis-generation.

7.2 How to Solve It: Then and Now

Answering the philosophical question how to produce *new* knowledge, one should first answer the question of how knowledge progresses and has progressed throughout history. But the answer to this question has changed and progressed throughout history (!). This progress and change are visible especially in the last few decades, when we have dismissed an account prominently anecdotical and psychological (*then*) and we have developed an inferential and 'logical' one (*now*).

The present, most influential, and actionable answer to this question holds that the advancement of knowledge takes place through the solution of problems (see e.g. Russell, 1912; Popper, 1961, 1999; Simon, 1977, 1987; Lakatos, 1976, 1978; Gillies, 2014, 2016; Cellucci, 2013, 2017; Ippoliti, 2015; Weisberg, 2006). In effect, the development and refinement of a theory of problems is one of the major contributions of the philosophy of science and its heuristic tradition. This tradition has produced a better and better theory of problem-solving that rationally accounts for both its aspects (see Simon, 1987), that is, how new and relevant problems are formulated (*problem-posing*) and how they are solved once formulated (*problem-solving*).

The development of an actionable theory of scientific problem-solving and discovery has provided us with a conceptual framework showing that:

1. Problem-solving cannot be achieved (solely) by deductive reasoning.
2. We need to *naturalize* scientific discovery and problem-solving.
3. Models of scientific discovery and problem-solving do exist.

Moreover, the development of an actionable theory of scientific problem-solving and discovery has been pursued in different ways and degrees by the main approaches to problem-solving theory over the past few decades, which are essentially three:

(i) the Popper-Laudan approach.
(ii) the Poincaré-Simon approach.
(iii) the Lakatos-Cellucci approach.

In more detail, these three approaches:

(a) Have progressively removed one of the main obstacles to producing a rational account for hypothesis-generation (i.e., the core of problem solving), the idea of the *romantic genius,* which maintains that the production of new ideas and hypotheses is subjective and inscrutable. This tenet has been shared by famous philosophers and scientists (e.g. Frege, 1960; Einstein, 1958), who supported an account for hypothesis-generation based on notions like intuition, creativity, or insight, which are conceptual black-boxes and do not offer actionable explanations.

(b) Have provided us with models and conceptual frameworks for scientific discovery and problem-solving. The main models are (see Sect. 7.4):

- The **computational model** (e.g. Simon, 1977, 1987).
- The **evolutionary model** (e.g. Campbell, 1960; Kantorovich, 1989, 1993, 1994).
- The **analytic model** (e.g. Cellucci, 2013, 2017).

The *computational model* maintains that we can automatize a great part of scientific discovery by means of algorithmic procedures. The *evolutionary model* maintains that the mechanism of evolutionary process, namely blind variations and selective retention (BV + SR), provides us with an explanation for discovery. The *analytic model maintains* that discovery can be guided by the analytic method supplemented with a set of heuristic rules (ampliative inferences).[1]

These approaches and models have given us a more refined and detailed understanding of the methods and tools available for generating new knowledge. This represents a significant contribution from the philosophy of science to science itself, positioning it not only to interpret scientific practices but also to offer and enhance tools that scientists can actively use in their work.

[1]There are other approaches like abductive inference and serendipity. However, I won't explore them here due to space limitations and because they do not offer comprehensive 'models': abductive inference doesn't specify how to generate a hypothesis (it assumes the hypothesis is already known before applying the inferential rule), while serendipity largely depends on chance.

It is worth noting here that while (i)–(iii) are historically and conceptually interwoven, their outcomes differ in many ways (see for example Ippoliti, 2018a, 2018b; Ippoliti & Cellucci, 2016).

The position held by Popper and Laudan (Popper, 1999; Laudan, 1977, 1981) argues that while knowledge advances by problem-solving, hypothesis-generation is a subjective and idiosyncratic process and cannot be rationally or logically investigated, reconstructed, or transmitted. Basically, this position argues that a solution to a problem is worked out by random trial and error and under this respect, this position does not dismiss the romantic tradition. This specific thesis is challenged by the other two approaches, which argue that rational procedures to solving problems and generating new hypotheses can be provided.

As a matter of fact, the Poincaré-Simon tradition argues that knowledge progresses through problem-solving, and unlike the Popper-Laudan approach, it asserts that we can rationally explain how hypotheses are generated. Poincaré (1908) suggests that it's possible to express and somewhat describe the subconscious processes involved in seeking solutions. He proposes that when conscious efforts to solve a problem fail, the unconscious takes over, generating various combinations based on specific ideas. The subconscious first mixes these ideas and then selects a smaller subset. Wallas (1926) and Hadamard (1958) further developed Poincaré's view with Hadamard's well-known four-stage model: (i) preparation → (ii) incubation → (iii) illumination → (iv) verification. According to Poincaré, during incubation, the unconscious works in parallel (unlike conscious thought, which is serial) and generates a wide range of ideas in a combinatorial explosion. While many of these combinations are irrelevant, a few offers potential solutions to the problem at hand.[2]

Simon (1977, 1987) refines this account by offering an explanation and a formalization of incubation and illumination in hypotheses-generation. Simon's well-known outcome is a theory of problems that (a) conceptualizes the 'problem space' into the initial state, the end state, the operators, and the constraints, and (b) argues for the 'mechanization' of scientific hypothesis-generation.

The third perspective, represented by Lakatos and Cellucci embodies the modern 'heuristic view' (see Cellucci, 2013; Ippoliti, 2015; Jaccard & Jacoby, 2010). In this context, heuristics pertains to the study of the methods and principles of discovery and invention (see Polya, 1954), which diverges significantly from the common psychological interpretation, where heuristics are understood as mental shortcuts used by humans in decision-making (see, for example, Gigerenzer & Todd, 1999).

The Lakatos-Cellucci position contends that there is a method for generating hypotheses that, as such, may be examined, transmitted, and learned in a rational and logical manner.[3] But this is not to say that we must accept that this method is algorithmic, at least not in all parts. In addition, this view provides a way to account

[2] Nevertheless, Poincaré's account does not offer a cogent methodology for analysing how these combinations are formed and selected in a rational way. In effect, Poincaré's solution employs unclear criteria to make this selection like mathematical 'beauty'—intended as 'harmony', 'symmetry', or 'simplicity' on a case-by-case basis.

[3] See also Petroni (1992) on the logic of discovery.

for hypothesis-generation that reunites problem-solving and problem-finding, as well as discovery and justification, within a single coherent framework by showing that they are not separate as the orthodox philosophy of science maintains.

The most recent version of the heuristic view argues that a method for hypothesis-generation and discovery is the 'analytic method' (appropriately revised), which must be supplemented with a set of heuristic rules, namely inferences such as induction, analogy, metaphor (see e.g. Cellucci, 2013, chapters 20–21; Jaccard & Jacoby, 2010; Ippoliti, 2018a, 2018b). The heuristic view maintains that a hypothesis, once introduced, must satisfy the requirement of 'plausibility', and not 'truth', that is, "the arguments for the hypothesis must be stronger than the arguments against it, on the basis of experience" (Cellucci, 2013, 56). This of course implies that solutions to problems are not certain, but only plausible and revisable.

The method to work out solutions to problems is a revised version of the analytic method as put forward by Plato, and it states that to solve a problem, one looks for a hypothesis that is a *sufficient* condition for solving the problem, that is, such that a solution to the problem can be deduced from it. The hypothesis is obtained from the problem, and possibly other already available data, by non-deductive rules and it must be plausible. But in turn the hypothesis is a problem that must be solved, and it is solved in the same way. That is, "one looks for another hypothesis that is a sufficient condition for solving the problem posed by the previous hypothesis; it is obtained from the latter, and possibly other already available data, by a non-deductive rule, and it must be plausible. And so on, ad infinitum" (*Ibid.*, 55). Thus, this version of the analytic method explicitly draws on an 'infinitist' solution to the regress argument and argues that knowledge is provisional and can be revised as new data come out.

7.2.1 The Naturalization of the Science of Problem-Solving

As I have underlined, a key step in achieving the science of problem solving and scientific discovery was the gradual *naturalization* of the process underlying it—hypothesis generation—and the removal of the idea of Romantic Genius, that is, the tenet that discovery and problem solving are fundamentally black boxes. This is one of the main features telling apart not only the orthodox view (e.g. Frege, 1960; Einstein, 1958; Popper, 1961) and the heterodox view (e.g. Cellucci, 2013, 2015; Ippoliti, 2018c), but also the *first* approaches to problem-solving and discovery (e.g. Popper, 1999) and the *present* ones, especially the heuristic view.

The view that supports the idea of genius can be seen as a type of supernaturalism, meaning it explains certain aspects of the world by invoking supernatural entities or events. Given that various forms of naturalism exist (see e.g. Cellucci, 2015), it's essential to clarify the specific type of naturalism being referred to here. The form used here asserts that "the world does not include supernatural entities or events, and no aspect of the world, including ourselves, is to be explained in terms of supernatural entities or events – entities or events that stand outside the world and whose doing cannot be understood as part of it" (Cellucci, 2015, 219). This means that we can explain problem-solving and discovery without using opaque concepts as *insights*, *genius*, but rational and inferential means.

In this perspective, the genius, or the insight, is a natural phenomenon: she/he is simply one using these means mostly unconsciously and who is well trained in doing this. Thus, real of fictional examples of genius like Ramanujan, Will Hunting, do no possess any unique set of special or unfathomable qualities: on the contrary, they apply techniques that can be transferred, learnt, and trained. Thus, this form of naturalism contrasts Quine's naturalism, according to which epistemology "is contained in natural science, as a chapter of psychology" (Quine, 1969, 83), and it maintains that that psychology is not enough to deal with problem-solving and scientific discovery.

7.3 New Knowledge and Breaking Constraints

As I have stressed, the central epistemic and methodological question of the science of problem-solving and discovery is how to produce *new* knowledge. Now, there are two main ways of characterizing the concept of *'new'*, or *'novelty'*, that is what can be considered as new knowledge. First, we can qualify 'novelty' *strictu sensu* (or strong novelty) as the inferential production of information in the conclusion that is not contained in the premise of a reasoning (Cellucci, 2013; Ippoliti, 2015). This characterization has consequences for the methods and tools for advancing knowledge. For instance, deduction is *useful* for our knowledge, but not logically ampliative: it remains an inference that does not extend our logical knowledge, but it is useful because a reformulation of the premises can make explicit information that is only implicit, thus facilitating the comparison of the premises with the data and experience.

Second, we can qualify weak novelty as the inferential production of new logical relations between known entities. In this sense, deduction can provide us with 'weak new knowledge' and a typical deductive activity, such as re-proving (or re-deducing) theorems, can be regarded as *weakly* ampliative because it establishes new logical relations between known elements (like axioms or properties) and theorems, providing a new justification of it.

Thus, under the 'weak' characterization, not only a proposition establishing a new theorem counts as a genuine form of ampliating our knowledge, but also the production of new logical relations between known findings counts as a form of ampliating knowledge, for example re-proving known mathematical theorems (see e.g. Ippoliti, 2024).

Moreover, at least theoretically, novelty is a *positional* concept. Knowledge can be defined as 'new' only in relation to something already known at a certain time. Therefore, what we know is the basis for exploring the 'unknown', and something is unknown only in relation to the known. This could seem a trivial point, but it is important to remember it since it tells us what the main drive of solving unsolved problems and discovery is, that is, the attempt to break conceptual constraints posed by what we know (i.e. present theories). In effect, this is one of the most powerful heuristic procedures that scientists follow in the attempt to advance at the frontier of knowledge (see Nickles, 2015; Ippoliti, 2015). The heuristics of *constraint-breaking* (see Ippoliti 2016) identifies one or more constraints (or solvency conditions) of a problem, i.e. properties that the solution of the problem includes now, and which it violates deliberately by positing that they do not hold. Many important scientific problems have been solved precisely by violating the constraints until then considered firm: for example, Kepler breaking the constraint that the orbits of the planets are circular is one of the most important heuristic procedures that led to the eventual formulation of Newtonian celestial mechanics. One difficulty of this procedure is that often these constraints are implicit, and to make them explicit is an essential part of solving problem and discovery.

7.4 How Scientific Problems Are Solved: A Heuristic Pint of View

As I have claimed, the present science of problem-solving and discovery has improved and outdated the old version of it by producing several models to account for scientific how hypotheses are generated. The main model three are:

- The **computational model.**
- The **evolutionary model.**
- The **analytic model.**

Let us examine them in more detail.

7.4.1 The Computational Model

The computational model posits that a substantial portion of scientific discovery can be automated since heuristic procedures can be applied to generate hypotheses within the problem-space. This means that hypothesis generation can generally be directed by heuristic techniques: "the search for a solution to a problem is not carried on by random trial and error, but it is guided in the direction of a goal situation (or symbolic expressions describing a goal) by rules of thumb, called heuristics" (Simon, 1987, 4). Simon identifies two types of heuristics, weak and strong, which "make use of information extracted from the problem definitions and the states already explored in the problem space to identify promising paths for search" (*Ibid.*).

A key aspect of Simon's approach is the concept of mechanizing discovery, as demonstrated in the various iterations of the BACON software: "our method of inquiry will be to build a computer program (actually a succession of programs) that is capable of making nontrivial scientific discoveries and whose method of operation is based on our knowledge of human methods of solving problems-in particular, the method of heuristic selective search" (*Ibid.*, 5). The central example Simon offers is the discovery of Kepler's law using the BACON software (version 1), which employs a set of straightforward mathematical rules, that is, recursive procedures that identify regularities in datasets.

7.4.2 *The Evolutionary Model*

This model naturalizes scientific discovery and incorporates evolutionary epistemology through an explicit analogy with biology, specifically "the metaphor of an organism as a problem-solver" (Kantorovich, 1993, 224)—it also integrates the concept of serendipity. This perspective relies on the mechanisms of evolutionary processes—namely, blind variation and selective retention (BV + SR)—as a framework for explaining discovery. According to this view, the development of a theory is analogous to the growth of an organism: "both a research program, and an organism are engaged in problem-solving. A research program solves, for instance, problems arising from the need to adjust its basic ideas [...] to new observational data. Similarly, an organism adapts to new environmental conditions on the basis of its genetic makeup" (*Ibid.*).

In fact, BV + SR mechanisms appear to be "incredibly creative problem solvers" (Nickles, 2009, 193) and "biological evolution can be interpreted, metaphorically, as an innovative problem-solving process" (*Ibid.*, 182). A notable advantage of this approach is that it seems to enable generating more knowledge from less knowledge: the recombination and blind variation of a few simple elements can yield something more complex and sophisticated.

A key aspect of the evolutionary approach to scientific discovery is that science does not pursue a predetermined goal. Additionally, the human mind serves as the 'host' for the process of blind variation and selective retention, and "these processes can be categorized as unintentional or involuntary. The incubation process is well known but is not well understood. It can be explained by the model of natural selection which is applied to mental elements created quasi-randomly in the discoverer's mind; the discoverer hosts the process in his mind, so to speak" (Kantorovich, 1994, 20). Tellingly, this conception of the mind as a mere vessel for a combinatorial-selective process supports a computational take on the evolutionary approach, which employs genetic algorithms (see Koza, 1992, 1994; Koza et al., 1999).

7.4.3 The Heuristic Model

This approach naturalizes scientific discovery by demonstrating that knowledge advancement is guided by methods, even if not in a strictly algorithmic manner. Specifically, knowledge progresses through analytic methods combined with an open set of ampliative rules.

This version of analytical methodology refines the approach used by Plato: it follows a bottom-up path from problems to hypotheses that are non-deductively derived from them. The idea is that to solve a problem, one seeks a hypothesis that serves as a sufficient condition for the solution. The hypothesis, which must be plausible, is obtained from the problem and potentially other available data using a non-deductive (ampliative) rule.

The set of ampliative rules includes inductions, analogies, metaphors, what-if inferences, and more (see Ippoliti, 2018a, 2018b, 2022; Ippoliti & Cellucci, 2016; Cellucci, 2013, 2017; Jaccard & Jacoby, 2010 for detailed discussions of these ampliative procedures).

An example is the famous Fermat's Last Theorem. Ribet addressed it by showing that the Taniyama-Shimura conjecture is a sufficient condition for its solution. For, let E be an elliptic curve over Q: the Taniyama-Shimura Conjecture (also known as the Weil-Taniyama Conjecture) states that E is modular. Ribet demonstrated that the Taniyama-Shimura conjecture implies Fermat's Last Theorem. Wiles (and Taylor) then solved the problem posed by the Taniyama-Shimura hypothesis by demonstrating that certain other hypotheses are sufficient conditions for its solution, and so forth.

This model, therefore, offers a variety of rational tools and rules for advancing scientific knowledge. By highlighting the role of ampliative inferences in scientific discoveries, it underscores the importance of interdisciplinary and intradisciplinary research (see Darden, 2006) in the quest for new knowledge. Emily Grosholz (2007) also illustrates this point with examples of discoveries enabled by what she terms 'productive ambiguity.'

This model also has the advantage of integrating a psychological perspective, such as Poincaré's (1908). In fact, the heuristic model reveals that concepts like 'intuition' or 'insight' are merely unconscious non-deductive inferences. Why do only certain scientists experience these insights? Because they possess adequate background knowledge and have intensely considered the problem. This provides them with data to make competent conjectures through unconscious non-deductive reasoning. In fact, many scientific concepts and results defy intuition (hence they are often labelled as 'monsters').

Moreover, the heuristic view overcomes the limitations of the evolutionary approach, which relies on 'discovery by chance.' It demonstrates that so-called serendipitous discoveries, such as those attributed to Fleming, Columbus, or Poincaré, were achieved through numerous inferences, strategies, and plans—all rational tools. While the final outcome may have been unexpected, it emerged after many possibilities were explored and dismissed by the discoverer or others researchers (as in Poincaré's mathematical discoveries), or after several controlled experiments (as in Fleming's penicillin discovery), or because the final discovery was not ruled out from the outset (as with Columbus, who could not exclude the possibility of encountering new lands while attempting to circumnavigate the earth).

Naturally, these three models of scientific discovery have differences and limitations. For instance, the evolutionary model only explains certain aspects of scientific discovery, and the notion of serendipity as a general model is untenable. First, such discoveries are rare and mostly epiphenomenal. Second, biology and evolution can offer a more rational explanation for scientific discovery. In fact, it can be argued that evolution shows that biological processes involve inferential problem-solving even at very basic and primitive levels (see von Helmholtz, 1866; Gregory, 1970, 1980). Vision serves as a prime example in this regard. Our visual apparatus can be understood in inferential terms, as it constantly formulates hypotheses to solve problems—it is not a 'camera' but an inferential engine (see, e.g., Cellucci, 2013, 2017). These inferences are ampliative, unconscious, non-propositional, and built into us by evolutionary adaptation: they process data (like light beams) and generate hypotheses using non-deductive rules based on past experiences.

7.5 Procedures to Solve a Problem: The Heuristic Way

The heuristic perspective has enriched the science of problem-solving and discovery by analysing various rational, inferential, and 'logical' methods for generating hypotheses to address problems. Here, heuristics refer to the study of methods and rules for discovery and invention (see Polya, 1954), which differs significantly from the common psychological definition—namely, the mental shortcuts people use for decision-making (see, e.g., Gigerenzer & Todd, 1999). The primary aim of a heuristic, therefore, is to direct the search for a hypothesis to solve a problem. Specifically, it is a method for constructing the problem space and navigating it by identifying which paths are rational to pursue and which should be disregarded. Broadly speaking, and in line with Lakatos (1978), heuristics can be categorized into positive and negative types. Additionally, the study and classification of heuristics can be further refined by distinguishing between primitive heuristics and derived heuristics.

The distinction made by Lakatos defines two main aspects of research, namely the methodological rules that tell us which paths of research to avoid—negative *heuristics*—and which paths to follow—positive *heuristics*. Both obviously have a great impact on problem solving because not only do they greatly reduce the problem space and the combinatorial explosion that can take place within it, but because they literally contribute to shaping it.

Moreover, Lakatos observes an important relationship between heuristics and metaphysics (which cannot be explored in detail here).[4] He argues that metaphysics (where metaphysics means the basic notion of assertion that does not admit potential falsifiers) can be a heuristic device, i.e. it provides a tool to guide research through flexible principles, which allow problems to be posed as well as solved. As an example of such heuristic metaphysics, Lakatos brings Cartesian metaphysics, i.e. the mechanistic theory of the universe according to which the universe is a huge clock (and system of vortices), where thrust is the only cause of motion.

As concerns the second distinction, *primitive* heuristics are analogies and inductions, and from these we can obtain more articulated and complex heuristics (the *derivative* ones), which can intervene in the steps of the search for a solution to complex problems. Examples of them are the analysis of paradoxical cases, the analysis of a deviant case, the change of scale, the focus on processes, considering abstractions or specific instances, the use of remote associations, role-playing, definition as analysis, metonymy, metaphor, the use of simulations, the use of the figure and so on.

7.5.1 Problem-Finding

As already mentioned, these heuristic procedures can be used also to pose new problems. For example, analogy is a way to generate a new problem. A famous example of problem generation through analogy is the Poincaré conjecture, formulated in 1904 by the famous French mathematician and then solved by the Russian mathematician Grigori Perelman in 2002. Poincaré had developed a mathematical tool that distinguished, and therefore allowed for the topological classification of, all 2-dimensional manifolds. At this point, he conjectured that a similar phenomenon would hold in dimension 3, namely that this tool would distinguish at least the 3-dimensional sphere from other manifolds. In this way, he had constructed a new problem, analogous to one he had solved.

Naturally, also induction can be used to generate new problems. A famous example of problem generation is the Goldbach conjecture, which states that any even integer greater than or equal to 2 can be written as the sum of two prime numbers (not necessarily distinct). Goldbach examined various ways to represent integers as sums, looking for properties that might emerge. To this end, he examined many examples of integers. For example, if we consider the number 2, we can observe that it can be expressed as 1 + 1. Examining the number 4, we can see that it can be written as 2 + 2. Then we might observe that the number 6 can also be represented as 6 = 3 + 3. Moreover, 10 can be represented as 3 + 7, or 5 + 5. Also, 14 can be expressed as 3 + 11, or 7 + 7. What do these sums have in common? They are sums

[4] See Franck (1948) and Wartofsky (1965) for two seminal works on the relationship between metaphysics and heuristics.

of prime numbers, except for the number 2. At this point, we can ask by induction whether this property holds for all numbers, and in this way, we obtain the problem known as the Goldbach conjecture.

7.6 Mechanizing Problem-Solving and Discovery

As we have seen, a key question in the science of problem-solving is whether problem-solving and scientific discovery can be automatized, leading to the creation of algorithms capable of achieving this task. This issue is especially relevant today, given its implications for the Big Data Revolution and machine learning.

Various research programs and algorithms have been developed to pursue this goal. One of the earliest and most notable computational models of discovery is the BACON software created by Simon et al. (1987), which has yielded valuable insights (see e.g. Hesse, 1976; Ippoliti, 2018a, 2018b; Gillies, 2014, 2016; Cellucci, 2013; Weisberg, 2006). The primary conclusion from this effort, which is representative of the computational approach in many ways, is that only a specific, more straightforward aspect of scientific discovery can be explained computationally—namely, the least complex part.

First, BACON and similar 'discovery' algorithms are unable to perform a critical step in hypothesis generation, that is, identifying relevant variables and data (a point we will explore further in the next section). These variables and data are mostly supplied by the programmers, and the software merely applies a set of predefined heuristic rules to them. Consequently, the actual problem these algorithms address is limited to estimating parameters from a given dataset. While they offer an alternative approach to solving this basic problem, they do not generate new generalizations. Additionally, the process of identifying relevant variables through algorithms has significant theoretical limitations (see Sect. 7.6.1).

Secondly, it serves as a reconstruction of a historical process. This reconstruction gains an advantage from having a *backward perspective*—the knowledge that a particular issue exists, that its data can be approached using specific heuristics, and that the problem has a known solution. Therefore, at best, this software uncovers an alternative solution to an already solved problem rather than addressing an unsolved one.

Moreover, it lacks the fundamental capability of producing genuine new concepts or properties, or, more broadly, of building a conceptualization that is required to solve and pose problems. The impossibility of production of new concepts by means of algorithms is a highly debated topic (see e.g. Cellucci, 2017; Coveney et al., 2016, Kantorovich, 1993). For example, Simon argues that the software BACON can produce a new concept (a new variable), corresponding to a new theoretical concept (see Simon et al., 1987). By 'concept,' Simon refers to a theoretical term with physical significance. Simon defines the new concept, termed 'gravitational mass,' as follows: in a given planetary system, the quantity P^2D^3 remains constant, represented by a value K. If the BACON algorithm is applied to various planetary systems, like the satellites of Neptune, different K values will emerge. Through this process, the concept of gravitational mass is revealed because, according to Newton's theory of universal gravitation, K is proportional to the gravitational mass of the central body within the planetary system. This seems like a genuine discovery, as the algorithm has identified a new physical quantity.

However, Kantorovich (1993) argues that this is not truly a discovery. If a machine is provided with two physical quantities, such as P and D, and then produces a new combination that fits within a theory like Newtonian mechanics, this doesn't equate to the machine discovering a new concept. For a discovery to be genuine, BACON would need to uncover a theory or law in which 'gravitational mass' plays a crucial role, not just generate a combination of the variables P and D. The concept of gravitational mass has content beyond being a mere combination of these variables. Fundamentally, what BACON does is not inference. The recursive heuristic rule (REC) programmed into BACON is not equivalent to a rule or set of inference rules. REC guides a process that identifies hidden regularities in the data. When a mechanical procedure finds regularities in data, it only indicates that the statistical rules are appropriate. Kantorovich rightly suggests that an important aspect of discovery lies in inventing effective statistical rules. In this limited sense, BACON itself could be considered a discovery: it is a machine that, when given suitable data, uncovers hidden regularities within them. But it has not produced a new concept.

Bottom line: BACON, just like any software, helps and boosts human beings in computational tasks such as finding regularities and making calculations, but cannot replace them in the conceptualization of phenomena. In addition, a critical part of problem solving, that is finding the relevant variables to solve a problem, suffers strong computational limits. Let us examine this issue in more detail.

7.6.1 *Automatizing the Process to Find Relevant Variables to Solving Problems*

Solving the problem of finding the relevant variables to solve a problem in an inductive and mechanical way suffers strong limits.

A common inductive-algorithmic approach to solving this problem involves procedures that attempt to determine relevance by tracking statistics through simple trial and error. For example, an algorithm[5] might begin by assuming that no characteristic or variable is relevant to the problem at hand (e.g., deciding which action to take or assigning a numerical value). Essentially, it starts with the assumption that the same action or value can be applied across all situations. Then, for each potentially relevant characteristic of a situation, the procedure monitors statistics to see what happens when the characteristic takes on different values—typically

[5] See e.g. Chapman and Kaelbling (1991).

observing the outcomes when the characteristic is either 'present' or 'absent'. If the statistics indicate that the presence or absence of the characteristic has a significant impact on actions or outcomes, it is deemed relevant. As more characteristics are identified as relevant, the description of the situation becomes increasingly accurate. However, this procedure has significant limitations. Three main drawbacks can be identified.

First, a characteristic may not be relevant on its own but can become relevant when combined with another characteristic. For instance, in soccer, being left-handed may not be important when deciding how to defend against a forward if the defender is effective against left-handed players. But if the forward is also shorter than 1.70 meters, being left-handed might become relevant because the defender struggles more against left-handed players of that height, whereas the same issue does not occur with right-handed players. In this case, left-handedness only becomes relevant when combined with another characteristic like height. To address this, the algorithm would need to track statistics on combinations of characteristics, leading to an overwhelming number of potential combinations to analyse.

Second, this approach assumes that the relevance of a characteristic is fixed across the entire domain, measuring its importance across all observed situations. But a characteristic might be relevant in some situations and irrelevant in others. This would require collecting separate data on relevance for each situation, once again leading to a combinatorial explosion in the number of statistics needed, making this method impractical for determining relevance computationally.

Third, there is no inherent limit to the number of characteristics that could be relevant in different situations. Starting with all potentially relevant characteristics and then collecting statistics to identify those that can be ignored is not feasible. Additionally, if the process begins with a predefined set of potentially relevant characteristics, there is no mechanism to introduce new ones if the existing set proves insufficient for adequately understanding the situation.

An interesting example to illustrate these limitations is provided by sports, where algorithms are increasingly used to make important decisions at crucial moments in competitions, e.g., the batting order in baseball or the penalty takers in soccer. Imagine the following situation. The owner of a soccer team provides the coach with an algorithm containing all the data on each player's performance in penalty kicks under various conditions. One day, after consulting the computer before the penalty shootout in the Champions League final, the coach decides to replace the player suggested by the algorithm, Cristiano, with another player, Diego, for the final penalty kick. Diego scores, and the team wins the match. However, the owner is still annoyed and accuses the coach of not using the computer correctly and claims that he was just lucky in his choice, as the algorithm clearly shows that Diego has a lower scoring average than Cristiano. The coach counters that the computer also showed that Diego has a higher average in night matches, and this was a night match. The owner then responds that Diego also has a lower average against goalkeepers taller than 1.83 meters, and the opposing goalkeeper was 1.91 meters tall in that game. And so on.

The point is that a coach's experience, and expertise in general, lies in being able to identify, evaluate and respond to relevant facts. A computer can help by providing more facts than the coach could remember, but only experience allows the coach to understand the current state of affairs and determine what is relevant. This know-how of the coach cannot be expressed by a computer simply by adding more facts, as the problem here is which perspective is the fruitful one to establish which facts are relevant.

There are also other reasons why the theory of relevance proves to be an inadequate solution for accounting for the analogical process. In fact, allowing an algorithm to identify which characteristics of a situation should be treated as a relevant subset and use them to determine actions and values is not only difficult, but it is also a circular process. Indeed, it is possible to discover which characteristics of a problem are relevant only by determining what type of situation the current situation or problem is. But this, in turn, requires determining the relevant past situations, and, as Dreyfus (1992) points out, this triggers a regression that is difficult to avoid: the circularity of relevance and the circularity of similarity and relevance.

Since an algorithm must specify the notions of similarity and relevance, this raises the logical problem of similarity and relevance. Establishing what similar or relevant characteristics are between objects and situations is not a problem that can be solved in a purely 'inductive' or probabilistic way. In fact, we have at least four fundamental and unavoidable problems:

(i) The underdetermination of similarity: every object is similar to another under one or more points of view.
(ii) The combinatorial explosion of potentially relevant statistics.
(iii) The circularity of relevance: determining which characteristics of a situation or a problem are relevant can only be done by determining what type of situation or problem the current one is. But this requires referring to relevant past situations. And so on.
(iv) The circularity of similarity and relevance: determining a degree of similarity between objects requires the notion of relevance; determining the relevance of certain characteristics requires the degree of similarity.

These problems can be resolved computationally only by trivializing the notions of similarity and relevance, that is, reducing them to deductions by adding background knowledge. But this is possible only for simple problems and become increasingly difficult as we move forward problems at the frontier of knowledge or for unsolved problems.

Moreover, since these properties, i.e. similarity and relevance, cannot be derived purely bottom up, they must be appropriately defined in the algorithms through specific metrics. These metrics are contextual, their effectiveness is local, and they require having in mind what counts as data and acceptable outputs. So even if domain knowledge is not directly embedded in the data (as it is in supervised algorithms), it is still encapsulated in the construction of the algorithm (even in unsupervised machine learning, see Ippoliti, 2023).

These limitations of inductive procedure to find relevant variables for solving problems are well-known after all. For instance, the five methods of experimental inquiry by Mill (1843) provide us with a method to identify relationships between phenomena and find relevant variables for a target problem. Now, these methods (the method of agreement, of difference, the joint method of agreement and difference, the method of residues, and concomitant variations) cannot solve the problem of finding such variables in a definitive and purely inductive way. They can help us but cannot substitute our theoretical work. For instance, the method of agreement identifies a relevant variable by looking for a common factor in multiple instances where a particular effect occurs. If a certain factor is present in all cases where the effect is observed, that factor is likely to be the cause (or at least a part of the cause) of the effect. For example, if people fall ill after eating at a restaurant, and the only common dish they all consumed was a particular fish, the method of agreement tells us that the fish is the likely cause of the illness. Of course, the procedure is defective and has several limitations. The main ones are two:

(i) This method assumes that the common factor identified is the cause, but it may overlook other variables that could also be causal factors (overlooked variables)
(ii) If multiple causes are involved, this method might fail to isolate the true cause.

Similar drawbacks affect not only all Mill's methods, but the inductive-statistical procedures in general, and automatizing them, of course, does not solve the problem.

Moreover, Mill's methods, along with other inductive-statistical methods, at most offer a rational, though not definitive, way to solve part of the problem of searching for relevant variables, such as selecting among existing variables, i.e., those already observed or known. The search becomes even more challenging when such variables are not known or observed (i.e., at the frontier of knowledge), and in this case, the inductive-statistical approach is even more limited and needs to be accompanied by a theory: identifying the relevant variables can hardly be done without some initial theory or at least well-motivated hypotheses.

These limits are not surprising at all and have been explored extensively by the philosophical literature on the subject, which has clarified that we cannot ignore at least four factors that affect the automatic search for relevant variables.

The first is the role of theory: as Popper (1999) at other have shown, scientific progress relies on pre-existing theories that guide observation and experimentation. We don't just collect random data: we look for evidence that supports or refutes existing ideas.

The second is negative instances: since just one counter-example can potentially disprove a seemingly universal pattern derived from observations and data, how do we account for the exceptions within a inductive framework?

The third is the role of biases: observations and data can be influenced by our existing beliefs and expectations and our prior knowledge might shape what we see and how we interpret it.

The fourth is the underdetermination of theory by data: multiple theories and hypotheses can potentially account for the same set of data. How do we deal with all these hypotheses?

In effect, computational tool and machine learning cannot solve the problem of identifying and selecting relevant variables, as a theory is still needed to do this. To say otherwise would be to say that machine learning has solved the problem of induction and ampliative reasoning, when it is nothing more than a way, however evolved and refined, of implementing some inductive-statistical procedure. Moreover, this limit becomes increasingly marked when we move to the frontier of knowledge, where the characteristics of a problem (initial state, final state, operators, constraints and available data) can be unknown and become the real object of research.

Of course, this does not mean that algorithms (and machine learning) cannot help in this task, but it is necessary to remember that it is, in fact, an aid in those processes that can be reasonably automated, and not completely replace the human being. They can assist in identifying regularities and patterns that enable the formulation of conjectures that would be difficult to achieve without algorithms capable of analysing large datasets (a notable example being number theory).

In this sense, a paradigmatic example is the four-colours theorem and its solution (see e.g. Ippoliti, 2023). In fact, this theoretical situation closely resembles a debate already present in the philosophy of mathematics in discussions surrounding the solution to the four-color problem. In this instance, the entire process of conceptualizing the problem was developed by humans. Computers effectively handle part of the problem-solving process by executing the calculations encoded in the instructions within algorithms ad we trust tits outcome. We can, in principle, trace back all the steps according to the rules and instructions embedded in the software code. Although these steps can be so numerous that reconstructing them may be difficult or impractical, they are still comprehensible to a human mind because they involve straightforward recursive processes. The fact that the reasoning can be lengthy (e.g., 500 pages for the four-color theorem) doesn't imply that it is 'obscure' or 'alien'. Furthermore, algorithms can make mistakes (like code errors) just as humans do, which makes them not so different from us.

7.7 Conclusion

The issue of producing genuine new knowledge and making scientific discoveries has increasingly become a focus of both philosophical and scientific inquiry, evolving into a distinct field of study. This shift is largely due to the advancement of the heuristic view, which argues that these processes can be understood through 'logic,' 'rationality,' and inference. This contrasts with the traditional view, which holds that logic and rationality are sufficient only for justifying knowledge.

The heuristic view has showed that the problem of knowledge acquisition and scientific discovery can be framed as a problem-solving exercise, leading to the development of more sophisticated problem-solving science. By addressing the question of how to produce new knowledge through the lens of how to solve scientific problems, this view has provided rational and replicable methods for generating and addressing new and unresolved issues.

The crucial question of 'how to solve it' has seen evolving answers, with new methods emerging to address this challenge. Specifically, recent advancements have focused on naturalizing the problem-solving process, offering fresh approaches to the question of 'solving it'. In recent decades, the heuristic view has been significantly developed by scholars such as Popper, Simon, Poincaré, Polya, Lakatos, Nickles, Cellucci, Gillies, Grosholz and Magnani, who have contributed valuable insights into knowledge acquisition.

Acknowledgments I would like to express my deepest gratitude to Lorenzo Magnani for the generosity, openness, and liberality that have always defined both his academic and personal life.

I also would like to thank the two anonymous referees for their valuable comments and suggestions.

References

Campbell, D. T. (1960). Blind variation and selective retention in creative thought as in other knowledge processes. *Psychological Review, 67*, 380–400.

Cellucci, C. (2013). *Rethinking logic*. Springer.

Cellucci, C. (2015). Rethinking knowledge. *Metaphilosophy, 46*(2), 213–234.

Cellucci, C. (2017). *Rethinking knowledge*. Springer.

Chapman, D., & Kaelbling, L. P. (1991). Input generalization in delayed reinforcement learning: An algorithm and performance comparisons. In *Proceedings of the international joint conference on artificial intelligence. Sydney, Australia*, pp. 726–731.

Coveney, P., Dougherty, E., & Highfield, R. (2016). Big data need big theory too. *Philosophical Transactions of the Royal Society A Mathematical Physical and Engineering Sciences.*

Darden, L. (Ed.). (2006). *Reasoning in biological discoveries: Essays on mechanisms, interfiled relations, and anomaly resolution*. Cambridge University Press.

Dreyfus, H. L. (1992). *What computers still can't do: A critique of artificial reason* (revised ed.). The MIT Press.

Einstein, A. (1958). A testimonial. In J. Hadamard (Ed.), *The psychology of invention in the mathematical field* (pp. 142–143). Dover.

Frank, P. (1948). The place of logic and metaphysics in the advancement of modem science. *Philosophy of Science, 40*, 275–286.

Frege, G. (1960). *The foundations of arithmetic. A logic-mathematical enquiry into the concept of number*. Harper.

Gigerenzer, G., & Todd, P. M. (1999). *Simple heuristics that make us smart*. Oxford University Press.

Gillies, D. (2014). Why do scientific revolutions begin? In E. Ippoliti (Ed.), *Heuristic reasoning*. Springer.

Gillies, D. (2016). Technological origins of the Einsteinian revolution. *Philosophy and Technology, 29*, 97–126. https://doi.org/10.1007/s13347-015-0191-1

Gregory, R. L. (1970). *The intelligent eye*. Weidenfeld & Nicolson.

Gregory, R. L. (1980). Perceptions as hypotheses. *Philosophical Transactions of the Royal Society of London Series B, 290*, 181–197.

Grosholz, E. (2007). *Representation and productive ambiguity in mathematics and science*. Oxford University Press.

Hadamard, J. (1958). *The psychology of invention in the mathematical field*. Dover.

von Helmholtz, H. (1866). Concerning the perceptions in general. In *Treatise on physiological optics* (Vol. III, pp. 1–37). Dover.
Hesse, M. B. (1976). Truth and the growth of knowledge. In F. Suppe & P. D. Asquith (Eds.), *PSA 1976* (Vol. 2, pp. 261–280). Philosophy of Science Association.
Ippoliti, E. (2015). Reasoning at the frontier of knowledge. In E. Ippoliti (Ed.), *Heuristic reasoning* (pp. 1–10). Springer.
Ippoliti, E. (2016). Ways of advancing knowledge. A lesson from knot theory and topology. In F. I. Sterpetti & E. T. Nickels (Eds.), *Models and inferences in science*. Springer.
Ippoliti, E. (2018a). Building theories. The heuristic way. In D. Dank & E. Ippoliti (Eds.), *Building theories*. Springer.
Ippoliti, E. (2018b). Heuristic logic. A kernel. In D. Danks & E. Ippoliti (Eds.), *Building theories*. Springer.
Ippoliti, E. (2018c). Scientific discovery reloaded. *Topoi, 39*(4), 847–856.
Ippoliti, E. (2022). On the heuristic role of mathematical representations. *Synthese. 200*, 407. https://link.springer.com/article/10.1007/s11229-022-03883-9
Ippoliti, E. (2023). *Guida critica alle intelligenze artificiali*. Egea.
Ippoliti, E. (2024). Deduction and ampliativity: A critical appraisal. In A. Piccolomini D'Aragona (Ed.), *Perspectives on deduction: Contemporary studies in the philosophy, history and formal theories of deduction, synthese library* (Vol. 481, pp. 233–250). Springer.
Ippoliti, E., & Cellucci, C. (2016). *Logica*. Egea.
Jaccard, J., & Jacoby, J. (2010). *Theory construction and model-building*. Guilford Press.
Kantorovich, A. (1993). *Scientific discovery: Logic and tinkering*. State University of New York Press.
Kantorovich, A. (1994). Scientific discovery: A philosophical survey. *Philosophia, 23*(1–4), 3–23.
Kantorovich, A. (1989). Serendipity as a source of evolutionary progress in science. *Studies in History and Philosophy of Science, 20*(4), 505–529.
Koza, J. (1992). *Genetic programming: On the programming of computers by means of natural selection* (Vol. I). MIT Press.
Koza, J. (1994). *Genetic programming II: Automatic discovery of reusable programs*. MIT Press.
Koza, J., Bennett, F., III, Andre, D., & Keane, M. (1999). *Genetic programming III: Darwinian invention and problem solving*. Morgan Kaufmann.
Lakatos, I. (1976). *Proofs and refutations: The logic of mathematical discovery*. Cambridge University Press.
Lakatos, I. (1978). *The methodology of scientific research programmes*. Cambridge University Press.
Laudan, L. (1977). *Progress and its problems*. University of California Press.
Laudan, L. (1981). A problem-solving approach to scientific progress. In I. Hacking (Ed.), *Scientific revolutions* (pp. 144–155). Oxford University Press.
Magnani, L. (2003). Euristiche abduttive per la scoperta. *Sistemi Intelligenti, 15*(1), 81–87.
Magnani, L., Nersessian, N., & Thagard, P. (Eds.). (1999). *Model-based reasoning in scientific discovery*. Kluwer/Plenum.
Mill, J. S. (1843). *A system of logic, ratiocinative and inductive: Being a connected view of the principles of evidence, and the methods of scientific investigation*. Cambridge University Press (2011).
Nickles, T. (Ed.). (1980a). *Scientific discovery: Logic and rationality*. Springer.
Nickles, T. (Ed.). (1980b). *Scientific discovery: Case studies*. Springer.
Nickles, T. (1981). What is a problem that we may solve it? *Synthese, 47*(1), 85–118.
Nickles, T. (2009). Life at the frontier: The relevance of heuristic appraisal to policy. *Axiomathes, 19*(4), 441–464.
Nickles, T. (2013). Scientific revolutions. In: Zalta, E. (Ed.), *Stanford encyclopedia of philosophy* (Winter 2013 edition),
Nickles, T. (2015). Heuristic appraisal at the frontier of research. In E. Ippoliti (Ed.), *Heuristic reasoning* (pp. 57–87). Springer.

Nickles, T. (2016). Fast and frugal heuristics at research frontiers. In E. Ippoliti, F. Sterpetti, & T. Nickles (Eds.), *Models and inferences in science. Studies in applied philosophy, epistemology and rational ethics* (Vol. 25). Springer. https://doi.org/10.1007/978-3-319-28163-63

Nickles, T. (2018). Prospective versus retrospective points of view in theory of inquiry: Toward a Quasi-Kuhnian history of the future. In M. Beaney et al. (Eds.), *Aspect perception after Wittgenstein: Seeing-as and novelty*. Routledge.

Petroni, A. M. (1992). On some problems of the logic of scientific discovery. In W. Newton-Smith, C. T'ien-chi Chiang, & E. James (Eds.), *Popper in China* (p. 77). Routledge.

Poincaré, H. (1908). L'invention mathématique. *Enseignement mathématique, 10*, 357–371.

Polya, G. (1945). *How to solve it: A new aspect of mathematical method*. Princeton University Press.

Polya, G. (1954). *Mathematics and plausible reasoning* (Vol. I e II). Princeton University Press.

Popper, K. (1961). *The logic of scientific discovery*. Science Editions.

Popper, K. (1999). *All life is problem solving*. Routledge.

Quine, W. (1969). *Ontological relativity and other essays*. Columbia University Press.

Russell, B. (1912). *The problems of philosophy*. Home University Library.

Simon, H. (1977). *Models of discovery*. Reidel.

Simon, H. (1987). Is scientific discovery a topic in the philosophy of science? In N. Rescher (Ed.), *Scientific inquiry in philosophical perspective* (pp. 1–15). University Press of America.

Simon, H., Langley, P., Bradshaw, G., & Zytkow, J. (1987). *Scientific discovery: computational explorations of the creative processes*. MIT Press.

Wallas, G. (1926). *The art of thought*. Cape.

Wartofsky, M. (1965). Metaphysics as heuristic for science. In M. Wartofsky (Ed.), *Models* (pp. 40–90). Reidel.

Weisberg, R. (2006). *Creativity: Understanding innovation in problem solving, science, invention, and the arts*. Wiley.

Chapter 8
The School of Milan and the Problem of a New Critical Rationalism

Fabio Minazzi

Abstract Through a loose narrative, the Author reconstructs the history of the Insubric International Centre, which preserves the Archival and Library memory of the Milan School. The Centre currently preserves some thirty philosophical Archives, together with seven Author Libraries. By the work and thought of Piero Martinetti and Antonio Banfi, through to Banfi's great pupils of the 1930s—Enzo Paci, Remo Cantoni, Giulio Preti and Dino Formaggio—the Milan School took on a breadth that is unique in the cultural landscape of the twentieth century. This school was not just *philosophical*, because it also included poets such as Antonia Pozzi, Vittorio Sereni and Daria Menicanti, musicologists, educationists, intellectuals, and filmmakers. Today, the legacy of this school constitutes a research programme with a rationalist-critical and phenomenological framework that still displays notable critical relevance.

Keywords Critical rationalism · Epistemology · Philosophical archives · The international Insubric Centre

8.1 The Genesis of the International Insubric Centre

Lorenzo Magnani has made important contributions to the study of scientific philosophy, not only within the context of the Italian culture at the turn of the century, but also on a European and international level. His original research programme devoted to 'scientific philosophy' was thus inevitably entwined with my studies on the problem of critical rationalism and also critical realism developed in the context of the Milan School. There are many contributions by Magnani in which the theme

F. Minazzi (✉)
Dipartimento di Scienze Teoriche e Applicate, Università degli Studi dell'Insubria, Varese, Italy
e-mail: fabio.minazzi@uninsubria.it

S. Arfini (ed.), *Scientific Cognition, Semiotics, and Computational Agents: Essays in Honor of Lorenzo Magnani - Volume 2*, Synthese Library 506,
https://doi.org/10.1007/978-3-031-96688-0_8

of 'scientific philosophy' is placed in direct relation to the magisterium of philosophers such as Giulio Preti and Ludovico Geymonat. But the specific merit of Magnani's thought lies above all in his study of the function of rationalism within Giulio Preti's epistemology, a crucial theme to which he has devoted two important contributions, namely Magnani (2001a) and (2001b).

Moreover, in carrying out this original reflection—at once historical-critical, but also intensely theoretical—Magnani had the ability to deal with these topics on different critical levels, placing the work of Geymonat and Preti in the precise context of the first half of the twentieth century. Hence he set it in a historical and cultural period in which these authors necessarily had to cope with both declining nineteenth-century positivism and also the dominant Italian neo-idealism (Magnani, 2009a). It was precisely by studying these movements of thought that Magnani was compelled to place these Italian debates in the broader and more highly articulated panorama of European philosophical research in those years. This allowed him to highlight the European value of the contributions made by philosophers such as Preti and Geymonat. It was thus inevitable that my research programme devoted to critical rationalism and critical realism should have intersected, quite naturally, with Magnani's study and research programme. In this way, not only did a deep friendship arise between the two of us, but my own syllabus often drew on many of Magnani's contributions (see in particular Magnani, 2009b).

As you know, at the Università degli Studi dell'Insubria (Centro Internazionale Insubrico), we are collecting the Archives of the School of Milan. Our Centre was founded in the spring of 2009. To date we have collected some fifteen of such Archives and in addition we have also collected, whenever possible, the Libraries of the Authors together with their Archives.

We started with the Preti Collection, which contains Giulio Preti's complete Archive. In this case it includes no letters, because he had a habit of destroying all the letters he received every evening. However, as happens with these somewhat bizarre characters, many letters Preti sent to his correspondents have been preserved, so we also have his very one-sided correspondence. However, another important bequest that is always closely bound up with the cultural history of Milan was given to us, again in the spring of 2009: the Archive of the manuscripts of Carlo Cattaneo (1801–1869), those from his exile until his death, hence from 1848 to 1869.

This bond between Cattaneo and Preti is related to my *Leben.* I was teaching in Lecce (2001–2008) when Preti's collaborator, who had preserved and saved all of Preti's papers, died. This was Professor Ermanno Migliorini (1924–1999),[1] who taught in Florence, where Preti held the chair of Philosophy until his death (in Djerba, Tunisia on 28 July 1972). Migliorini, through his wife, decided to leave me the whole collection of papers on which I had worked as a student at the Università Statale in Milan. I had been sent by Mario Dal Pra, my thesis supervisor, to conduct, as stated by the mandate of the Department of Philosophy—which had only recently been instituted—an "exploratory mission" to report on Preti's papers in Florence. I

[1] On Migliorini's thought cf. Peruzzi (2011).

catalogued them (cf. Minazzi, 1984), and saw that they were a heritage of great importance, and this was the fund that I then inherited from Migliorini. I should also say, and this is interesting, that offers from abroad had arrived to purchase this collection, in particular from the University of Italian Switzerland, where I had worked for 5 years (1999–2003). I refused because I believed that it should remain in Italy.

By putting together these two core collections of Cattaneo and Preti, our dream was to trigger a "magnet effect", which did in fact happen, with the result that we began in 2011 to organize a conference on a major figure of the School of Milan. In 2011 we began with a conference on Giulio Preti, because it was the centenary of his birth (cf. Minazzi, 2015a; Magnani, 2009b). This was followed the year after by a conference on Carlo Cattaneo; then a conference on Banfi (2013); and in 2014 one on Dal Pra (cf. Minazzi, 2018). Then followed one on European critic rationalism in 2015; on Piero Martinetti, the so-called philosophical "grandfather" of the School of Milan in 2016; on Husserl in Milan in 2017; and the last on Ludovico Geymonat, in 2018 (Magnani, 2009a, 2009b).

In fact, while we were busy with these initiatives and starting to file this material and work on it, other archives came to hand, including the extraordinary Archive and Library of Antonia Pozzi (1912–1938), which has recently arrived. In 2011 we received the *Secret Archive* of Antonio Banfi (1886–1957), containing some five thousand letters, ranging from his stay in Berlin (when he was majoring in philosophy), until his death. This is the so-called *Secret Archive*, not because it contains secret information, now that history has already brought it to light: it is secret mostly because the letters of the period when he was a senator of the PCI reveal his fairly radical dissent with the Secretary of the Italian Communist Party. When we archived the letters, we found that many of them contained the recurrent term that we were unable to decipher, because Banfi wrote: "Today *Canopo* did…", "Today *Canopo* said…", etc. When we asked the philosopher's grandson, "Does anyone in the family know who this *Canopo* was?", he told us it was Togliatti. So, they were "*secret*" for this very reason. But the correspondence is extraordinary. For example, there are letters in which he relates when he met Husserl, when he entered Simmel's home, the impression made on him by reading the *Philosophie des Geldes*, etc.

I have borrowed Fulvio Papi's acute comment, when he saw this Archive and commented: "Banfi's intellectual biography will have to be rewritten, because now we can really reconstruct it *day by day*." Because Banfi was a graphomaniac, and wrote three letters a day, every day, to his fiancée, later his wife (Daria Malaguzzi), and so we have *five thousand* of them. But there are precedents for this. In short, there were no telephones, and the postal service was efficient…

And then there is the other Archive, of which the salient facts are as follows. The volumes number about 16,000 in 30 archives. We have the Archive of Evandro Agazzi, who studied at the Università Cattolica under Gustavo Bontadini, and then went to work with Ludovico Geymonat's group at the CNR, because his early work of 1961 was devoted to problems of axiomatics (cf. Agazzi, 1961). We also have his library, a specialised library of mathematical logic and the philosophy of science, a large library in which all the key texts are in many languages, testifying to his international outlook.

Then there are also other smaller holdings that are no less significant. There is the small Archive of Vittorio Sereni (1913–1983). Another small Archive of Daria Menicanti (1914–1995). We have materials of Guido Morpurgo Tagliabue (1907–1997). All Giovanni Vailati's papers on the Royal Commission for the Reform of the Italian School System, Bruno Widmar's Archive and Library, Augusta (Lella) Monti's Archive and Library, Guido Bersellini'Archive and Library, Silvio Ceccato's Library and Archive, etc. I will stop here: those of you interested can go to the site of our online Centre, which has a film clip about our Archives and Libraries. The reference is to the Centro Internazionale Insubrico "Carlo Cattaneo" and "Giulio Preti", of course.

An important point is that we received all these Archives and Libraries from *donations*. I wish to emphasize this key fact because in this Italy, so devastated and ravaged, there still exists a part of *civil society* that is essentially *healthy*. Then there is a second factor: whenever one makes a contact to obtain these libraries and archives, those who have been involved in the process know that it is never a relationship that ends with the single donation, but always remains open to further relations, so that certain Archives are continuously augmented, for a reason that is perfectly comprehensible. The owners of these papers also have an emotional tie with them, so when a donation is made there is always a remainder, which then gives rise to a subsequent donation. Behind these Archives there are always *human relationships*, so that I have to say that around these Archives, this Centre, a network of study of human relationships has grown up that was established many years ago, when I was a young man. How is it that all these Archives relating to the School of Milan were not collected here in Milan? Actually, even in Milan something that is very important has started to develop, perhaps since the creation of our Centre. The Library of Philosophy at the Università Statale has now set to work, and it has already collected some very important Archives. But even earlier, it already contained several archives; it had Giuseppe Rensi's Archive, Giovanni Vailati's Archive, the partial photocopy of Preti's Archive (which I had collected when I was graduating). And now it contains the whole of Dal Pra's Archive, then Guido Neri's, Mondolfo's, and finally a group of very interesting papers by Antonio Banfi.

8.2 The School of Milan

I speak as a former student of the Università Statale in Milan. I was born in 1955 and I enrolled in the Statale in 1975, so I attended courses in the second half of the seventies. To establish a time frame, I followed Paci's last, extraordinary course (*Leibniz and Husserl on Monadology and Relationism*), and I was in time to take the courses held by Remo Cantoni, as well as those by Ludovico Geymonat and Mario Dal Pra. Now that university, which was my university, no longer exists. I was in time to follow the first course given by Sini (who had just arrived in Milan) on Nietzsche. The last philosophical voice of the School of Milan, as we will now explain, was Sini's. Why did Milan fail to collect this material? Following my time at the Università

Statale something happened. I am anticipating something that I was not in time to record, but this is my impression: it did not happen by chance and a primary responsibility goes to these same masters. The responsibility for the decline of the Statale of Milan, in my opinion, is a mistake that has its root in certain serious errors that were made by these masters. In particular I am referring to the academic and institutional management of two figures with whom I worked most: Geymonat and Dal Pra did everything they could do to choose people in ways that were not intelligent and therefore they caused this tradition to self-destruct.

So, we are now, in cultural terms, faced with a repression of the School of Milan around that extraordinary event which happened in the 1930s. When I consider what happened in the thirties in Milan, I am astonished at what happened. Why? Meanwhile, there was a transition that has not (in my opinion) been sufficiently investigated. Banfi, when he arrived in Milan, came on the recommendation of Martinetti, and Martinetti was the only philosopher who did not swear allegiance to Fascism, the only philosopher out of the twelve among some 1200 university professors, and it was Martinetti who persuaded Banfi to come to Milan, when in reality he did not want to come, because his civil position was the same as Martinetti's. He did not want to swear loyalty to fascism, but Martinetti persuaded him to do it. Yet Banfi was a great master. Besides Banfi was very fortunate. Those who teach know it well: when one finds a good student it is a gift from God, a divine grace. But he was also extremely fortunate, because within a few years he had a group of students, each of whom was better than the last.

So if I look at the School of Milan, strictly speaking the Banfian school, from this point of view I see it as a series of epicycles in the sense that there is the hard disc made up of philosophers, and then there are the great students of the thirties, the nucleus, then there are all these epicycles that extend to aesthetics, to literature, poetry, musicology, epistemology, history of science, cinema, etc. Hence Banfi's good fortune was to have these students and also to be a great Master. I believe this was already the root with Banfi's phenomenological approach, which meant that each of his students could freely follow their own *demons*. And this, of course, opens a historiographical problem: what is the relationship between the evolution of the great students of the thirties and Banfi's imprinting? Historiography, however limited, is now configured with two theses: there is the argument of those who say that the School of Milan—Cantoni, Paci, Preti, Rognoni, Formaggio, etc.—gradually lost this Banfian critical position (this is the thesis supported by Franco Cambi in a very good book (*Rationalism and Praxis in Milan (1945–1954)*[2] and Magnani, 2007). By contrast, there is the thesis that is advanced by other authors, among whom I count myself, who say that the pupils developed this Banfian criticality along highly original and creative paths. So, there was some quality precisely in Banfi's approach that made his students fruitful.

I am interested in another question. Why was this School of Milan so fruitful and interesting? Was there something specific, on the level of its theoretical formation,

---

[2] Cf. Cambi (1983).

that Banfi gave his students? Did he endow these students with an edge, which I think I can find in his student poets and student *littérateurs* and his student philosophers, and so forth? I think so and now, briefly, I would say that the critical approach that was the edge that this School had—which is, however, a problem in cultural terms in the pages that I read of the recent book by Sini on Paci (Sini, 2015 and Sini, 2013, pp. 105–115)—is that Banfi had a very original way of structuring this school, because behind him he had his original path of development.

So, evidently Banfi has to be periodised. It is certain that the action that he carried out here in Milan in the thirties had behind it his masterpiece, *I principi della teoria della ragione* of 1926. But what is the philosophical operation which Banfi performed in this work, seen in a theoretical terms?

There is an essay by Dal Pra, which I believe is very fine, and which he later published in *Filosofi del Novecento* (cf. Dal Pra, 1989; Dal Pra, 1984; Magnani & Li, 2012), which basically states: what is the core of the tradition of critical rationalism which Banfi drew on? The answer is very clear: the *critical rationalism* of Kant and his *trascendentalism*. Within this rationalism, *the soul of the soul* is *transcendental*. But how was the transcendental read by Banfi? *It was read with an effort to Hegelianise this Kantian transcendentalism, without slipping into a philosophy of history*. And then, we can say, the problem of *how* he read this relationship between Kant and Hegel was to take account of the *formal* character of this rationality, but seeing it in its *historical genesis*, hence in its *internal dynamism*, always keeping these two poles in tension. Dal Pra says exactly this: "For Banfi, only Kantian reason is *form*, but in Hegelian terms it is also a *structure* constructed in time" (cf. Dal Pra, 1984; Dal Pra, 1985). And these two theoretical poles are truly the innovative key, as I see it, of Banfi's lesson. In the thirties, even if one reads the texts of his published lectures, the famous courses collected by Formaggio and other students, one continually rediscovers this key. In particular, in the course on Plato in '37, in the final part, when Banfi sought to reconstruct the fecundity of Plato's teaching, in reality he wrote pages that applied to the School of Milan. When he speaks of the relationship between Plato and his pupils, he is actually speaking about the relationship between himself and his pupils. This point, however, is also the point in discussion at the School of Milan, because the School certainly developed in this critical rationalism, this *problematicism*; but at a certain point one realises that there is something that creates a problem, and this, at least as I interpret it, was clear to Remo Cantoni, to Enzo Paci, to Giulio Preti (and also to Antonia Pozzi),[3] each with his own formation and following his own path.

Now, I followed Preti's path because it seems clearer to me from this point of view. In an essay that was published posthumously, *In principio era la carne* (cf. Preti, 1983), but dating from the forties (1948), he saw that there was a very close similarity between the theoretical thought of Banfi and that of Giovanni Gentile. And here one should open a long discourse, because Paci also argued something like this, stressing the importance of Gentile. Here, again, I have to open a

[3] Cf. Minazzi (2015b) and Minazzi (2018)

parenthesis: some of these students of Banfi's had a direct relationship with the two famous neo-idealists. Of Paci we know that some of his letters to Croce have been published, but perhaps it is less well known that Preti himself went to see Gentile. We have published two letters from Preti to Gentile which are very clear from this point of view (cf. Minazzi, 2015a, vol. II, pp. 1229–1233). Meanwhile, what did it mean for a student of Banfi's to have relations with Gentile in the thirties? In the thirties Gentile was the "Excellency", and not just in the philosophical sense. There is a letter of invitation to Preti which I urge you to read (we published first it in *Il Protagora*), in which Preti says this very clearly. (Preti had the merit of thinking that in theoretical life and philosophy one should seek to tell the truth, always, without mediation, including social mediations, which created a lot of problems, naturally.)

He wrote this letter saying that he had sent out for reading an essay for publication in the *Giornale Critico della Filosofia Italiana*, an essay—among other things—of Hegelian historiography, in which he maintained the relation between *continuity* and *discontinuity* in the history of philosophical thought, a very interesting topic that is Banfian (cf. Preti, 1983; Magnani, 2009a). And he says, since he had to introduce himself, "You cannot consider me one of your students because I am not, nor am I an idealist, but you cannot consider me either—as I am not one—to have been part of the tradition of Catholic thought"; nor of all those alliances—this was in '37—that were created in idealism from the practical point of view, after the Lateran Pacts (which Garin describes so well in his famous *Cronache di filosofia*).[4] He says "the only person who listens to me is Banfi, who gave me the opportunity to publish and with whom I work. And then I have only one destiny which I feel is mine, and that is to follow My Lady Philosophy." But it is significant that these students—Paci with Croce and Preti with Gentile—had a relationship with the famous neo-Hegelian philosophers, and this is an interesting point.

I think that Preti, in particular, at a certain point realized that the problem he found in Banfi was the same theoretical problem there was in Gentile. And what was this problem? It is the problem that at a certain point there was created a sort of rift between the *structures of rationality*, very *formal* but very *empty*, and the *richness of life*, of experience, which stands as another pole, and there is *no critical mediation* between the two poles: the pole of the *formality of reason* and the *pole of experience*. Bear in mind that what is true of all these authors was also true of Preti; namely, Banfi for them meant Husserl, he meant *phenomenology*. Preti graduated with a thesis (which we have been unable to find because he destroyed it) on the *Historical significance of Husserl*. It was a thesis some of whose chapters were published in the *Archivio di filosofia* (1934) and in *Sophia* (1935),[5] but in its entirety Preti retrieved it from the Archive of the University of Pavia and ensured that it disappeared. The Archive contains the whole of his personal file, yet the thesis by Preti is missing. Preti graduated in Pavia under Guido Villa (1867–1949) and then later, thanks to Paci (1911–1976), he met Banfi and as soon as he graduated, he

[4] Cf. Garin (1975, vol. II, pp. 345–488).

[5] Cf. Minazzi (1984), pp. 113–114.

moved, with his family, to Milan, and went to live in the famous Via dei Cinquecento (cf. Pozzi, 2015, p. 439), where Dino Formaggio (1914–2008) lived. And I must also say that this emerges clearly from the papers. In the Banfian group, why did this harmony arise, that would last a lifetime, between Formaggio and Preti? I think it was due to the fact that in Banfi's circle there were those who came from a humble social background: Formaggio from a working-class family, he himself a worker, and Preti came from a very humble and very poor family, to the point where his mother had actually worked as a laundress in Pavia. I remember his wife, Daria Menicanti, who told me that Preti, because of his mother's insistence, had gained a teaching diploma from the *Istituto Magistrale* (4 years of study), and then enrolled at the University. However, he wanted to study philosophy and so the extra year that he "gained" he spent studying alone, as an autodidact, for the matriculation examination, presenting himself as a private candidate. He was the same age as the children of Pietro Vaccari (Enrico and Giovanni), and a friend of theirs. Vaccari (1880–1976) was the *podestà* of Pavia and a famous professor at the University of Pavia, and he had two children, one of whom was a contemporary of Preti's, who attended the "Foscolo" Classical secondary school in Pavia. So Preti, to study on his own and prepare himself as a private candidate, went to the home of the Vaccari brothers, who gave him notes, especially on their Greek lessons. In this respect there is a charming episode: he wanted to listen to the Greek lessons and so they came to an agreement that whenever there was a Greek lesson the Vaccari brothers would open the window of the classroom on the first floor of the school, and he would creep secretly under the window and try to listen to what the teacher was saying…

This fine episode explains what lies behind Zambelloni's anecdote of 1971 (cf. Minazzi, 2015a, vol. II, pp. 1105–1106), in a speech here at the Università Statale by Preti. (This was in '71 and we were discussing education in the university.) At one point, in the midst of a great debate, in the midst of the student movement and everything else, Mario Dal Pra (1914–1992) at the presidential table saw Preti among those present, and called on him to make a contribution. After he had discussed the whole universe, Preti made this simple comment: "I think culture has to form people with butts as tough as leather." So here, behind these words lay this experience, of one who had such a deep love for culture that he would go and listen to the Greek lessons outside the window. To this I should add, to illustrate his poverty, that he used to call at the Vaccari brothers' home to study because they were the children of the *podestà* and the President of the Faculty of Jurisprudence at University of Pavia. This meant that there was always some afternoon tea and something to eat there, because in Preti's house there was never much to eat.

8.3 Banfi's Great Pupils of the 1930s

Faced with the problem of *formal reason*, which fails to comprehend the richness of experience and so creates a dichotomy which opens on the one side to an irrationalism seen in all its richness, but not dominated by reason, and on the other side to a reason so *abstract, formal* and *empty* that it does not have a content, what did Banfi attempt to do? To unite the lesson of epistemology (and he looked especially at developments in neo-positivism in all its various phases) as that space of construction of *structures of the intellect* (*Verstand*, in the Kantian sense), which can perform a very special operation in relation to the object, in this case, of knowledge, which *constructs an objectivity within a theoretical framework with the mediation of experimental control*. But this, I believe, is also what Paci did with existentialism, with relationalism and then with phenomenology, Cantoni with anthropology, and Formaggio with artistic practice. In short, this seems to me the point that enabled these great students to supersede a critical point in Banfi's teaching. From the properly cultural perspective, one who accepted with great finesse a rather difficult relationship with her master was Antonia Pozzi (1912–1938). We have seen her *Diary*, which was published (cf. Pozzi, 2015), but as a diary it is an infinitesimal part of the authentic diary that was then torn up. Anyway, from that *Diary* you understand that poetry in this case was an opportunity for Pozzi to understand that there was something that failed to function on the theoretical level in Banfi's system (cf. Minazzi, 2007).

I must also say—and the human aspects have emerged inevitably—that in Banfi these problems did not explode because perhaps also because of the man's character, which was capable of holding it all together, *à la* Goethe. And this was to some extent Banfi's optimistic key, his smile, for he also had a harmony that did not exist among his students (cf. Minazzi, 2013). I should add that it was lacking among his students because they had experienced the darkest moment in Italian history, namely the 1930s. They had perceived what Formaggio expressed very clearly, the *leaden weight* that rested on their future. And this of course also had a political dimension, in the sense that most of these students were then active in the Resistance. And there was a problem that exploded first with Preti, because already in the spring of '46 he decided not to renew his membership in the Italian Communist Party because he dissented with the party line, but he also dissented with Banfi (cf. Minazzi, 2011b). But if you go and check for yourself, you will see that to the detriment of Preti we have to say that he made this choice *before* gaining a university chair, and so he paid the price, whereas the others did the same *after* gaining a chair, and so they then paid a smaller price. In this respect Preti appeals to me more, in the sense that if he had something to say he said it. His wife, who tried to keep her feet firmly planted on the ground, told me that she said to him: "Don't say these things to Banfi, in public, in front of everyone! You'll ruin yourself academically!" But Preti would not listen to her because he was a *free spirit*.

Now, in the paper you will find that in this respect, superseding Banfi in this way also posed a problem for Preti, but here there are two points I would like to make. The first point is this. We talk about the School of Milan, but I think we should do so in *extensive terms*. Now we are in 2015, and we have to look at these events, I believe, with an eye that allows us to see all the impulses at work. In my opinion, the person who, better than anyone else, has reconstructed the unity around the critical rationalism of the School of Milan was Dal Pra, when he wrote that wonderful essay on critical rationalism in Milan in the volume published by Laterza in 1985,

La filosofia italiana dal dopoguerra ad oggi. In it he managed to see how all these thinkers actually responded to a common horizon, and what was the novelty of this horizon. I feel it was precisely the *transcendental* (cf. Dal Pra, 1985; Magnani, 2001b). This was pretty well the added value that came from Banfi. And here you can see this clearly in relation to another voice that we cannot help but consider as belonging to the School of Milan, although it followed a different trajectory, yet it is still a voice that was important in the School of Milan, namely the School of Geymonat. I am impelled to ask myself a question. What is the different image of rationality that is present in Geymonat? Geymonat was a neo-rationalist, a neo-Enlightenment thinker. All these authors identify themselves in neo-Enlightenment. But the image of rationality that an author such as Preti had—who was part of the neo-Enlightenment itself—was very different from Geymonat's image of rationality. The difference was profound, because in Geymonat's image of rationality *there was no such dimension of the transcendental*. There was not, and there never would be.

And I must say that on this point the cultural deafness has remained. And here I should say that in terms of *conceptual traditions* Geymonat remained always tied to the great tradition of positivism. But the criticism that Preti made of him was precisely this: not to see the functioning of the plane of the transcendental within reason. And here, in Geymonat's activities there is also a very interesting movement. It appears in that little volume published by Garzanti in the "Saper Tutto" series on scientific thought (cf. Geymonat, 1954), a volume that is valuable for Geymonat's intellectual biography, not only because it opens at the birth of this famous scholastic Manual of Philosophy, which then formed the basis of his greater *History*, the one in seven volumes (Geymonat, 1970-1976). But in that little book, when Geymonat reflects on the concept of reason, of critical rationality, at a certain point there is a wonderful page—he writes a very fine discourse on rationality in Galileo—in which he declares: "We can never confuse the structures of rationality with reality. We cannot short-circuit the structures of rationality with the structures of reality, because otherwise we would slip into a metaphysical position which is based on a philosophy of identity." When he then wrote, in 1977, his finest metaphysical book, *Scienza e realismo* (cf. Geymonat, 1977), he returned to exactly the same position that he had criticized in the fifties, namely that there is no identity between structure and rationality and structure and reality in the key of dialectic. And this is very significant: he returned for a reason that now we cannot consider, which is the defence of realism; however he returned to a position that was truly metaphysical.

This aspect, then, enables us to ask: does this tradition of critical rationalism in the School of Milan have its own importance, its more general value? And at this point I will enlarge the discourse. Here we have to engage in a lengthy reflection, and be able to understand what was the source of critical rationalism within the School of Milan, which often remained rather obscure: Piero Martinetti (1872–1943). Because this will enable us to understand Martinetti and Banfi (and we should also invoke another philosopher: Giovanni Emanuele Barié, 1894–1956). In Banfi's essay on his *Three Masters—Husserl, Simmel and Martinetti* (cf. Banfi,

1946)—what does he say about Martinetti? He uses an expression that for us, as we reflect on the image of reason in the School of Milan, is important. I quote Banfi's own words: "Martinetti represented the commitment to reason in his life." I think that this typical legacy of Martinettian rationalism is what Cantoni, in a fine essay published in *Studi filosofici*, first series, calls "the mythology of reason", present also in Banfi (cf. Cantoni, 1943). When the great students of the thirties criticised this lack of critical mediation between the *formal structures* of rationality and the *richness of experience*, they grasped this "mythological" aspect of reason, that Banfi brought from the lesson of Martinetti.

This, then, enables us to say that this School of Milan in the broadest sense, namely the school of Banfi, of Martinettian derivation and including Geymonat, built an interesting tradition in the debate on neo-Enlightenment in the fifties, which proved fruitful and powerful in the history of Italian philosophical culture. True, it is a tradition that has been defeated, because this tradition has been repressed, effaced, hidden, despised and unrecognised (and this explains why so many archives have ended up in our Centro Internazionale Insubrico in Varese). If you take the usual tools of the trade, for example the *Enciclopedia filosofica* published by Bompiani in 12 volumes (cf. Centro di Studi Filosofici di Gallarate 1968–69), or a good dictionary of philosophy such as Abbagnano's (cf. Abbagnano, 1971), under the entry *Critical Rationalism*, what do you find? You find only references to Popper's critical rationalism; the critical rationalism of Milan does not appear. And this is a theoretical problem. Why does critical rationalism not appear? Because it was defeated, in the first place. And in the second, what nexus exists between the critical rationalism of the tradition of the School of Milan and the critical rationalism of Popper? None. In fact, if you read Popper, you see a world of difference. Where is the huge difference? In the image of reason, and here Preti had seen very clearly, because when he discusses the various phases of the Vienna Circle and of empirical logic, he precisely captures this limit to the neopositivist image of rationality.

Hence in more general terms (now I am looking at the history of European critical rationalism in the Western tradition, so please forgive the "barbarism" of what I am about to say), I want to say that we have a rationalism that runs from Socrates and constantly traverses the Western tradition, with several modulations, in karstic fashion, with all the richness that we know. And then the turning point in this rationalism—namely its complication, which is precisely the Kantian Copernican revolution—and the rationalism of the School of Milan, with all its Husserlian, phenomenological derivation, is within this other history, namely the one that begins from Kant. And what is the difference? Once again, the concept of the *transcendental*. If I had to find a synthesis of this, I would say that it is an expression that is found in Paci's *Diario fenomenologico*, a truly extraordinary book (cf. Paci, 1973, pp. 84–88). There is a passage that has always struck me, a famous page, in which he says: "… today I went to see Banfi and asked him 'What is phenomenology?' And instead of answering, he said, 'Describe these flowers that are on the table.'" (Pozzi also devotes a page of her *Diary* to this "horrid" vase of plastic flowers.[6]) However, at one point Paci makes a statement that cannot fail to strike one, because he writes, "the transcendental is man". Now, I think that with this formulation of Paci's we really have the heart of what was the problem, the value and the limit of Banfi's teaching. Because "the problem is man" means that the problem has been understood: *How do I grasp the object of the comprehension of the world?* I grasp it through the operation that intelligence performs on the world to impoverish this world by building an *objectivity*, which I know is *relative*. Preti asks: "Relative to what?" To the *regional ontology* within which I have *constructed* it and in the *practices* with which this regional ontology *measures itself*. And I think this is an important issue, because it is the *theme of objectivity*, raised in the fifties. In Preti, then, it is interesting because, in the sixties, when he again reflected theoretically on this, he returned to Banfi; and the curvature of Preti's return is evident in *Retorica e logica* (cf. Preti, 1968; Preti, 2018). This is a "Banfian" book: you can feel this both in the theoretical chapters and in its historiographical aspects.

But this also opens up a more interesting problem, because I read the School of Milan in this key (cf. Minazzi, 2018; Magnani, 2012, 2016). The School has been characterized by this bond with the Kantian *transcendental*, which means at least bearing in mind the importance of the *structures of intentionality*, which relate these *structures of intentionality* with the *time of history*. Preti followed what was his most mature path, and at one point he returned—through his studies of medieval logic (cf. Preti, 1976, vol. II, pp. 3–194)—to debate late Scholasticism, because in it he identified the emergence of what he called "logical neo-realism" or "phenomenological realism". He then sought to develop this by the theme of *critical ontology*, which would be consigned to the unpublished part of his thinking. He died in July 1972 while he was collecting a series of essays on this very subject of "phenomenological neo-realism" or "logical neo-realism" (cf. Preti 1969–1970 and Minazzi, 2011a). The names and labels changed, but the issue was this. And this is an interesting issue because it leads back to the *transcendental* concept of *intentionality*. I will stop here: it seems to me that the important thing about this School is also to explain its repression.

I will close with a comment, a true one: I graduated here in Milan. My supervisor was Geymonat with Dal Pra as the co-examiner. This was in 1978. Geymonat then retired and was replaced by one of his students, who became my supervisor. I had presented the thesis on Preti's historical epistemology, and naturally started from Banfi's critical rationalism and problematicism. I am recounting this anecdote because it struck me then, and it shows that there was a problem and what it was. The new supervisor, having read a single chapter of my thesis, remarked: "Hum, '*problematicism*'? I do not know *problematicism*, I only know *problems*." This remark sums up the repression of the School of Milan.

[6] Cf. Pozzi (2015), p. 39.

8.4 Relevance of the Enlightenment Lesson

I very much agree that in their lessons Paci and the other Philosophers of School of Milan educated to think. That being said, there were not just the different paths taken by individuals; there was something more, and now the Archives are showing this. This appears, for example, in the correspondence of Geymonat with Dal Pra (cf. Minazzi, 2010, pp. 263–292), in which they discussed philosophical problems. Among other things there is a letter from Geymonat on the history of the School of Milan, written in '78, in which he said: "I am retiring, I can finally devote myself full-time to philosophy." Dal Pra replied that he was very disappointed at the changes taking place in philosophy at the *Statale*. In turn Geymonat replied: "Don't worry," and quoting Vico, wrote: "[I]t's a matter of recurrent cycles, of historical recurrences." Then he mentioned the person he regarded as his ideal teacher, Piero Martinetti, writing "of course, we are no longer in the glorious phase of Martinetti", with respect to which the season of the School of Milan to which he belonged already seemed to be a phase of decadence, "so that now we are going towards a period of great decadence, but then you have to be optimistic about the future". This is interesting, because they reasoned in terms of Schools, not in the sense of the Frankfurt School, but a *tradition of thought* in the broad sense, in which each pursued his own *demon*. They had different methods of approach, problems and concerns, yet they recognised themselves as the offspring of a common tradition, that of *critical rationalism.*

I would add that it was not only an *education to thought,* because in '75, when I attended the lectures given by these figures, behind certain words I felt the weight and the strength of a *lifetime*. I will try to explain this. The operation that Banfi had carried out in the 1930s was an interesting task, and it explains very clearly the repression, of which the problematicism mentioned was an unwitting epigone of this history. And it consisted of saying: "We are in a time of serious crisis, although fascism is triumphant. But it is not enough to fight against fascism and declare oneself against it, but we must fight it by *building a new culture*." It was this to which that group of philosophers felt committed, and their participation in the Resistance was conducted on a high level, fighting fascism because, said Preti, of the "dead weight of tradition", a backward tradition compared to the cultural debate in Europe and America. Now, philosophy today is not meant to be, as it were, "the grand narrative", but if it becomes that analytical discussion on the problems taken from analytic philosophy, it ends up, according to Evandro Agazzi, to put it metaphorically, *like someone who takes up a shotgun to go out hunting butterflies*, for which a butterfly net would be more suitable.

Philosophy lives to the extent that it has a social presence. This is the teaching of Socrates, who was a philosopher because he influenced the society of his time. For the School of Milan, there was not a tenuous position in the intellectual tradition of critical rationalism in which these thinkers recognised themselves, albeit with different and even conflicting positions. Dal Pra in his essay (cf. Dal Pra, 1985) emphasises the ability to bring these thinkers with their autonomous theoretical concerns within that tradition, to take part in a *common dialogue* with their voices, even in conflict with each other. Now I would like to contest the common statement that they are no longer read. It is not true. Abroad they are read, and I am referring, for example, to the book by a philosopher of mathematics and an epistemologist, Jean Petitot, *Per un nuovo illuminismo* (cf. Petitot, 2009), which is emblematic, because he, a scholar immersed in French culture, says that the only reference that helps his epistemological problem is in the reflections of the School of Milan. In fact he is amazed that it is not more highly regarded in Italy, and yet he believes that "it is a jewel", and he reads Banfi and Preti. For Petitot the dialogue continues only with the School of Milan, and he reads them because the theoretical problem in historical epistemological reflection is the one that those thinkers had already raised in the 1960s, and which international culture came to *later.* How can we relate the *formal structure* of the transcendental, the *a priori,* with the *historical dimension of the conceptual dynamic internal to scientific thought.* This is the problem of Preti.

The counterproof: if you read an author who passed through Milan, Paolo Rossi Monti (1923–2012), but who then had a conflicting relationship with this school, if you take his biography (cf. Rossi Monti, 2013), or the various earlier essays devoted to members of the School of Milan written in a polemical spirit that sometimes goes too far, you will see that they are not correct. For example, Paolo Rossi Monti (cf. Rossi Monti, 1986, pp. 244–247) undertook to reconstruct Geymonat's book *Filosofia e filosofia della scienza* (cf. Geymonat, 1960), in which he posed the problem of conceptual movements in scientific theories, and there was the famous controversy with Giulio Preti, who had seen these things more profoundly in his *Storia del pensiero scientifico* (cf. Preti, 1957), which, not coincidentally, looked back to a position of Gentile's. Paolo Rossi related him to Kuhn, who is completely irrelevant to the matter, because Kuhn's theory is that of the *incommensurability* of theories, while to Geymonat the problem is exactly the opposite: "How do I construct *continuism with discontinuity*."

And then Paolo Rossi Monti's operation does not help us to understand this School, but rather its repression. This part of the will of the Milanese school to construct a *positive thought* against a profound Italian tradition of three centuries, or since the seventeenth century, that of *counter-revolution*. Moreover, there was in them the admonition to *think for those who could not have thought,* and behind it was the partisan experience, hence a philosophy that *served society.* They were in deep accord with the Enlightenment (cf. Luzzini, 2018). It is clear that if we reduce philosophy to a series of technical problems, for example as Diego Marconi maintains (cf. Marconi, 2014), we have killed philosophy, because philosophy has a value of its own to the extent that it disturbs us, as Socrates did with his questions. See Banfi's *Socrate* of 1943 (cf. Banfi, 1943), an extraordinary book. It was reviewed by Preti in '46 in *Studi Filosofici*, where he presented a critique of the book. Banfi maintained that the philosophy of Socrates coincided with his life, taking into account the testimony of Plato in the *Theaetetus* and of Aristotle in the first book of the *Metaphysics,* while Preti says that in Socrates there was also a *theoretical dimension* that was the problem of the concept.

Beyond this, ours was a School that did not recognise itself in a *manifesto*; it was not the Frankfurt School, but a School that recognised itself in a tradition that is critical rationalism, and that came into marked conflict with what Preti called "the dead weight of the Italian tradition". Consequently he wrote: "There has always been a minority in Italy that looked at culture more alive wherever it developed internationally, that did not recognise itself in tradition, and that sought to construct a European Italy." This is also our problem today. For example, it means that in the historiographical instruments, there is no reference to *critical rationalism*, not just of the School of Milan but in general, because this orientation, which was engendered from the Kantian "Copernican revolution" onwards, did not put an end to the movement of thought such as neo-positivism (which had a manifesto, a journal, a school, and an organisation to disseminate it), since by its internal orientation it is made up of people who, in principle, *think for themselves*. So that in Poland we have epistemologists who go their own way, in France, Switzerland and Italy… But if we look at all these voices today, we find among Gonseth, Piaget, Bachelard, the school of Lvov, the Milan group, Enriques, all voices that were part of a very composite and rich tradition, which made its weakness, because it was not a philosophical party (cf. Papi-Minazzi 1995 and Magnani, 2007, 2017).

That said, I see also the sign of a cultural defeat, which from this point of view has a name, Federico Enriques (1871–1946), for example in the collocation of philosophy at universities, about which he argues that philosophy ought to be transversal to the various degree courses. Benedetto Croce (1866–1952) won by the combination of Literature with Philosophy, which persists today. Finally, regarding art, apart from the group around the magazine *Corrente di vita giovanile* (1938–1940),[7] there was a relationship with art, the osmosis was continuous. Preti was in relationship with an artist of the next generation like Piero Manzoni (1933–1963). They wrote about the *Galleria Il Milione*, they engaged in debates about exhibitions. But this is also the lesson of Banfi; there were his *Lectures on Aesthetics*, and he recommended not judging life a priori, but *understanding the reasons* that were developing in every sector. For this reason Preti, for example, in *La fiera letteraria,* wrote more articles about art than philosophy (cf. Preti, 1970). Then there was the *Minuziano* series edited by Rognoni. Moreover, Banfi's students wrote works about aesthetics, like Hume's *Essay on Taste*, about Fracastoro, Spalletti, Schelling, etc.

In the end there is the point of *schematism.* I think it is a misleading point; rather it is a confirmation of what I described as a problem with the School of Milan. This is that, when Preti equated Gentile (1875–1944) with Banfi, he did so on a decisive aspect, that both Banfi and Gentile found the "nullification of the intellect", the nullification of that intermediate plane between the *empty formal structures* and the *richness of reality* that has to be taken into account. I will just give a quote from Preti: "What is epistemology? Epistemology is that which studies the *functional structures* of the *intellect*, analysing the *construction of objects* in the light of those

[7] Cf. *Corrente* (1978).

same structures" (*In principio era la carne*, 1948).[8] And then, if I were to translate these indications, philosophy must be placed in a meta-reflection, philosophy no longer has its purpose, its content, but it should go and learn from other disciplines. That is what, in very different terms, Geymonat also said ("I'm going *to learn*, I'm not going *to teach*"). Hence there is not a "grand narrative", and I within the problem, but not as a technician of the specific problem; rather with the Socratic philosophical key that is *critical rationalism*. From this point of view there is a document that we want publish, a manuscript that dates back to 1945–46, concerning the seminars that were organized in Banfi's house, with Enzo Paci, Remo Cantoni, Giulio Preti, Giovanni Maria Bertin and Dino Formaggio. A theme would be assigned to a speaker and there was a person who kept the minutes (Angela Colantoni Stevani). Cantoni would correct the minutes, which have come down to us in this form. They were discussions that lasted for 2 years on the problem of reason and the concept of *a-priori*, with wonderful contributions that take up the movements that we have mentioned here. They were never in agreement with each other. Banfi intervened very little, and when he did so it was to give an impulse to further discussion, and from this comes the confirmation that the problem was this, namely: "How can we relate the structures of rationality to experience?" Banfi insisted strongly on the fact that the system which comes from phenomenology is the idea of *correlation*, which comes from two terms, the objective and the subjective, which, being translated into the *Leben* of the School, meant the relationship between *Geist* and *Leben*. This relationship came from Simmel: *life—more life—more than life*. This was the problem, because Paci, Preti and Formaggio were aware that thought was an emergence from the vital that is rooted in a body, a physicality… So that even the outcome of death is not there unless you have this life. This is a sign that also in the biographies of these philosophers and the relationship with artists stemmed from such a sensibility. The artist is seen in his *praxis*, whence a discourse on practice, on objects and on techniques. It is a highly relevant orientation, because they raised the relationship we have with the practices that Sini has investigated, but the term was already theirs. On Hegel, this point does not bother me, because they do not carry the weight of Hegelian metaphysics, but it is a Hegel that is analytically disarticulated and brings the richness of Hegelian phenomenology (it is not the Hegel of the *Encyclopaedia*). In this sense, the dialectic is devised as a *heuristic instrument* of rational transcription of that polarity between *subjective* and *objective*. And this speaks to our culture of today.

References

Abbagnano, N. (1971). *Dizionario di filosofia*. Utet.
Agazzi, E. (1961). *Introduzione ai problemi dell'assiomatica*. Vita e Pensiero.
Banfi, A. (1943). *Socrate*. Garzanti.

[8] Cf. Preti (1983)

Banfi, A. (1946), Tre Maestri (Simmel, Husserl, Martinetti). *L'Illustrazione Italiana*, LXXIII, 44, 284-285 (reissued in Banfi 1970, pp. 245-251).
Cambi, F. (1983). *Razionalità e prassi a Milano (1945–1954)*. Cisalpino-Goliardica.
Cantoni, R. (1943). L'illuminismo religioso di Piero Martinetti. *Studi Filosofici, IV, 2-4*, 216–233.
Dal Pra, M. (1984). Kantismo ed hegelismo in Banfi. In M. Dal Pra, D. Formaggio, & P. Rossi (Eds.), *Antonio Banfi: le vie della ragione* (pp. 21–35). Edizioni Unicopli.
Dal Pra, M. (1985). Il razionalismo critico. In *La filosofia italiana dal dopoguerra ad oggi* (pp. 31–92). Laterza.
Dal Pra, M. (1989). *Filosofi del Novecento*. Franco Angeli.
Dal Pra, M., Formaggio, D., & Rossi, P. (1984). *Antonio Banfi: le vie della ragione*. Edizioni Unicopli.
Garin, E. (1975). Cronache della filosofia italiana 1900-1943. In *Appendice Quindici anni dopo 1945–1960* (Vol. 2 Vols). Laterza.
Geymonat, L. (1954). *Il pensiero scientifico*. Garzanti.
Geymonat, L. (1960). *Filosofia e filosofia della scienza*. Feltrinelli.
Geymonat, L. (Ed.). (1970-1976). *Storia del pensiero filosofico e scientifico* (Vol. 7). Garzanti.
Geymonat, L. (1977). *Scienza e realismo*. Feltrinelli.
Luzzini, F. (2018). Bibliographical Distortions, Distortive Habits: Contextualizing Italian Publications in the History of Science. *Isis*, 109, SI, 3-13—Italian Version: Luzzini, F. (2019). Distorsioni bibliografiche e storture etiche. Le pubblicazioni di storia della scienza nel contesto italiano. *Il Protagora*, XLVI, 31-32, 327–340.
Magnani, L. (2001a) Philosophy and geometry. Theoretical and Historical Issues, Kluwer Academic.
Magnani, L. (2001b). Abduction, reason and science. In *Processes of Discovery and Explanation*. Kluwer Academic/Plenumn Publishers.
Magnani, L. (2007). *Morality in a technological world. Knowledge as duty*. Cambridge University Press.
Magnani, L. (2009a). *Abductive cognition. The epistemological and eco-cognitive dimensions of hypotetical reasoning*. Springer.
Magnani, L. (2009b). *Matematica e Morale in Ludovico Geymonat* in Minazzi (2009), a cura di, pp. 335-344.
Magnani, L., & Li, P. (Eds.). (2012). *Philosophy and cognitive science*. Springer.
Magnani, L. (2012). Scientific models are not fictions. In *Model-Based Science as Epistemic Warfare in Magnani-Li* (pp. 1–38). Springer.
Magnani, L. (2016). Scientific models are distributed and never abstract. A naturalistic perspective. In E. Ippoliti, F. Sterpetti, & T. Nickles (Eds.), *Model and inferences in science* (pp. 219–240). Springer.
Magnani, L. (2017). The abductive structure of Scientifc creativity. In *Am essay on the ecology of cognition*. Springer.
Marconi, D. (2014). *Il mestiere di pensare*. Einaudi.
Minazzi, F. (2007). La pura teoreticità banfiana e il pensiero scientifico. In: S. Chiodo-G. Scaramuzza, Ad Antonio Banfi cinquant'anni dopo, pp. 257–269. : Edizioni Unicopli.
Minazzi, F. (2010). *Geymonat epistemologo. Con documenti inediti e rari (un inedito del 1936, il carteggio con Moritz Schlick, lettere con Antonio Banfi e Madio Dal Pra)*. Mimesis.
Minazzi, F. (2011a). *Suppositio pro significato non ultimato. Giulio Preti neorealista logico*. Mimesis.
Minazzi, F. (2011b). *Giulio Preti: le opere e i giorni. Una vita più che vita per la filosofia quale onesto mestiere*. Mimesis.
Minazzi, F. (2013). Nel sorriso banfiano. In: Minazzi (2013a), pp. 19–147.
Minazzi, F. (Ed.). (2015a). *Sul bios theoretikós di Giulio Preti, 2 Vols*. Mimesis.
Minazzi, F. (2015b). L. Geymonat. Scegliere la ragione contro la barbarie nazifascista. *Humanitas, LXX, 1*, 83–94.

Minazzi, F. (Ed.). (2018). *Mario Dal Pra nella "scuola di Milano"*. Mimesis.

Paci, E. (1973). *Diario fenomenologico*. Bompiani.

Peruzzi, F. (2011). *L'estetica di Ermanno Migliorini. La militanza e la misura*. Leo S. Olschki.

Petitot, J. (2009). *Per un nuovo illuminismo, edited and transl. From French by F. Minazzi*. Bompiani.

Pozzi, A. (2015). In G. Bernabò & O. Dino (Eds.), *Parole. Tutte le poesie*. Centro Nazionale Insubrico.

Preti, G. (1957). *Storia del pensiero scientifico*. Mondadori.

Preti, G. (1968). *Retorica e logica*. Einaudi.

Preti, G. (1970). *Que seà, serà*. Il Fiorino.

Preti, G. (1976). *Saggi filosofici* (Vol. 2). La Nuova Italia. (English selected transl. Preti 2011).

Preti, G. (1983). In principio era la carne. In M. Dal Pra (Ed.), *Saggi filosofici inediti (1948–1970)*. Franco Angeli.

Preti, G. (2018). *Retorica e logica, new amended and enriched edition, edited with an introduction and notes by F. Minazzi*. Milan.

Rossi Monti, P. (1986). *I ragni e le formiche. Un'apologia della storia della scienza*. il Mulino.

Rossi Monti, P. (2013). *Un breve viaggio e altre storie. Le guerre, gli uomini, la memoria*. Raffaello Cortina Editore.

Sini, C. (2013). *Incontri. Vie dell'errore, vie della verità*. Jaca Book.

Sini, C. (2015). *Enzo Paci*. Feltrinelli.

Chapter 9
Horror Vacui: Characterizing the Experience of Paradigm Crisis Through Magnani's EC-Model of Abduction

Alger Sans Pinillos

Abstract This work pays tribute to the work of Professor Magnani by analyzing and applying his Eco-Cognitive Abduction Model (also known as the EC-Model) to introduce axiological values' role in the generation of scientific hypotheses. For this purpose, the cognitive dimension of the researcher is proposed in this work as a complementary factor to better understand the scientific paradigm shift. In particular, it is assumed that the experience of discovering a surprising fact that neither current theories nor methodologies can account for generates a state of radical ignorance that can provoke in the agent the emotion of anguish in the face of the imperative need to find a solution. Faced with ignorance as the only available evidence of the fact that triggers anguish, the agent's abducted hypotheses will be much more fragile than usual because they contain very distant elements, which will imply a more significant commitment to the interpretation of the scientific discipline involved, as well as to the agent's own vision of how the paradigm should be understood. This fragility, however, does not undermine the process but revitalizes the discipline in crisis with novel proposals that motivate suggestive new lines of investigation.

Keywords Wild abduction · Radical ignorance · Phaistos disc · Existential anguish · Inert affordance

A. Sans Pinillos (✉)
Department of Computer Applications in Science and Engineering (CASE), Barcelona Supercomputing Center (BSC), Barcelona, Spain
e-mail: alger.sanspinillos@bsc.es

S. Arfini (ed.), *Scientific Cognition, Semiotics, and Computational Agents: Essays in Honor of Lorenzo Magnani - Volume 2*, Synthese Library 506,
https://doi.org/10.1007/978-3-031-96688-0_9

I got a head with wings.
A head with wings.
Now I'm floating around up here way above the clouds.
So high, above the ground.
And the only thing that holds my head to the ground.
Is this one little skinny string.
(Morphine, 1993, A Head with Wings).

9.1 Introduction to the Horror Vacui

Partly an academic publication, partly a tribute to a person who has been my mentor, colleague, and friend, this essay aims to exalt Lorenzo Magnani's work, firstly, by presenting the relevance of his research career and, secondly, by showing how his proposals are alive and can contribute to future investigations.

By *horror vacui* or fear of emptiness, I refer to the other side of the coin when talking about abduction. It is usual to understand this form of inference or reasoning as the one that allows the hypothesizing of possibilities, generally associated with anomalous situations where other forms of inference or reasoning cannot operate satisfactorily. In sect. 9.3, a detailed explanation of what this means is offered. However, for now, what is important to introduce the object of this work is to point out that this conception assumes that abduction emerges as a safeguard against those moments in which the world is "interrupted" for the agent. As will be seen, surprise and bewilderment are two psychological conditions commonly used to characterize the circumstance in which the agent must change cognitive strategy to resolve the disruption caused by the ignorance of one or more facts that will ultimately give the world a new turn. However, to borrow Feyerabend's very apt words:

> The world we inhabit is abundant beyond our wildest imagination. There are trees, dreams, sunrises; there are thunderstorms, shadows, rivers; there are wars, flea bites, love affairs; there are the lives of people, Gods, entire galaxies (2001: 3).

If this sentence is considered, then we should revise the ontological assumptions about this world because, instead of a comfortable reality for us in which interruptions are few, it could well be that our condition is an eternal conflict to reconcile ourselves with this reality:

> History generally, and the history of revolution in particular, is always richer in content, more varied, more many-sided, more lively and subtle than even the best historian and the best methodologist can imagine (Feyerabend, 1993: 9).

With these two quotations from Feyerabend, I intend to approach the classical problems of incommensurability and paradigm shift to propose a re-reading based on the aforementioned *horror vacui* based on Magnani's Eco-Cognitive Model of Abduction (hereinafter, EC-Model). Therefore, it is not intended to enter into the classical debates around scientific realism or anti-realism or to address the debate on the contexts of justification and discovery. However, as is obvious, these are issues to which I will necessarily refer in order to situate and develop my approach.

On the contrary, I intend to present a characterization of abduction by assuming the ontological understandings and assumptions of Pragmatism, a theory that is based on this form of inference (Sans Pinillos, 2022):

1. Pragmatism assumes that the experience configuring reality is structured from a continuous flux of information (world of pure experience).
2. Continuous experience maintains knowledge in a certain degree of provisionality, which obliges us to revise and adapt our beliefs to the changes we experience (fallibilism).

My primary motivation is to analyze these debates from the perspective of the agent experiencing the rupture that causes incommensurability and may eventually lead to a paradigm shift. Taking one of Feyerabend's acute observations again, one could say that my approach takes as its point of reference the questions of the participant (those raised by scientists in scientific practice) and leaves aside those of the observers (those which philosophers and historians of science can ask, precisely because they analyze what has already been discovered and, because of this—and, often, also despite this—, they have data that scientists cannot count on) (cf., Feyerabend, 1978: 18).

Since the participants do not have a historical perspective, their questions can only be aligned with the current state of the research and, therefore, those events considered paradigmatic on paper will be surprising, chaotic, even dangerous, and compromising (cf. Feyerabend, 1967: 401). In other words, it does not matter whether the event is paradigmatic or not since it will always cause surprise to a greater or lesser degree (Sans Pinillos, 2022), but whether the participant can really ask the observer's questions when scientific theory and methodology enter a crisis. I do not close my mind to an affirmative answer. However, I believe that it will then acquire nuances that only a cognitive approach can account for since it is not the same to accept an open and unfinished conception of the world from the idea of progress and scientific success as to do so from the experience of defeat, usually accompanied by the risk of losing one's job, funding, recognition and all those personal and social factors that are so rarely considered in philosophical research.

In the same way that the success stories commonly known as discoveries can be characterized as abductive processes, this work intends to delve into the abductive dimension of a kind of hypothesis that depends almost exclusively on the participant's determination to continue in the face of results that do not offer opportunities to propose a possible course of action. Because of their condition of being sustained almost exclusively by the will of the agent to find sufficient indications to continue an investigation initially declared dead from the point of view of methodology and the scientific community, the hypotheses that will sustain future actions will be a kind of wild abductions because of their apparent disconnection, awaiting reconciliation, which will always imply a change in the methodology or conception of the discipline.

Usually, ignorance and its mitigation are part of an epistemological system, where the agent is situated about various types of affordances (in short, opportunities for action, see Gibson, 1979) that provide him with qualitative and procedural information about the distributed cognitive system in which he finds himself (environment). However, there are extreme cases in which ignorance about something is absolute. I refer precisely to radical ignorance because its novelty challenges the methodology so that the crisis of the discipline will be lived by the agent from anguish. In short, anguish is taken from its existentialist philosophical meaning that points to what Kierkegaard called the "dizziness of freedom": freedom as a possibility of action. Taking responsibility for our actions (jumping or not into the abyss to which I have leaned out) leads to self-awareness (Kierkegaard, 1980: 61).

Anguish occurs when faced with the hypothesis of an unknown affordance that does not offer opportunities for action. Therefore, the researcher will be responsible for abandoning or generating a hypothesis that allows them to rethink the research. For this, they will surely resort to meta-theoretical resources, even axiological ones, to support their decision, even if only subjectively. This decision is made before the attentive gaze that the abyss fixes on us moments before we jump into it. In other words, the horror vacui may come to challenge the worldview of the discipline and propose a new methodology that, in the long run, could be identified in the books as a case of Paradigm Shift. For example, Copernicus' sixteenth-century heliocentric theory faced great resistance due to the lack of empirical evidence and the entrenchment of the geocentric model. However, his decision to go ahead and rethink astronomy with new observations and mathematical methods sparked the Scientific Revolution, radically changing the worldview and laying the foundations of modern astronomy.

For this reason, the anguish prior to proposing a possible line of action and its establishment as a new paradigm is an essential scientific moment to consider in order to understand scientific practice better. In this paper, I will focus mainly on the Phaistos Disc. However, there are other enigmas, such as the Voynich manuscript (cf., D'Imperio, 1978) or the debate on the artificiality (manufacture) or naturalness of the Klerksdorp spheres (cf., Heinrich, 2008). Archaeology offers a good conceptual framework to address the subject of this work because it is a discipline that depends heavily on abducting hypotheses that make sense of concrete facts on an ongoing basis.

Then, in cases such as those described above, radical ignorance is concentrated in concrete artifacts that become inert affordances because they do not offer any opportunity in the distributed system where they have been found. Since the artifact provides no inherent opportunity in this context, the cognitive agent hypothesizes the affordance by contextualizing it based on the sociocultural values that define their worldview (cf. Sans Pinillos, 2023). In other words, the artifact is not truly an affordance by itself. However, it becomes one to the extent that we associate some action with it, thereby generating a hypothesized opportunity for further investigation.

In cases of radical ignorance, the agent's relationship between self-awareness of their ignorance and their subsequent judgments is critical to understanding the apparent paradox of one who becomes aware of their own ignorance. Thus, agents' self-awareness of their ignorance is the process of understanding that one is ignorant of something, and it can be characterized as a cognitive tool with which agents can manage both the particular problem about which they are ignorant and the distress they feel in this cognitive situation (Sans Pinillos & Magnani, 2022).

Section 9.2 presents scientific research as a social activity that integrates technical and social elements influenced by professional goals and institutions. Research success and failure depend on dynamic collaboration and decisions based on researchers' informed intuition. Lack of progress can cause existential angst, which motivates philosophical reflection and abductive reasoning. Although not all scientists share a uniform understanding of science, their ability to cope with crises can lead to unexpected discoveries.

Section 9.3 presents the EC-Model of abduction as a non-arbitrary process of embodied and situated hypothesis generation and selection aimed at exploiting available resources to the maximum to suggest courses of action to cope with the uncertainty inherent in research. Among these resources, emotional and axiological factors are highlighted, which allow the reorganization of the data to propose new objectives in situations of paradigmatic crisis.

In sect. 9.4, self-awareness of ignorance of being ignorant about something in Nature is distinguished as a cognitive state precursor of philosophical questions that contribute to handling situations of radical ignorance that occur when, for example, an object does not offer opportunities (affordances) for action. For such an exposition, the case of the Phaistos Disc is taken as an example; the lack of conclusive answers is especially relevant in abduction, where hypotheses emerge as plausible solutions to unresolved questions. This abductive approach is essential in disciplines such as archaeology, where ignorance becomes a catalyst for exploration and discovery even when the objects studied seem inert and offer little opportunity for action.

Finally, in the conclusions (sect. 9.5), we explore the paradigmatic crisis lived from the existential anguish as a responsibility and commitment to think about this radical ignorance and thus create the opportunity to abduce wild hypotheses (detached from the scientific institution) that propose a revision of the discipline that cannot give an answer to this crisis.

9.2 The Social Nature of Scientific Research: Navigating Crises and Discoveries

Scientific research is a social activity in the profound sense that there is no clear distinction between the technical and the social, between the natural fact and the artifact (e.g., a scientific model), and, therefore, neither between the researcher's thinking and his scientific practice developed within a particular worldview. By this, I mean to highlight that successes, failures, and discoveries are closely linked to the research's professional objectives and the institution that sustains it. This is nothing new (cf., Latour & Woolgar, 1986). However, it is important to point it out in order to emphasize and, with it, better introduce the idea that there are professional circumstances during research that, despite not being strictly linked to scientific methodology, have a significant weight for its proper development. In fact, following Latour's perspective, it is crucial to understand that research is a dynamic process of partnerships between various actors (human and non-human) whose partnership movements can be suspended or restarted according to the researchers' decisions (Latour, 2005).

From this perspective, when the methodology enters an impasse, it is the decisions based on the informed intuitions of the researchers (Schön, 1983; Klein, 1999) that will tip the balance toward a new line of action that will allow progress toward a definition shared by the community, thus ensuring the adequate and successful development of the research. However, it is not realistic to consider all people equally informed. Professionals often only know their small parcel of expertise, even when the project in which they collaborate pretends—to place at the end of the ideal—to discern the meaning of the Universe. Therefore, it is essential to remember that researchers do not need a clear idea of what science is or share the values generally associated with its practice beyond their training and professional background.

It is also important to realize that there is a link between professional goals that are not closely linked to these values, such as the discovery of reality or contributing to the progress of humanity, but that, in their effort to confront and resolve crises in more local research, new and interesting truths are ultimately arrived at. A common way of referring to these occasions is serendipity, often misunderstood as "fortuitous luck," but which instead aims to be a kind of capability (de Rond, 2014) to generate creative explanations (Ross & Arfini, 2023) that, among other things, anticipate the new and interesting results referred to above like Fleming's discovery of penicillin, Antoine Henri Becquerel's finding of X-rays, Watson and Crick's reflection on the alpha-helix structure of DNA, and a long etc. (Bennet & Chung, 2001; Magnani et al., 2022).

As is known, these examples are successful cases, i.e., situations in which the novel, surprising fact offered opportunities for interaction to hypothesize a line of research. However, as has been said, this work focuses on those situations where the surprising fact has not yet offered opportunities, and it seems that it never will. Therefore, the hypothesized new line of action will depend on several parallel factors, such as financial factors, economy of effort and time, and the commitment of the researchers to the project. Because of its condition, this line of action will have to be sustained almost exclusively by the agent's will to find sufficient evidence to continue an investigation initially declared dead from the methodology and the scientific community's point of view.

For this reason, the hypotheses that will sustain future actions will be a series of radical abductions, wild by their apparent disconnection, awaiting reconciliation, which will always imply a change in the methodology or conception of the discipline. Both are cases that can be characterized by abduction. However, the latter requires some considerations that, although they may be present in the former, become essential here to understand a genuine part of scientific research.

Let us leave out the discussion for a moment about cost-effectiveness as a factor for canceling research and assume a case in which the scientists (participants) must make the decision. We have, then, the real picture of people committed to their decisions and, therefore, with mixed feelings at the prospect of canceling the course of action set without first answering why their project has failed.

The fact that in many cases, there is no response can be attributed to a variety of circumstances, such as the diversity of people's abilities, a certain degree of luck, and, of course, the fact that in most of the world's investigations, there is more than one way to achieve the stated objective. However, it is indisputable that experiencing a crisis in an investigation because of an anomaly will always be daunting, conflicting, and ultimately distressing. A way of dealing with radical ignorance can be introduced through self-awareness of this ignorance from the notion of existential distress, ultimately leading to philosophical reflection on the reasons for this situation through abductive reasoning.

9.3 Unveiling the Different Faces of the EC-Model of Abduction

The current concept of abduction is a reinterpretation of Peirce's (1958, CP 5.14–40) recovery of the Aristotelian apagōgē (ἀπαγωγή) (Aristotle., 1964, An. Pr. II 25, 69a20–35), intended to ground the acquisition of experience (Peirce, 1958, CP 5.348; Magnani, 2015). Then, it can be stated that the function of abduction is to handle the experience of novel facts by generating hypotheses or conjectures that indicate new courses of action (Sans Pinillos, 2022). In this sense, abduction is defined as a non-arbitrary process of hypothesis generation and selection (Magnani, 2015). It is activated in situations or phenomena perceived as surprising or puzzling due to their novelty because classical epistemic resources cannot address them (Thagard, 1988).

For this reason, abduction requires a non-classical and non-monotonic formalization, as abductions are problematic judgments with a high degree of expectation (Peirce, 1958, CP 5.192). It must, therefore, be understood as a deviation from classical logic (Gabbay & Wood, 2005; Aliseda, 2006). More specifically, abductive reasoning adds nuances to perception through hypothetical inference, which, as will be seen below, is a conclusion that can be reformulated in ecological-adaptive terms whose validity lies in its plausibility for proposing courses of action for research (Magnani, 2001).

It is important to consider the distinction between Peirce's early and later conceptualizations of abduction when discussing its application in contemporary contexts (Minnameier, 2017), so this paper starts, on the one hand, with the classification of the essential features of Kapitan's abduction and in Magnani's multimodal cognitive proposal. On the first point, Kapitan distinguishes four essential features that define early pragmatic abduction (1997: 447–448):

1. *Inferential Thesis*. Abduction is, or includes, an inferential process or processes (Peirce, 1958: CP: 5.188–189,7.202).

2. *Thesis of Purpose*. The purpose of "scientific" abduction is both (i) to generate new hypotheses, and (ii) to select hypotheses for further examination (Peirce, 1958: CP: 6.525); hence, a central aim of such abduction is "to recommend a course of action" (MS 637:5).
3. *Comprehension Thesis*. Scientific abduction includes all the operations whereby theories are engendered (Peirce, 1958: CP: 5.590).
4. *Autonomy Thesis*. Abduction is, or embodies, reasoning that is distinct from, and irreducible to, either deduction or induction (Peirce, 1958: CP: 5.146).

Although nowadays, we tend to distinguish between what Aliseda rightly identified as common sense, static reasoning (i.e., medical diagnosis), and scientific reasoning (i.e., Kepler's discovery of the elliptical orbits of the planets) (Aliseda, 2006: 29–31), it should also not be forgotten that "Peirce not only developed a philosophy of possible experimental verification and the 'experimental community of interpretation' that it presupposes, but" […] "Peirce's semiotic uses the social interrelationship between the regulation of behavior and possible experience to interpret the problem of understanding meaning" (Apel, 1981: 7). In this sense, focusing what has been said on the subject at hand, it could be said that scientific reasoning is introduced by what Aliseda identifies as common sense based on abduction (Aliseda, 2006: 29):

I. *Explaining observations with simple facts*: I see smoke, so (I hypothesize) there will be fire.
II. *Laying causal connections with facts*: It is very windy today. In the morning, I see that the pots of my plants are broken and scattered on the ground, so (I hypothesize) the wind must have blown them over.
III. *Facing a contradiction*: I know the wind often blows my plants to the ground, but today it has not blown a bit, so how are my plants scattered on the ground?
IV. *When something does not work*: No wind blew today, but the plants are on the ground. I ask my partner if he/she has been on the balcony, but that is not the case. In addition, I live on the tenth floor, and there is no access (i.e., for someone who hates my plants) to my balcony. So, (I hypothesize) my neighbor's cat will have managed to get from her balcony to mine.

Reading these cases from Kapitan's four theses, abduction is characterized as reasoning to cope with the uncertainty that ignorance about novel facts causes through generating and selecting hypotheses that suggest possible courses of action to handle the situation (Woods, 2013). Hypothesis setting is subject to the epistemic virtues of minimalism, i.e., the entrenchment (Gärdenfors, 1988) arising from preference-based ordering that aligns a state of belief (Aliseda, 2006: 181), and of economy of inquiry, i.e., the inclination to take those hypotheses that a) do less violence to the prior cognitive state (the vital starting point) and b) require fewer resources for their fixation (Aliseda, 2006: 37).

A seasoned reader of the contemporary debate on abduction will have noticed my continued avoidance of specifying the theory under which these two virtues are ascribed. The reason for this is simply that this paper cannot and does not pretend to encompass the problem of explanationism in abductive inference, and, as such, my choice of references should be taken exclusively expository of the subject matter of this paper and not as a theoretical positioning. Moreover, as has been introduced, this research starts from the EC-Model, so that the notion of tentativeness characteristic of the explanationist model (AKM) and the preservation of ignorance of the non-explanationist model (GW), are taken from a multimodal perspective which understands that it is the demands of cognitively situated action that determine the conditions in which a hypothesis arises and is chosen.

9.3.1 The Role of the EC-Model in Understanding Abduction

The perspective offered by the EC-Model makes it possible to understand the challenge that hypothesizing poses for the agents in an environment that is not necessarily regular for them in their efforts to find answers. For this reason, in addition to their novelty, the hypotheses triggered should suggest a line of action to navigate an ignorance mainly lived from strangeness (Sans Pinillos & Magnani, 2022: 198). In the following section, the notion of ignorance is addressed in more depth, so for the moment, it will serve to say that I refer to the condition of epistemic limitation because the available knowledge is insufficient to account for the situation.

Thus, as will be seen, ignorance is also a vital epistemic element since it constitutes the cognitive niche, namely the natural and sociocultural cognitive human activities (cognitive representations—methods devised by the mind to understand and navigate the outside world—(Magnani et al., 2022)) that transform the natural environment into a cognitive one (Laland et al., 2001; Odling-Smee et al., 2003), which encourages the agents to investigate in order to overcome surprise and perplexity and try to reduce their ignorance of the situation. The agent will then constitute three possible scenarios (Woods, 2013; Barés Gómez & Fontaine, 2023):

A) *Subduance*: The puzzling fact yields information that eliminates ignorance (e.g., empirical scientific discovery).
B) *Surrender*: The fact remains unknown (persistence of ignorance).
C) *Abduction* (ignorance mitigation): A hypothesis that suggests new courses of action is established.

All three scenarios presuppose that (a) the ignorance of the ignored fact or circumstance is not total, for when faced with the annoying but acceptable circumstance that there are facts we do not know, the ignorance is immediately signified with familiar emotions and feelings and that (b) the ignorance is always local (not general), indicating that the agent has resources, often heuristic, tangential and divergent, to deal with the ignorance.

Abduction arises as an auxiliary process of tentative (without the property of certainty, Aliseda, 2006: 91) and provisional (Aliseda, 2006: 36–37) reconciliation of extraneous data with the solid epistemic baggage of the cognizing being (the assumed and reasonably variable world). Therefore, the agent needs to take advantage of all available information to generate and then consider plausible (select) a hypothesis that allows the agent to cope with surprise, bewilderment and, in extreme cases, to manage anguish in his efforts to mitigate ignorance in the course of the investigation (Sans Pinillos & Magnani, 2022: 195).

In this sense, both tentativeness and provisionality are due to the impossibility of responding to ignorance (subduance) together with the agent's persistence in not surrendering to which abduction arises as a hypothesis that mitigates ignorance: Abduction preserves ignorance so that the hypothesis will have a tentative epistemic value or, in other words, as a conjecture to be verified (Gabbay & Wood, 2005: 47). The advantage of taking the EC-Model as a reference becomes even more evident here since from this approach it is assumed that the hypotheses are reified and externalized through action so that the "explanations" are the local successes that contribute to the objective set precisely to guide the course of the research.

The emphasis falls on cognitive abduction (Magnani, 2009), which means the naturalization of the logic of hypothesis generation and selection, which comprises both the minimal cognitive expression of micro-processes of hypothesizing in perception (Vallverdú & Sans Pinillos, 2023: 1292) and complex cognitive processes (i.e., scientific discovery or medical diagnosis). It is possible to identify the naturalistic project of logic to transfer the epistemic issues of scientific research to its cognitive dimension (Magnani, 2017: 1-1n). In this sense, agents are considered as embodied distributed cognitive systems, which emphasize the knowledge-enhancing property of considering all information as a valuable resource to manage facts (Magnani, 2017: 207).

From this perspective, abduction is, as far as truth is concerned, a knowledge-enhancing inference, which means that it is a process by which reliable experience is transformed into new and exciting suggestions in the form of hypotheses aiming at new precisions for reasons parallel to epistemological ones (Magnani, 2017: 90) considering all information as valuable resources for the handling of facts (Magnani, 2017: 207), such for example, its axiological value (considering that a hypothesis is not a fact) or its conformity with empirical data that contradict "what should be the case".[1]

In this sense, abduction is the form of reasoning that integrates the interactive predispositions of agents (experience) with their environment (Vallverdú & Sans Pinillos, 2023: 1301). Therefore, the structure of abduction is conceived as multimodal in terms of the different cognitive strategies that define it:

[1] It is important to always keep in mind that with abduction, Peirce (1958: CP: 5.348) tried to find a satisfactory answer to Kant's claim about the impossibility of demonstrating the existence of things external to consciousness (Kant, 1998: B, Preface, xxxix; Dilman, 1973: 2; Sans Pinillos, 2022). Magnani considers that despite the differences between Kantian synthesis and Peircean abduction, the connection between both concepts lies precisely in the fact that Peircean research on the characterization of Kantian processes and rules of ampliative inference requires abduction (Magnani, 2001: 42). The reason for this is that the difficulty pointed out by Kant suggests a gnoseological limit that invites a revision of the very notion of experimentation ("the lock upon the door of philosophy" (Peirce, 1958, CP 5.348)).

(a) It must optimize the different situations we live in (optimization of situatedness).
(b) There is a mutability (permutation) between the roles (input/output) of the elements (maximization of changeability).
(c) Abduction is sensitive to absorbing information from what is presented to us continuously (information-sensitive).
(d) It is necessary to enrich the inferential system we have to acquire all these requirements (Magnani, 2017: 138–139).

These requirements are oriented to represent the means that organisms have to relate to the information presented to them, which implies considering also the possible new information generated through these interactions (Magnani, 2017: 139). Then, the non-monotonicity of abductive inference is subsumed from the cognitive perspective with selective and creative strategies (Hintikka, 1998) that allow the agent to retract in the light of new evidence (Aliseda, 2006: 53).

From this perspective, the EC-Model is defined as a hybrid system that will prioritize the requirements that will define the form of the abducted hypothesis depending on the confronted case, being more creative in the strangest cases (Magnani, 2017: 206) or more selective when there is more conformity between the novelty and the agent's initial condition (Magnani, 2017: 201). Likewise, the EC-Model considers manipulative and theoretical abductions and sentential and model-based abductions in addition to selective and creative abductions (Fig. 9.1).[2]

Considering the information expressed in Fig. 9.1, it can be understood that theoretical abduction is articulated through an internal relationship between the knowledge of the cognizing being and the cognitive schemes and strategies aimed at experimenting and acquiring new information. In this way, theoretical abduction is considered to exploit the agent's creative resources to generate and select hypotheses. In this way, a distinction can be made, on the one hand, between theoretical-sentential abductions (relating to logical and verbal/symbolic inferences) and

Fig. 9.1 The pattern of theEC-Model of abduction

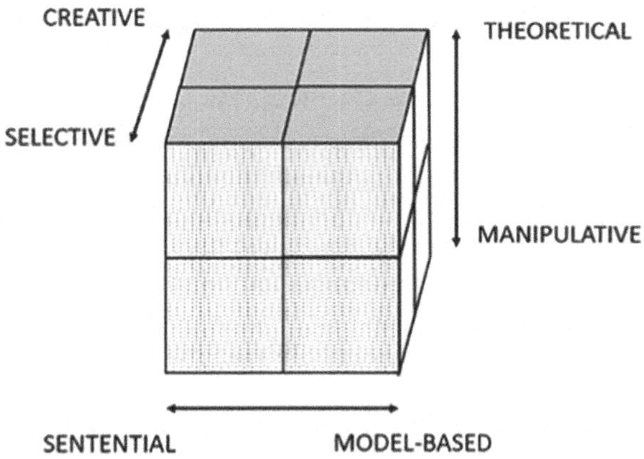

[2] Figure inspired by Park's work (2017: 12) (Cf., Vallverdú & Sans Pinillos, 2023: 1302).

model-based abductions (the exploration of the environment through the externalization of models, diagrams, and images) (Magnani, 2009: 11) (Magnani, 2009: 11). On the other hand, manipulative abduction predominates as a tacit process of knowledge.

Likewise, this perspective can be extended in such a way that it also contemplates situations in which emotions, feelings, and axiological values determine a form of unreflective hypothesis but with heuristic value. To argue this, I take into consideration the epistemological distinction between off-line (theoretical) and on-line (manipulative) cognitive processes (Magnani, 2009: 11). This means that the distinction between theory and action does not make sense in all situations. This statement represents an interesting gap to explore for abduction since it allows us to understand theoretical cognitive processes (off-line) as the anguish before a disconcerting fact that does not offer opportunities for action and their embodiment through actions as manipulative processes (on-line).

In this way, abduction becomes a crucial tool not only for the generation and selection of hypotheses in scientific and rational contexts, but also in the more everyday and emotional contexts that are also involved in research. Integrating elements such as emotions and axiological values in the abductive process highlights the importance of considering the totality of the agent's cognitive experience, including his or her affective and evaluative responses to novel and puzzling situations.

For example, in a situation where an individual is faced with an unexpected and emotionally disturbing event, such as a case of radical ignorance, abduction can help generate hypotheses that not only explain the event but also guide actions to mitigate the emotional impact and reorient knowledge in such a way that options for action emerge based on the free choice (anguish) that only conviction can endow our actions in situations where we have only ourselves to go on.

9.4 Self-Awareness and Existential Anguish

From the Delphic sentence "*know thyself*" (Greek: γνῶθι σεαυτόν) to the Socratic sentence "*I know that I know nothing*" (ἓν οἶδα ὅτι οὐδὲν οἶδα), self-awareness of ignorance is one of the fundamental pillars of philosophy, in the sense so beautifully expressed by Colli of "not knowing and wanting to know" (1978). As is well known, both sentences draw the dialectical foundations of Socratic maieutics. However, it was Aristotle who best embodied the philosophical way of thinking focused on physis (φύσις) and, therefore, who had to classify the ways of inferring (syllogisms) in order to grasp the essence of reasoning. From this, two forms of ignorance are distinguished: ignorance in investigating nature and ignorance of philosophical questions. While the former is characterized by what was later called the hypothetico-deductive method, the scientific method par excellence, the latter is characterized by not having a conclusive answer.

One way to understand this is the way Aristotle shapes Socrates' last words in the Platonic dialogue Menon. As is well known, Socrates ends his inquiry into whether virtue is teachable in aporia. However, Socrates indulges in a very suggestive epilogue referring to a "dispensation awarded by the gods, without any knowledge," for making expert politicians that presumably a politician could come to have (Plato, 1924: Men. 100a-c). To this conclusion, Aristotle confronts the form of political knowledge (differentiated from knowledge of physis) as hypothetical inference, i.e., abduction as a plausible solution that only attains the status of "possibility". Aristotle concludes that if we assume the rule that "whatever is knowledge, can be taught" and take the fact that "virtue is knowledge," then we can hypothetically infer (abduce) the case that "virtue can be taught" (Aristotle., 1964: Anal. Pr.: II 25 20–35).

It is important to note that, although the aporia is not resolved, abduction allows us to characterize the answer as a hypothesis, which gives it a specific epistemic role: to make sense of broad-spectrum and unfinished questions that tend to refer to fundamental problems of life. In the specific case analyzed in this paper, the self-awareness of ignorance is an abductive process that predisposes us to continue investigating blindly because we do not foresee a resolution in the short term (Sans Pinillos & Magnani, 2022). The reason for this is that, in these situations, the agent reconsiders the facts, assuming that he does not have the necessary resources to deal with it and, therefore, freely undertakes his efforts to solve the enigma without thereby—wanting—to compromise the essential values of science and his discipline. I identify this new approach with the philosophical view of the scientist, which allows not only the question of the facts under investigation but also the judgment of the assumptions that define the branch of knowledge that constitutes the point of reference for the agent.

This philosophical view is not only abductive because of its hypothetical condition but also because it needs to combine elements of a different ontological order to operate, even when such relations are considered erroneous from a logical point of view. I am referring to the relation of facts and values, of logic and psychology (Putnam, 2002), etc. (cf., Sans Pinillos, 2022), dichotomies that, although they have had an essential theoretical meaning within, let us say, Paradigms, are differentiations that are blurred in the practical dimension, where "exploring to the maximum the available resources" does not imply a single methodology.

A clear example of this can be found in scientific models during the paradigm shift or, in other words, during their constitutive role in the process of discovery. In these cases, these scientific models are epistemologically productive in an abductive sense: while they are inert models in the face of the fact being observed (e.g., Phaistos Disc, sect. 9.5), they play a causal role in producing new knowledge. Thus, scientific models abductively connected with novel facts may be empirically false and, in turn, potential knowledge by allowing the agent to revise his assumptions in the light of new unknown information (Sans Pinillos & Magnani, 2022: 199).

9.4.1 *The Role of Ignorance in the EC-Model of Abduction*

This makes more sense within the EC-Model of abduction because, as stated in the preceding section, the tentativeness and preservation of ignorance are approached from a cognitive-situated perspective so that the ignorance caused by the scientific model, methodology, etc., is not the same as that which persists in empirical research, although they are certainly related. Thus, taking Arfini's definition of ignorance as:

> a cognitive condition that can be either passively (and unconsciously) bore by an agent or actively nurtured by her and a condition that entails epistemic limitations (which can be any lack of knowledge, belief, information, or data) that affects the behavior, the belief system, and the inferential capacity of the agent (Arfini et al., 2020: 185).

we can also identify the second-order ignorance that links the psychological aspect of research with the empirical research that states that all knowledge acquisition requires a certain degree of ignorance, so that ignorance, as introduced, is part of the event that provides opportunities to act: to explore and discover (Arfini et al., 2020: 102–103). In other words, we know that we are ignorant about something because of how we can interact with it (Sans Pinillos & Magnani, 2022: 199).

This notion of ignorance fits perfectly with the EC-Model of abduction and allows for the analysis of how ignorance enables knowledge through interaction with objects understood as affordances, in which contextualized action produces abductive hypotheses through the meaning that occurs during manipulation in terms of opportunities (Varela et al., 2017). In the next section, we delve deeper into the notion of affordance to introduce cases in which the properties of an object do not provide opportunities. For now, however, this is sufficient to understand how action can be motivated by an abductive hypothesis about the opportunity triggering ignorance. Then, archeology is introduced as a scientific discipline to delve into this topic and, finally, to introduce how the fact of being aware of one's own ignorance (whether by surprise, bewilderment, or anguish) can be a trigger for the philosophical look that allows to approach situations of radical ignorance through hypotheses alleviated about an object that does not offer opportunities (inert affordance).

9.4.1.1 The Role of Abduction in Archaeological Investigation

An archaeological investigation requires hypothesizing a plausible scenario to make sense of the remains discovered. Thus, archaeology is a discipline in which abduction plays a fundamental role because of having to deal continuously with the uncertainty caused by the ignorance of crucial data, so archaeologists have to be creative in imagining and proposing hypotheses in their research (Shelley, 1996). In particular, archaeologists must make sense of the objects they study and their distribution in the place where they have been found (Vallverdú & Sans Pinillos, 2023: 1300). For this reason, Shelley (1996) considers that visual abduction predominates in archaeological research: there are patterns that are applied as processes of recognition and identification of the ignored as plausible properties that encourage—provide opportunities for—action.

Here, Shelley seems to understand recognition patterns as a refined version of the classical sense of conscious inference in an epistemological sense of carrying out a particular inquiry (i.e., Hanson, 1958; Popper, 2002) through abduction (Peirce, 1958, CP: 5.14–40; Kowalski, 1994). From this perspective, Shelley (1996: 280) identifies four types of visual abductions:

1. recognition of objects as of theoretical significance (classical abduction);
2. abduction on shapes of objects (which can provide insight on the causal history capable to situate the object under investigation in a context);
3. abduction on structure of objects (which is the representation of the object-based exclusively on its structure);
4. analogical inference, which refers to a novel hypothesis when constructed analogously to a known one.

Nevertheless, although certainly, in most cases, the hypotheses in archaeology operate according to the four types of abductions that Shelley classifies, it is important to keep in mind that this is so because it is usual to be ignorant of something and to know many other things. When this happens, it could be said that the puzzling fact is "respecting the rules of the game" and, therefore, the abduction acts as an under-coded process (Meyer, 2010), that is, the fact offers opportunities to raise a plausible hypothesis. Following Shelley's (1996: 293–294) suggestive study, it can be seen that the characterized situations are those in which, faced with the impossibility of directly recognizing something by, for example, historical dating, the rule of analogy is activated, allowing the generation of novel hypotheses. In these cases, symmetry governs the analogy to relate and compare similar structures with similar solutions from generalizations and particularizations (cf., Longino, 1983) oriented to a defined and localized problem of ignorance.

However, as has been said, there are also cases in which the disconcerting fact surprises them by its persistence in not offering opportunities, thus remaining as an inert affordance (an artifact that lacks inherent utility in its context but gains meaning when interpreted through sociocultural values, suggesting a potential use). In cases like this, it becomes clear that there are cases of ignorance that can only be managed through action-oriented by the agent's certainty of knowing himself to be ignorant. Then, instead of under-coded abductions, the agent will rely on over-coded ones (Eco, 1983): unconscious (Schurz, 2008: 207) and immediate (Magnani, 2001) that occur during manipulation when cognitive processes manifest themselves in the form of interaction, which human beings adopt in response to suggestions offered by an object or a fact through factors such as memory, sensations, emotions, narratives and feelings that will contribute to predisposing the way in which the inert affordance is perceived (Vallverdú & Sans Pinillos, 2023: 1301) through interaction built on the knowledge of our absolute ignorance about it.

I call these hypotheses wild abductions, which are introduced in the following section, using the Phaistos Disc case as an argumentative context. With them, I intend to try to capture the genuine mechanisms for managing the commitment that some agents may suffer in the face of existential anguish: choosing to continue despite the psychological impossibility of stopping in the face of an enigma that puts both the discipline and them in check and, precisely for that reason, they cannot stop investigating.

9.4.2 *The Enigmatic and Still Unknown Phaistos Disc*

The Phaistos Disk (Fig. 9.2) is one of those paradigmatic phenomena about which we know practically nothing. It is classified as a clay tablet located in the middle Minoan period, approximately 1400 BC. However, nothing is known about it. "Tablet" is synonymous with the disk form, whose terminological use refers to the inscriptions or sequence of signs engraved on this mysterious artifact (Balistier, 2000). Therefore, it is obvious that many hypotheses try to account for this surprising fact. However, a detailed study of these failed attempts to find the key to reveal what this artifact is shows us that most theories focus on the idea that the disc supports a recorded code (Godart, 1990 and, more recently, Duhoux, 2000) or musical writing (Younger, 2007). In the face of the feeling of defeat, some opt for the hypothesis that the artifact is a fraud (Eisenberg, 2008).

However, the fundamental issue is that there seems to be a limit to the number of hypotheses within the framework of a given archaeological theory (Sans Pinillos & Magnani, 2022: 204–205). This means that, even without opportunities offered by the object, the researcher is subject to the logical constraints of the methodology (the trained form of, for example, minimalism, economy, coherence, etc.). The possibility to affirm this comes precisely from the EC-Model of abduction, which defines hypothesizing as the situated action-oriented to exploit to the maximum the available resources as an adaptive strategy of constant openness in an indeterminate world (Magnani, 2017: 15).

Fig. 9.2 Phaistos Disc, sides A and B (C messier, edit by Bammesk, n.d.)

This points to the idea that our state of indeterminacy is based precisely on the fact that there are de facto unlimited opportunities for an object to offer us (Vallverdú & Sans Pinillos, 2023: 1301). Drawing on Chemero's terminology (Chemero, 2003), from the fact that an affordance-property of an artifact does not offer options for action, we cannot conclude that no affordances-characteristic of our incessant interaction-arise.

Therefore (see Fig. 9.2), although triggers (i.e., surprise, disconcert, distress) and constraints (i.e., minimality, economy) are cognitive structures that hierarchize and mediate the way in which we interact with the environment (Gabbay & Wood, 2005: 84), it is also true that we have the phenomenological capacity to bracket (*Einklammerung*, Husserl, 1960: First meditation) them or adapt them theoretically to achieve other objectives, just as they are translated into rules for various theories with different presuppositions and purposes. By this, I mean that faced with a fact of radical ignorance, as is the Phaistos Disc, the decision to continue the investigation is abductive-based. This can only operate assuming the inertness of the artifact under investigation and, therefore, allowing some flexibility for hypotheses based on our own characteristics, among which are imagination, desires, etc., all of them cognitive elements of our actions. As explained below, the existential point in these cases is to assume one's own ignorance from the anguish caused by being unable to resolve it and, with it, explore other options based on what is unknown (Fig. 9.3).

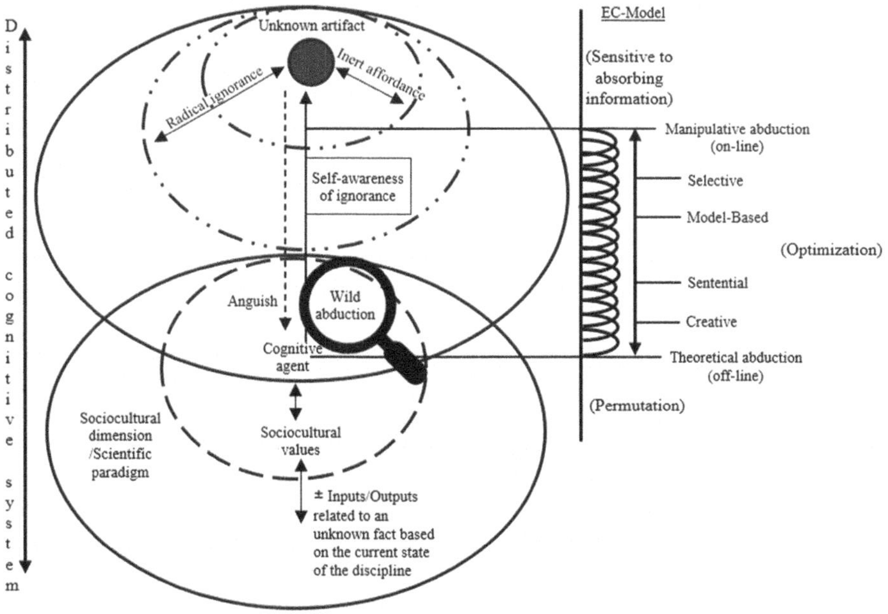

Fig. 9.3 Diagram of wild abduction

9.5 Conclusions: Radical Ignorance and Wild Hypotheses

The case of the Phaistos Disc is an artifact of which we know nothing directly (empirical identification) or indirectly (possibility of making analogies), so the agent is devoid of any opportunity for action in terms of the properties the object offers. In general terms, it can be said that affordance has been lost from the point of view of properties and, therefore, becomes inert for our cognitive characteristics. However, considering that we are still in front of an unknown fact, we can turn the inert into fruitful through reorganizing available resources and objectives. In this case, the archaeologist can replace the objective of knowing what this object is with the objective of analyzing why it is so unavailable for the methodology he is using and, with this, to promote a revision of the discipline.

As I am sure the reader has already noticed, the Phaistos Disc is taken in this paper as an occasion to exemplify a facet of research little explored: the step prior to the paradigm shift as a crisis for researchers. This Minoan artifact is an object about which we know nothing directly (e.g., through its manipulation) or indirectly (it resists analogies, for example), so researchers are deprived of any opportunity for action. However, the fact that affordance has been lost from the point of view of artifact properties does not mean that we cannot generate opportunities through our cognitive actions. As has been said, our characteristics include imagination, desires, feelings, and emotions, all of which define our way of being in the world and, therefore, also intervene in our everyday and professional investigations.

We see their intervention in common cases, where even the most extreme hypotheses have an anchor point with which agents can determine plausible courses of action based on certain knowledge. Hesse (1954) proposed that the paradigm shift was based, among other factors, on the imagination defined by a universal scientific language (Fernández Buey, 2004: 70), whose articulation would be based on what a confirmed theory could predict unproven results by analogy with similar confirmed cases of the same theory (Hesse, 1966). Likewise, it is known that in the discovery of Neptune, Le Verrier performed an existential abduction (Hanson, 1958:124-1n; Thagard, 1988: 54) based, among many other scientific considerations, the belief of a stable universe (Grosser, 1979).

However, when there is no anchor to hold on to in order to support a hypothesis because, for example, we are faced with an unknown fact because the object under investigation does not offer opportunities for action, then the agent either gives up or tries to convert the inert into fruitful through a reorganization of the available resources and the setting of other objectives. For example, a researcher may replace the objective of discovering what this unknown object is with that of analyzing why this object resists the methodology he is using and, in so doing, revise the discipline. Such a reorganization must be understood under a trigger beyond mere curiosity: anguish. As is well known, this affective state is usually the reaction to the unknown, where the first step-knowing oneself to be ignorant, is the loss of innocence that provokes anguish (Kierkegaard, 1980: Chap. I, § 4). This idea can be contextualized in scientific research, assuming that the loss of innocence is the experience of the paradigm crisis caused by knowing oneself to be ignorant in the face of the impotence to account for a surprising fact.

This anguish can take the form of responsibility for our actions (Sartre, 1966: 33), when researchers maintain ethical and epistemic integrity towards the ideals they project on their discipline. Thus, for example, an archaeologist could commit himself to a line of research not supported by the scientific institution precisely because he considers that there are severe errors in its assumptions and methodologies, thus generating a "proto-research program" that can come to coexist as a scientific model at the same time (cf., Lakatos, 1978). Existentialism offers an ideal conceptual framework to understand the change of perspective as an abductive process to understand a situation of ignorance lived once the technical vision of the world that allows supposing a defined function for all the things of the world is saturated (Sartre, 1966: 18). In other words, the existentialist view implies the "realization" that reality is, synthesizing Feyerabend's words, from the beginning of this text, coarser and more complex than what methodologies and people can manage to think.

Thus, from knowing oneself to be ignorant, researchers can take the historical perspective to be in solidarity with all those who have found themselves in the same situation and understand that part of the discovery implies the collapse of outdated systems. Likewise, appropriating Sartre's famous sentence "l'existence précède l'essence", it follows from knowing oneself to be ignorant of an object that offers no expectations of the self-awareness step of criticizing the complete models of the world (innocence) and assuming the disconnection between our will and Nature (Sartre, 1966: 19) and, thus, to consider research as a human attempt to make sense of reality and our place in it (a thesis that, as we have seen, is also assumed from pragmatism and, I add, Marxism, cf., Apel, 1981: 1).

It is possible to characterize the experience and overcoming of the paradigmatic crisis through existential anguish through abduction because novel cases that cause radical ignorance, such as the Disc of Phaistos, require a very intense generation and selection of hypotheses. After all, it is done from the self-awareness of ignorance and the evidence of the disconnection between Nature and will. Therefore, it is clearly shown at that moment that the hypothesis will depend to a great extent on our will, which the responsibility of anguish will guide before the need to answer that unknown fact; now no longer out of mere curiosity, but as a commitment assumed with certain ideals attributed to the discipline that is practiced and, therefore, that constitute our existence.

This abduction is wild because the hypothesis does not focus on clearing the impossible but on discovering the causes of this unfortunate situation. In this case, solving ignorance means only taking advantage of it by transforming the chance of its appearance as an occasion to ask ourselves about the unknown: "Every point of departure in time is eo ipso accidental, an occasion, a vanishing moment" (Kierkegaard, 1980: Ch. I, A IV, 205). The abducted opportunity allows us to ask the question about the reasons for ignorance and to find a hypothesis that points to a methodological and epistemological investigation: the revision of the discipline as a possible cause of the crisis and, with it, to propitiate what in history books is known as a Paradigm Shift.

Acknowledgments This work has been possible thanks to the discussions I have had with my colleagues over the last few years, among which I would like to highlight those with Lorenzo Magnani, Selene Arfini, David Casacuberta, Jordi Vallverdú, Anna Estany, Mario Gensollen, Ariel Guersenzvaig, Agustín Adúriz-Bravo, Vicent Costa and Dafne Muntanyola. I would also like to thank Carla Tomas Kohlberg and Maria Nieva for their comments made from the perspective of the professional scientific eye that faces hard setbacks during their work and that has allowed the seminal ideas of this work to be put forward.

Funding Alger Sans Pinillos was supported by the "Generación D" initiative, Red.es, Ministerio para la Transformación Digital y de la Función Pública, for talent attraction (C005/24-ED CV1), funded by the European Union's NextGenerationEU program through the PRTR.

Ethics declarations Funded by the European Union - Next Generation EU. However, the views and opinions expressed are solely those of the author(s) and do not necessarily reflect those of the European Union or the European Commission. Neither the European Union nor the European Commission can be held responsible for them.

References

Aliseda, A. (2006). *Abductive reasoning: Logical investigations into discovery and explanation*. Springer.

Apel, K.-O. (1981). *Charles S*. Peirce. From Pragmatism to Pragmaticism (Trans. John Michael Krois). University of Massachusetts Press.

Arfini, S., Bertolotti, T., & Magnani, L. (2020). The antinomies of serendipity how to cognitively frame serendipity for scientific discoveries. *Topoi, 39*, 939–948. https://doi.org/10.1007/s11245-018-9571-3

Aristotle. (1964). In W. D. Ross (Ed.), *Analytica Priora et Posteriora*. Oxford University Press.

Balistier, T. (2000). *The Phaistos disc*. Verlag Dr Thomas Balistier.

Barés Gómez, C., & Fontaine, M. (2023). Medical reasoning and the GW model of abduction. In L. Magnani (Ed.), *Handbook of abductive cognition* (pp. 481–506). Springer.

Bennet, J. W., & Chung, K.-T. (2001). Alexander Fleming and the discovery of penicillin. *Advances in Applied Microbiology, 49*, 163–184. https://doi.org/10.1016/S0065-2164(01)49013-7

C messier, edit by Bammesk. (n.d.). Δίσκος της Φαιστού πλευρά A 6380.JPG [Photograph]. *Wikimedia Commons*. CC BY-SA 4.0. https://commons.wikimedia.org/w/index.php?curid=141943567

Chemero, A. (2003). An outline of a theory of affordances. *Ecological Psychology, 15*(2), 181–195.

Colli, G. (1978). *La nascita della filosofia*. Adelphi.

de Rond, M. (2014). The structure of serendipity. *Culture and Organization, 20*(5), 342–358. https://doi.org/10.1080/14759551.2014.967451

Dilman, İ. (1973). *Induction and deduction. A study in Wittgenstein*. Basil Blackwell.

D'Imperio, M. E. (1978). *The Voynich manuscript: An elegant enigma*. National Security Agency/Central Security Service.

Duhoux, Y. (2000). How not to decipher the Phaistos disc: A review. *American Journal of Archaeology, 104*, 597–600.

Eco, U. (1983). Horns, hooves, insteps. In U. Eco, U. Eco, & T. A. Sebeok (Eds.), *The sign of the three. Dupin, Holmes, Peirce* (pp. 198–220). Indiana UP.

Eisenberg, J. M. (2008). *The Phaistos disk: One hundred year old hoax? Minerva* (July/August), 9–24.
Fernández Buey, F. (2004). *La ilusión del método*. Ideas para un racionalismo bien temperado.
Feyerabend, P. K. (1967). On the improvement of the sciences and the arts, and the possible identity of the two. In R. S. Cohen & M. W. Wartofsky (Eds.), *Proceedings of the Boston Colloquium for the Philosophy of Science 1964/1966*. (Boston Studies in the Philosophy of Science, vol 3., pp. 387–415). Springer. https://doi.org/10.1007/978-94-010-3508-8_22
Feyerabend, P. K. (1978). *Science in a free society*. Lowe & Brydone Ltd..
Feyerabend, P.-K. (1993). *Against method*. Verso.
Feyerabend, P.-K. (2001). *Conquest of abundance*. Chicago University Press.
Gabbay, M., & Wood, J. (2005). *A practical logic of cognitive systems: The reach of abduction—Insight and trial*. Elsevier.
Gärdenfors, P. (1988). *Knowledge in flux: Modeling the dynamics of epistemic states*. MIT Press.
Gibson, J. J. (1979). *The ecological approach to visual perception*. Houghton Mifflin.
Godart, L. (1990). *The Phaistos disc: The enigma of an Aegean script*. Itanos.
Grosser, M. (1979). *The discovery of Neptune*. Dover Publications.
Hanson, N. R. (1958). *Patterns of discovery*. Cambridge University Press.
Heinrich, P. V. (2008). The mysterious "spheres" of Ottosdal, South Africa. The mysterious "spheres" of Ottosdal, South Africa. Reports of the National Center for. *Science Education, 28*(1).
Hesse, M. (1954). *Science and the human imagination: Aspects of the history and logic of physical science*. SCM Press. Return to ref 1954 in article.
Hesse, M. (1966). *Models and analogies in science*. Notre Dame University Press.
Hintikka, J. (1998). What is abduction? The fundamental problem of contemporary epistemology. *Transactions of the Charles S. Peirce Society, 34*(3), 503–533.
Husserl, E. (1960). *Cartesian meditations. An introduction to phenomenology*. (Trans. Dorion Cairns). Springer-Science+Business Media, B. V.
Kant, I. (1998). *Critique of pure reason (transl Paul Guyer & Allen W Wood)*. Cambridge University Press.
Kierkegaard, S. (1980). The concept of anxiety. *A simple pshychologically orienting deliberation on the dogmatic issue of hereditary sin* (R. Thompte & A. B. Anderson, Trans.). Princenton University Press.
Klein, G. (1999). *Sources of power*. MIT Press.
Kowalski, R. A. (1994). Logic without model theory. In D. M. Gabbay (Ed.), *What is a logical system?* (pp. 35–71). Oxford University Press.
Lakatos, I. (1978). *The methodology of scientific research programmes: Philosophical papers*. Cambridge University Press.
Laland, K., Odling-Smee, J., & Feldman, M. (2001). Niche construction, biological evolution, and cultural change. *Behavioral and Brain Sciences, 23*(1), 131–146. https://doi.org/10.1017/S0140525X00002417
Latour, B., & Woolgar, S. (1986). *Laboratory life*. Princeton University Press.
Latour, B. (2005). *Reassembling the social*. Oxford University Press.
Longino, H. (1983). Beyond "bad science". *Science, Technology, and Human Values, 8*(1), 7–17.
Magnani, L., Sans Pinillos, A., & Arfini, S. (2022). Language: The "ultimate artifact" to build, develop, and update worldviews. *Topoi, 41*, 461–470. https://doi.org/10.1007/s11245-021-09742-5
Magnani, L. (2001). *Abduction, reason, and science: Processes of discovery and explanation*. Kluwer Academic Publishers. https://doi.org/10.1007/978-1-4419-8562-0
Magnani, L. (2009). *Abductive cognition: The epistemological and Eco-cognitive dimensions of hypothetical reasoning*. Springer Science & Business Media.
Magnani, L. (2015). The eco-cognitive model of abduction: Ἀπαγωγή now: Naturalizing the logic of abduction. *Journal of Applied Logic, 13*(3), 285–315. https://doi.org/10.1016/j.jal.2015.04.003

Magnani, L. (2017). *The abductive structure of scientific creativity: An essay on the ecology of cognition*. Springer.

Meyer, M. (2010). Abduction—A logical view for investigating and initiating processes of discovering mathematical coherences. *Educational Studies in Mathematics, 74*, 185–205. https://doi.org/10.1007/s10649-010-9233-x

Minnameier, G. (2017). Forms of abduction and an inferential taxonomy. In L. Magnani & T. Bertolotti (Eds.), *Springer handbook of model-based science. Springer Handbooks*. Springer. https://doi.org/10.1007/978-3-319-30526-4_8

Morphine. (1993). A Head with Wings [Track 6]. On Cure for Pain [CD]. Rykodisc.

Odling-Smee, F. J., Laland, K. N., & Feldman, M. W. (2003). *Niche construction. The neglected process in evolution*. Princeton University Press.

Park, W. (2017). *Abduction in context: The conjectural dynamics of scientific reasoning*. Springer.

Peirce, C. S. (1958). In C. Hartshorne & P. Weiss (Eds.), *Collected papers of charles sanders peirce (Vol. 1–6)*. Harvard University Press. 1931–1935; (Vol. 7–8) (A.W. Burks Ed.).

Plato. (1924). *Meno*. In Laches, Protagoras, Meno, Euthydemus (W. R. M. Lamb, Trans.). Harvard University Press.

Popper, K. (2002). *The logic of scientific discovery*. Routledge.

Putnam, H. (2002). *The collapse of the fact/value dichotomy*. Harvard University Press.

Ross, W., & Arfini, S. (2023). The role of serendipity in creative cognition. In L. J. Ball & F. Vallée-Tourangeau (Eds.), *The Routledge international handbook of creative cognition* (pp. 46–64).

Sans Pinillos, A., & Magnani, L. (2022). How do we think about the unknown? The self-awareness of ignorance as a tool for managing the anguish of not knowing. In S. Arfini & L. Magnani (Eds.), *Embodied, extended, ignorant minds. Synthese library* (Vol. 463). Springer. https://doi.org/10.1007/978-3-031-01922-7_9

Sans Pinillos, A. (2022). Neglected pragmatism: Discussing abduction to dissolute classical dichotomies. *Foundations of Science, 27*, 1107–1125. https://doi.org/10.1007/s10699-021-09817-x

Sans Pinillos, A. (2023). Abductive irradiation of cultural values in shared spaces: The case of social education through public libraries. In L. Magnani (Ed.), *Handbook of abductive cognition*. Springer. https://doi.org/10.1007/978-3-031-10135-9_51

Sartre, J.-P. (1966). *L'existentialisme est un humanisme*. Les Éditions Nagel.

Schurz, G. (2008). Patterns of abduction. *Synthese, 164*, 201–234. https://doi.org/10.1007/s11229-007-9223-4

Shelley, C. (1996). Abductive reasoning in archaeology. *Philosophy of Science, 63*(2), 278–301.

Schön, D. (1983). *The reflective practitioner: How professionals think in action*. Basic Books.

Thagard, P. (1988). *Computational philosophy of science*. MIT Press.

Vallverdú, J., & Sans Pinillos, A. (2023). The foundations of creativity: Human inquiry explained through the neuro-multimodality of abduction. In L. Magnani (Ed.), *Handbook of abductive cognition*. Springer. https://doi.org/10.1007/978-3-031-10135-9_71

Varela, F., Thomson, E., & Rosch, E. (2017). *The embodied mind: Cognitive science and human experience*. MIT Press.

Woods, J. (2013). *Errors of reasoning: Naturalizing the logic of inference. (studies in logic, vol. 45)*. College Publications.

Younger, J. G. (2007). The Mycenaen bard: The evidence for sound and song. In S. P. Morris & R. Laffineur (Eds.), *Epos: Reconsidering Greek epic and Aegean bronze age archaeology: Proceedings of the 11th international Aegean conference/11e rencontre égéenne internationale, Los Angeles, UCLA–The J. Paul Getty Villa, 20–23 April 2006* (pp. 71–78). Université de Liège, Histoire de l'art et archéologie de la Grèce antique; University of Texas at Austin, Program in Aegean Scripts and Prehistory.

Chapter 10
Fair Play for the Machines: Turing on Charity Versus Skepticism on the Question of Thinking Machines

Cameron Shelley

Abstract When Turing considered the question, "Can machines think?" he anticipated many difficulties this claim would have to overcome before being considered seriously, much less accepted. The most basic difficulty was bias against the idea embedded in the English words *machines* and *think*. When considering this and other difficulties for his claim, Turing (1947/2004) urged that "… fair play must be given to the machine." As Magnani (2022) points out, Turing was working at a time when the concept of *computing* was changing fundamentally due to the development of digital computers, in which Turing played a significant role. Turing's ideas about thinking machines were part of his response to this profound revision in history, which both motivated and grounded his call for fair play. The aim of this article to discuss and develop Turing's idea of fair play as a reconsideration of the trade-off between skeptical and charitable strategies for the assessment of hypotheses like *machines can think*. Specifically, Turing's call amounts to the view that critics of the hypothesis are overly skeptical due to historical bias and partiality. His call for fair play is a call for a more charitable assessment of the hypothesis, one that neutralizes the prejudices embedded in his contemporary context.

10.1 Introduction

One of Alan Turing's many historic contributions was his exploration of the relation between automatic computing and thinking. Famously, he raised the question, "Can machines think?" and answered in the affirmative, in the sense that digital computers could be considered to be thinking in virtue of running some appropriate program (Turing, 1950). He proposed a test, which he called *the imitation game* (now often known as the *Turing Test* in his honor) to identify thinking machines. He

C. Shelley (✉)
Centre for Society, Technology & Values, University of Waterloo, Waterloo, ON, Canada
e-mail: cshelley@uwaterloo.ca

S. Arfini (ed.), *Scientific Cognition, Semiotics, and Computational Agents: Essays in Honor of Lorenzo Magnani - Volume 2*, Synthese Library 506,
https://doi.org/10.1007/978-3-031-96688-0_10

confronted and responded to several objections both to his claim and to the test that he proposed.

Turing's claim and his test have been examined from many perspectives (Copeland et al., 2017). One such perspective has been as an instance of abductive or explanatory reasoning (Magnani, 2010): Turing's claim, as he describes it, can be understood as a consideration of the hypothesis that a given computer is thinking in light of the evidence provided by its behavior, particularly in conversation (Shelley, 2010).

As Magnani (2022) points out, Turing was not applying a self-evident concept of computation to the issue of thinking as computing. Instead, he was engaged in applying his own general theory of computation and historical concepts of calculation to the novel matter of digital computation. A crucial aspect of this development was the concept of *learning machines*, that is, digital computers whose behavior arises not only from their own programming but from their interactions with the world (Magnani, 2022). Turing's conception was of a digital computer having experiences, its "brain" transitioning from a "disorganized" state to an "organized" one in a process comparable to an education. Such a computer, he contended, might rightly be said to be engaged in thinking (Turing 1948/1969).

In developing his ideas, Turing considered several objections to them (Gonçalves, 2023a, 2023b). Skeptics of his claim that digital machines can think could point to what they consider to be crucial differences between digital computers and humans, the prototypical thinking beings. Such differences, skeptics would argue, point out qualities that computers lack that thinking beings must have.

Turing made responses specific to each sort of objection but also, in his early work, insisted that "fair play must be given to the machine" (Turing, 1947/2004). The meaning of this plea has not been examined closely by scholars but, in fact, indicates an important characteristic of Turing's response to many of the objections. Roughly speaking, Turing was asserting that people are apt to be biased in their assessment of the hypothesis that machines can think and his demand for "fair play" was a plea for them to grasp and set aside this bias.

More specifically, I argue that Turing's plea for fair play in this context can be understood as a request for adjustment between *skepticism* and *charity* in people's assessment of Turing's hypothesis. For the purpose of this article, skepticism is well-known as an epistemic strategy on which people tend to withhold belief in hypotheses until they attain a high degree of support of the appropriate kind (Williams, 1999). There are good reasons to be skeptical in many cases, such as claims about astrology or ESP (Thagard, 1988). Skepticism helps to keep a set of beliefs free from falsehoods. However, skepticism can be biased and carried too far (Ditto & Lopez, 1992). An incurable skeptic may remain largely free of false beliefs but will also reject many true hypotheses. Thus, skepticism can lead to and reinforce ignorance.

The alternative strategy to skepticism is charity. In this sense, charity means being more open to hypotheses before accepting them as beliefs (Feldman, 1998). A charitable thinker may accept a large body of beliefs. The payoff of this strategy is that the practitioner would be less ignorant than otherwise. Of course, the price of charity is that the body of beliefs resulting from charitable assessment may well contain falsehoods.

Turing was certainly aware of this situation but felt that people were simply apt to be excessively skeptical when considering the matter of thinking machines.

The aim of this chapter is to examine Turing's plea for fair play for machines as part of his case for the claim that machines can think. The context in which Turing makes his plea for fair play for machines is reviewed first, followed by discussion of why he thought that critics might assess his claim in a way that is unfair or prejudiced. The general nature of Turing's argument is discussed, followed by a general characterization of objections to it. Turing's conception of fair play is then explored in its historical context. This conception is then developed into a general distinction between skepticism and charity as epistemological strategies. The mathematical objection and the Lady Lovelace objection to Turing's claim are then construed in light of the trade-off between these strategies. This leads to the conclusion that Turing's call for fair play is a specific case of a general tension between these two strategies for obtaining knowledge and, thus, that Turing had a valid reason for making his plea.

10.2 Fair Play for Machines

Turing (1950) began his famous article on intelligent computers with the disarmingly brief question, "Can machines think?" Interestingly, Turing did not try to answer the question in any direct way in this article. Instead, he proposed to replace the question with another one: Can a digital computer be programmed to pass itself off as a human being, under certain circumstances? Turing argued that the answer to this second question might indeed be "yes." Although he could not offer a program that was up to the task, Turing explained how one might well come to exist within 50 years or so. The possibility of such a program, he argued, also suggested a positive answer to the first question: A computer programmed to pass itself off convincingly as a human is also a computer that thinks.

Turing's contention and his arguments for it have remained the subject of robust discussion and debate (Epstein et al., 2008). Much commentary on the matter concerns either the cogency of Turing's move from the first question above to the second one, or the validity of his arguments for answering it affirmatively (Horn, 2008). These issues are significant ones but there is another aspect of Turing's proposal that has escaped general notice: whether the issue of thinking machines could be addressed fairly.

Turing's concern for fairness has gone unnoticed likely because he does not raise it explicitly in his 1950 article. Instead, he raises it in his first exploration of the same topic, where he says that "… fair play must be given to the machine" (Turing, 1947/2004, p. 394). The remark occurs in the context of what he later called the *mathematical objection* to thinking machines, which rests on the claim that human mathematicians have infallible access to mathematical truths, which computers do not (more below). Turing's response is to point out that human mathematicians often make blunders in practice and, thus, are not infallible, and that a suitably programmed computer can make similar blunders. So, no fundamental intellectual distinction between the two is evident on this basis.

The notion of fair play here seems to relate to bias in how people are apt to infer intelligence from the performance of machines versus people. "It is easy for us," Turing adds, "to regard these blunders [committed by a mathematician] as not counting and give him another chance, but the machine would probably be allowed no mercy" (Turing, 1947/2004, p. 394). Such prejudice is unfair, Turing implies, and the appropriate remedy is to apply similar standards to both mathematicians and machines.

Although Turing does not use the term *fair play* later on, his writings on thinking machines are shot through with implicit appeals to it. So, understanding it is important to understanding Turing's thinking on the matter and what might be its relevance for today.

The meaning and role of fair play in Turing's work has received little scholarly attention. Gonçalves (2020, pp. 44–47) construes this notion as a shift in the burden of proof: Rather than Turing having to prove that machines can think, critics have to prove that they cannot. This construal does capture the sense Turing clearly has that critics of his claim are apt to demand more from him than they do of themselves in assessing arguments on the topic. However, the burden-of-proof construal would seem to place the burden squarely on Turing's shoulders after all, since he is the one, by his own admission, who is making positive and novel claims about digital machines. It is conventional for anyone to bear the burden of proof in such a situation (Walton, 2008).

Instead, I propose that Turing's call for fair play should be understood as the view that critics of his claims about the possibility of thinking machines are simply too skeptical. To apply the sporting metaphor implicit in the notion of fair play, the playing field in the match of Team Machine versus Team Human is not level, making it impossible for the former to score, that is, have any hope of pressing an argument successfully against the latter. To convince skeptics of this view, Turing has first to identify the bias that is at the root of their perspective.

10.3 Prejudice

As noted above, the burden of proof of his claim that machines can think would seem to lie with Turing, since his claim is both positive and novel. Turing's call for *fair play* amounts to a denial of this perspective. While understandable, simply laying the burden of proof on Turing places excessive emphasis on accidental features of the English language and thus detracts inordinately from the substance of Turing's hypothesis. In other words, history resulted in prejudice against his claim, prejudice that should be removed in order for the claim to be evaluated as it properly should be.

In historical terms, Turing was in an unusual position on the topic of thinking machines. During World War Two, he had worked at Bletchley Park on the Ultra project of breaking the Nazi Enigma code (Copeland et al., 2017). Initially, this process was accomplished manually by teams of experts. Turing had led the design and construction of an electro-mechanical computer that partially automated the decoding of Enigma messages, thus greatly aiding the war effort. As a result, Turing had seen first-hand how the tasks of thinking human beings had been delegated, in part, to machinery. Very few people had witnessed such a process at the time and his experience was, in fact, a state secret, which he could not share directly.

Prior to the 1950s, the term *computer* typically referred to a human being, often a woman, whose job was to set up and carry out difficult, mathematical calculations (Grier, 2013). Through the 1950s and 1960s, programmable digital machines replaced human computers so thoroughly that the term *computer* came to refer to those machines instead. Turing foresaw that digital devices would take over individual tasks that required thinking in humans and extrapolated that, some day, a digital machine capable of thinking in a general way would be produced. However, in the late 1940s, he was one of the few people with the expertise to consider such a development in an informed way. Without his experience, critics were likely to dismiss his extrapolations without due consideration.

Turing raised this concern at the outset of his paper describing the imitation game. Upon considering the question, "Can machines think?" the obvious response is to consult definitions of *machine* and *think*. If the term *think* can be applied felicitously to the concept *machine*, then the question is answered affirmatively. Otherwise, the answer is negative. However understandable, this procedure is not satisfactory. After all, how do we establish the definitions involved? Turing observes (1950, p. 433):

> The definitions might be framed so as to reflect so far as possible the normal use of the words, but this attitude is dangerous. If the meaning of the words 'machine' and 'think' are to be found by examining how they are commonly used it is difficult to escape the conclusion that the meaning and the answer to the question, 'Can machines think?' is to be sought in a statistical survey such as a Gallup poll. But this is absurd.

Turing does not explain this absurdity explicitly but it is plain enough: It lies in the fact that the definition of *thinking*, such as you might find in a dictionary or discover through a poll, is the result of a contingent, historical process. Modern English happened to develop in an era in which there were no artifacts that could credibly be said to think. Only human beings were evidently capable of that. So, this method would establish only the historical absence of thinking machines and not their impossibility.

Suppose instead that modern English had developed during a more technically advanced era in which apparently thinking digital machines were commonplace. In that event, we might be perfectly comfortable in saying that machines can think. Looking ahead to an era in which powerful digital computers are plentiful, Turing (1950, p. 442) stated that, "One will be able to speak of machines thinking without expecting to be contradicted." So, the claim that machines can think seems wrong due to a historical accident unrelated to their capabilities. The result is an unfair prejudice against Turing's position.

The case is reminiscent of the remark by the computer scientist Edsger Dijkstra, winner of the 1972 Turing Award, who said, "The question of whether a computer can think is no more interesting than the question of whether a submarine can swim" (Richards, 2019). Dijkstra was expressing some exasperation with the whole question raised by Turing. However, the analogy he raises is instructive. Most people would likely say that submarines cannot swim. Why? Simply because the term *swim* applies to biological organisms only. Why is that? Simply because organisms were the only things that propelled themselves through bodies of water when the term *swim* took on its current meaning. Had modern English developed later or had submarines been invented earlier, we might well be happy to say that submarines can and do swim. The fact that they do not swim reveals nothing substantial about submarines but reveals instead a simple accident of linguistic history.

This point is the one that Turing has in mind when he says that it would be absurd to consult the established meaning of *think* when considering the matter of thinking machines. There is a clear and present danger that a negative answer would proceed not from any insight into humans or machines but to mere historical coincidence. To avoid reaching conclusions based on such prejudice, we need a procedure that removes it from consideration. For that purpose, Turing proposed the famous imitation game.

10.4 The Imitation Game

Turing was quite interested in the matter of programming computers to play games. Like many Western contemporaries, he viewed the ability to play a good game of chess as a sure sign of intelligence and sophisticated thinking. His first mention in writing of thinking machines takes the form of musing that, although digital computers of the mid 1940s could be programmed to play a poor game of chess, it would not take too long before digital computers could be programmed to play a very good game (Copeland, 2004, p. 374). Such a feat would be a harbinger of the arrival of thinking machines, which may have prompted Turing's efforts to develop chess-playing programs (Copeland & Prinz, 2017).

In the end, Turing did not stick with chess-playing ability as a test of the presence of thinking machines. Instead, he proposed the imitation game (1950). The basic idea of the game is that a human judge converses with two players, the first another human being and the second a digital computer. The conversation takes place strictly through teleprinters, like an exchange of text messages today. The object of the exercise is for the judge to see whether or not they can distinguish the human player from the digital one. Turing suggests that, if the judge cannot reliably make this distinction, then there is no fundamental distinction to make. On the assumption that the digital computer's conversational performance is due to an ability to think, it follows that the judge should ascribe the ability to think to both human and digital players (Copeland & Proudfoot, 2008).

The matter of the validity of this test is outside the scope of this article (Proudfoot, 2017). What matters here is how the imitation game deals with the matter of prejudice. The setup of the game has two main effects: (1) it largely screens off the judge from the players and (2) it promotes conversational ability as an indicator of thinking. On the first point, the game prevents the judge from detecting prejudicial information such as the computer's artifactual appearance and the humanness of the other player. The fact that the computer is a machine and not a human being is irrelevant—or, at least, prejudicial—in Turing's view. So, hiding its physical identity behind what might be called a veil of ignorance helps to provide fair play for machines.

On the second point, although the game should be designed to screen off prejudicial information, it has to provide the judge with some reasonable means of drawing conclusions on the matter of which of the players is thinking. Turing opts for conversational ability. There are many reasons for this choice but the most obvious one is simply that the ability to converse and the ability to think have always been intimately associated (Shelley, 2010). To the extent that the criterion of conversational ability is applied in the same way to both players of the imitation game, it also promotes fair play for machines.

10.5 Objections and Prejudice

Turing anticipated and responded to several objections to his claim that machines can think. Even if suitably programmed machines could succeed at the imitation game, critics would maintain that they cannot think because of other considerations.

Many of these objections take on the following rough, general form:

Beings can think only if they possess attribute X

<u>Computers lack X</u>

Therefore, computers cannot think

Turing's responses typically assert that the objections are merely ways of introducing through a back door the sort of prejudice that the imitation game screens off at the front door.

For example, the *theological objection* holds that computers lack souls. Taking it for granted that a being must possess a soul in order to think, it follows that computers cannot think. Turing responds to this objection by demanding evidence that computers lack souls, at least of necessity. It is not at all clear what it means to have a soul, so the assertion that computers lack them seems like mere prejudice (Turing, 1950, p. 443). The objector might refer to scripture, a move that Turing argues is no better.

The *mathematical objection* asserts that computers lack the ability to answer certain questions correctly. Since human beings possess this ability—as only thinking beings do, it is assumed—then it follows that computers cannot think. The questions at issue would be generated by application of Gödel's theorems to a description of the computer's program. Turing does not challenge the putative linkage between Gödel's theorems and thinking ability, though this assertion is dubious in itself. Instead, he challenges the claim that people have this ability: "… it has only been stated, without any sort of proof, that no such limitations apply to the human intellect" (Turing, 1950, p. 445). He observes that human beings sometimes give wrong answers to questions or fail to find answers at all. For all we know, then, a question could be generated by application of Gödel's theorems to a description of any given human being such that the human being could not work out the correct answer. Again, the crucial assertion that human beings have some attribute related to thinking that computers lack is merely an unjustified prejudice, in Turing's view.

Similar issues attend other objections that Turing considers. He rejects each one on the grounds that it begs the question: Critics assert, without sufficient evidence, that human beings are crucially different from mere machines in a way that enables the former to think while excluding the latter. Instead of clinching the case, these objections merely expose the sort of prejudicial notions that violate Turing's insistence on fair play for machines.

10.6 Fair Play in Context

Turing called out this sort of prejudice in his appeal for "fair play for machines." Turing's call for *fair play* should be understood in its historical context. The concept was a product particular to the development of amateur sport in Britain, culminating in the nineteenth century with the invention and practice of boys' school games like soccer and rugby (Duke-Evans, 2022). Turing likely absorbed it through his time at the Sherborne School for boys (1926–1931), where he engaged in golf and long-distance running, and was a "useful forward" in his house rugby team (Hodges, 1983/2014, p. 74), in which fair play would have been an important consideration. In addition, Turing could assume that the audience of his papers, featuring male members of the British intelligentsia, would understand his reference to fair play and its significance. However, it is a complex idea and, given the brevity of Turing's appeal to it, it cannot be applied in great detail. However, given the analysis above, two facets of fair play in English sport stand out.

The first facet is the *sporting chance*: no team or contestant should be denied the possibility of success, even if one side is markedly superior in prowess to the other. This aspect of fair play is manifested in Turing's responses to many of the objections against thinking machines that he considers. These objections rely on premises that assert essential and unbridgeable differences between digital computers and humans that preclude the possibility that both could be thinking beings. In effect, these objections *presume* that digital machines cannot think. Of course, such insistence begs the question at issue and does not give machines a sporting chance: Some possibility must be granted or else such objections are not fair.

The immediate reason to allow for sporting chances may be to ensure that a sport constitutes an appropriate test for the contestants (Renson, 2009). One of the goals of modern sport is to overcome obstacles, including one's own limitations or the efforts of competitors. If the losing side in a sporting contest did not have at least a sporting chance, then the winning side would prevail without overcoming an obstacle, meaning their victory is not truly an accomplishment. In terms of thinking machines, then, to not allow even the conceivability of thinking machines is to deprive objections to thinking machines of an important part of their significance.

The second facet of fair play relevant to this situation is *impartiality*: Fair play means applying the rules of the sport in a way that is not favorable to the interests of any particular team or contestant. This sense of fair play appears in Turing's plea that critics of his proposal about thinking machines should not simply appeal to the conventional meaning of *thinking*. This meaning is favorable to critics of Turing's claim for reasons that he considers implicitly prejudicial. The "veil of ignorance" built into the imitation game is one way that Turing responds to this issue and, thus, advocates for fair play.

Why be impartial? Duke-Evans (2022, pp. 9–10) points out that impartiality relates fairness in sport to justice in its broadly moral sense. In a just society, people are found innocent or guilty of transgressions without favor to particular interests, and benefits and disbenefits are distributed in society in a similar fashion. In this respect, fair play in sport reflects the ideal of justice in society.

Of course, the issue of thinking machines is not obviously one of justice in society. However, the notion of fair play for machines can be considered in an epistemic sense that does apply to Turing's claim for thinking machines and objections to it.

10.7 Epistemic Fairness

The connection between fair play in a moral sense and fair play in an epistemic sense invites further exploration. This section briefly elaborates on one way of framing this connection and how it can illuminate Turing's argument for the possibility of thinking machines.

One way of representing issue of fairness in a moral sense is through a graphic representation known as a Taylor-Russell (T-R) diagram, named after the social psychologists who introduced its use for this sort of purpose (Taylor & Russell, 1939). Taylor and Russell developed this diagram as a way of analyzing fairness issues that arise in cases of conflict of interest between social conflicts (Shelley, 2012) (Shelley, 2017). In particular, they were interested in issues arising from adoption of standardized tests to decide on admission offers at universities. The point of standardized tests was to predict how well students would perform in their studies. At a minimum, tests should predict whether students would complete their studies with a high enough score to graduate and receive a degree.

Table 10.1 T-R diagram for SAT scores and graduation

		Prediction: Student test score exceeds admission requirement	
		No	*Yes*
Event: Student exceeds degree requirement	*Yes*	Student would have graduated	Student graduates
	No	Student not admitted	Student should not have been admitted

The T-R diagram for this situation is given in Table 10.1. The right-hand column of the table ("Yes") represents students whose scores on a standardized test are high enough that they predict academic success and, thus, merit an offer of admission. Conversely, the left-hand side ("No") represents students whose scores do not merit an offer of admission. The top row of the table ("Yes") represents students whose academic performance exceeds degree requirements and so would merit awarding them a degree. Conversely, the bottom row of the table ("No") represents students whose academic performance would not merit a degree.

Divided in this way, the resulting diagram has four quadrants, each of which has a significant meaning for purposes of fairness. Each quadrant and its nomenclature are indicated here:

True positives (Upper-right): Cases of students whose test scores were positive (merited admission) and who did go on to graduate (the test result predicted the true outcome);

True negatives (Lower-left): Cases of students whose test scores were negative (no offers) and who would not have graduated if they had been admitted (the test predicted the true outcome);

False positives (Lower-right): Cases of students whose test scores were positive (merited admission) but who would not go on to graduate (the test result predicted the wrong outcome);

False negatives (Upper-left): Cases of students whose test scores were negative (no offers) but who would have graduated if they had been admitted (the test predicted the wrong outcome).

A fairness issue arises here because different social groups have conflicting interests in this case (Hammond, 1996). For example, parents of potential students have an interest in having admission requirements set low enough that their children can be admitted and, thus, have access to career enhancements that come with university education. In other words, they prefer to minimize false negatives. Where universities are publicly funded, taxpayer groups have an interest in having admission requirements set high enough that students are kept out who, if admitted, would not graduate, thus not consuming public resources fruitlessly. In other words, they prefer to minimize false positives.

At a given level of accuracy for standardized tests, the only way to reduce one sort of error is to increase the other sort. The result is a fairness issue: What balance of errors would fairly address the conflicting but legitimate interests present in this case (Shelley, 2017)?

Table 10.2 T-R diagram for Neyman-Pearson hypothesis testing

		Prediction: Test merits rejection of null hypothesis	
		No	*Yes*
Event: Alternative hypothesis is true	*Yes*	True hypothesis rejected	True hypothesis accepted
	No	False hypothesis rejected	False hypothesis accepted

If this sort of structure sounds familiar from a different context, this may be because it is identical with the structure for scientific hypothesis testing originated by Neyman and Pearson (Snijders, 2001). Briefly put, the Neyman-Pearson regime involves experimental testing to decide between adoption of a hypothesis (usually that some test result is the effect of a cause) or an alternative "null" hypothesis (usually that the result is due to chance). The structure of this decision can be laid out in a T-R diagram as in Table 10.2.

In the abstract language of hypothesis testing, a false positive is known as a "type I" error while a false negative is known as a "type II" error. One of the main concerns of the Neyman-Pearson regime in practice is to minimize false positives overall. In statistical terminology, the α-level of testing is standardly set at 0.05, meaning that no more than 5% of alternative hypotheses that are adopted as a result of testing are actually false. Ultimately, the reason is that, for scientific inquiry to serve its purpose as a knowledge machine that generates a body of predominantly true hypotheses, the number of false hypotheses that are adopted must be kept to a low level.

Also, it is important to minimize false negatives in some applications. In medicine, for example, it may be important that research should not reject true hypotheses where those hypotheses might be crucial to the formulation of helpful treatments.

Another way of understanding the trade-off between false positives and false negatives in scientific research is to view them as a trade-off between strategies of skepticism and charity. Skepticism is important to science in the rigourous avoidance of false positives. If the threshold for adoption of hypotheses is high, then most adopted hypotheses will be true ones. This emphasis on skepticism helps to make science the knowledge machine that its practitioners want.

However, charity is important to science in its concern for the avoidance of false negatives. As noted above, it may be important to conduct scientific research in such a way certain truths are not rejected out of simple devotion to rigourous skepticism. If the truth of a hypothesis is important for a given purpose, such as a medical one, then testing of it should reflect that issue.

This trade-off between errors in hypothesis testing may be viewed as a kind of epistemic fairness, a balance between skepticism and charity as appropriate in a given scientific situation. Turing's call for fair play for machines can be read in this light.

10.8 Epistemic Fairness and Objections

Turing's call for fair play in addressing the matter of thinking machines can be elaborated in light of the concept of epistemic fairness sketched above. To see how this works, the concept is applied to the *mathematical* and *Lady Lovelace* objections.

As noted above, the *mathematical objection* (also known as the *Gödel objection*) hinges on the existence of questions that the judge and the human participant in the imitation game may know the answer to but which the digital participant would not. The details of this objection and Turing's response to it have been much discussed (Abramson, 2008). It is important here because it is the objection that prompts Turing to make his plea for fair play for machines. In brief, Turing's reply in this case is to point out that the objection relies on the claim that the machine must be able to answer questions infallibly in order to be considered a thinking machine. Such a claim hardly gives the machine a sporting chance. Also, it seems rather partial, favoring human judges and their presumption that no such limitation applies to them.

It would be rather more fair to dismiss the notion that the machine should need to be infallible. Instead, it might make educated guesses, employ heuristics, and so on. As Turing points out, any human judge faced with the sort of question identified in the mathematical objection would behave similarly, making educated guesses, employing heuristics, and so on. Since such inferential practices can and do lead to mistakes, experience suggests that humans and machines remain on par insofar as this objection is concerned.

In terms of epistemic fairness, the situation arising from the mathematical objection can be represented as in Table 10.3, where *X* is either a human being or a digital machine.

In this scenario, the judge in an imitation game puts a Gödel question to the participants.

The objection takes the skeptical position that false positives should be reduced. In other words, it seeks mainly to reduce the possibility of the test identifying a non-thinking machine that answers its Gödel question correctly, perhaps due to an educated guess, and is thus misidentified as a thinking being.

Turing takes the charitable position false negatives should be reduced. In other words, he argues that the imitation game seeks to allow for the possibility that even machines that do not answer their Gödel question correctly may yet be properly identified as thinking beings.

Table 10.3 A T-R diagram relating to the mathematical objection

		Prediction: *X* answers a Gödel question correctly	
		No	*Yes*
Event: *X* can think	*Yes*	X thinks up the wrong answer	X thinks up the right answer
	No	X cannot think and doesn't answer correctly	X cannot think but answers correctly

The skeptical and charitable strategies for the imitation game are both valid in the sense that each one conduces towards the goal of those who pursue it. The skeptical strategy seeks to avoid misidentifying non-thinking things as thinking ones. The charitable strategy seeks to avoid misidentifying thinking things as non-thinking ones. At a given level of accuracy, the imitation game results in a trade-off between these kinds of errors, requiring judges to arrive at a fair balance between the two.

These considerations are paralleled in another objection to Turing's claim, that programmed machines do not think because they merely carry out instructions designed for them by their programmers. Turing (1950) calls this argument the *Lady Lovelace objection*, in honor of the woman who wrote first computer program and pondered its implications (Gonçalves, 2023a, 2023b). In her notes on programming Babbage's mechanical computer, Lady Lovelace stated that, "The Analytical Engine has no pretensions to originate anything. It can do *whatever we know how to order it* to perform." In a sense, the thinking has occurred during the programming phase and thus, it is argued, does not occur when the program is run by the computer.

Turing's initial take on this objection is to construe it as the claim that a digital computer cannot surprise anyone who understands its programming. Knowing how a computer is programmed implies knowing how it will behave when running its program. Turing points out that this claim is demonstrably false: He is often surprised by the behavior of computers whose programs he had written himself. This situation occurs because programmers sometimes (often?) do not expend the effort necessary to anticipate every possible outcome of their work.

This response leaves open a critical reinterpretation of the objection that a computer does not think when running a program simply because the outcome of running it is *determined* by the program. Since thinking is inherently creative or non-deterministic, then running a program cannot constitute thinking. Turing turns to this construal of the Lady Lovelace objection and argues that it may be overcome by digital computers that learn. As Magnani (2022) explains, Turing thought a great deal about how a computer might learn, that is, transition from an "ignorant" artifact to a thinking one. Since learning involves what Magnani terms "eco-cognitive abduction," it is inherently ampliative in nature: the computer learns things that are not deduced from an initial knowledge base but arise from interactions with its environment. In terms of the objection, the states of the program, once running, depend not only on its initial design but also on the condition of its surroundings. Thus, the performance of the computer is not determined by its programming.

As such, a learning computer can do things that surprise its programmers, no matter how well they understand the program it is running. Thus, learning computers overcome the Lady Lovelace objection.

In terms of epistemic fairness, the situation arising from the Lady Lovelace objection can be represented as in Table 10.4, where *X* represents a digital computer.

This scenario is different from the mathematical objection in the sense that it is not framed in terms of the imitation game. Instead, it is framed in terms of what conclusions might be drawn about digital computers if we could examine how they are programmed. On the Lady Lovelace objection, no programmed digital computer should be considered a thinking being, no matter how well it simulates one.

Table 10.4 A T-R diagram relating to the Lady Lovelace Objection

		Prediction: *X* acts surprisingly	
		No	*Yes*
Event: *X* can think	*Yes*	X did not learn original thinking	X learned original thinking
	No	X's program is fully understood	X's program is not fully understood

As such, the Lady Lovelace objection takes the skeptical view that false positives should be reduced. These instances are cases where digital computers behave in ways that surprise their programmers even though they cannot think. Such instances might be put down to failure of the programmers to work out the consequences of their instructions, as Turing explains.

As before, Turing takes the charitable view that false negatives should be reduced. These instances are cases where digital computers are capable of learning and original thought but have, perhaps, not had either the programming or the education needed to realize this potential. This concern seems to underlie Turing's substitution of the imitation game for chess playing as an exemplary manifestation of thinking. In the development of his arguments for thinking machines, Turing realized that, since a digital computer could be programmed to play chess "automatically" by following pre-programmed rules, chess-playing ability would not convince people that it was thinking (Gonçalves, 2023a, 2023b). By contrast, Turing assumed that conversation could not be conducted automatically but is a skill that must be learned. So, conversational ability would better demonstrate the thinking potential of digital computers.

Both the mathematical objection and the Lady Lovelace objection employ skeptical arguments against Turing's claim that digital computers can think. Turing counters both arguments by taking a more charitable approach to the matter. The difference between these approaches can be understood as instances of different strategies regarding epistemic fairness. Both strategies are valid in the sense that they are reasonable approaches to take in light of the epistemic concerns that their adopters have. These qualities are illustrated by the T-R diagrams representing each case.

10.9 Conclusions

Turing faced many challenges in supporting his claim that digital computers can think. Among these challenges were the conventional meanings of *machine* and *thinking* in the English language as well as a natural human partiality towards one of the essential attributes of their own species, the power of thought. Turing correctly anticipated that these facts would pose substantial obstacles to acceptance, or even consideration, of his claim.

To gain a hearing for his ideas, Turing called for "fair play" for the machine. This concept of fair play was grounded in the norms developed in school sports in nineteenth century Britain that still applied in Turing's day. The upshot of his plea is that people considering his contention that machines can think should recognize and set aside their prejudice about digital computers and at least entertain the notion that they can think. The imitation game was designed, in part, to prompt people to screen off "unfair" aspects of conventional thinking while examining relevant aspects of the issue.

Brief though it is, Turing's call for fair play relates well to a general trade-off between skepticism and charity in the assessment of hypotheses. Skepticism is a strategy focused on rejection of false positives, that is, false claims that pass tests for adoption. This focus serves the aim of keeping a body of accepted beliefs largely free of error. However, this strategy can also lead to ignorance through the occurrence of false negatives, that is, true claims that do not pass tests for adoption.

Charity is a strategy focused on rejection of false negatives. This focus serves the aim of acquiring true hypotheses that may be important for practical purposes. However, this strategy can also lead to credulity through occurrence of false positives, that is, untrue claims that pass relaxed tests for adoption.

In the case of thinking machines, the skeptical strategy is inherently conservative. That is, it regards the view that thinking is more-or-less unique to human beings as established, such that it is "fair" to be quite skeptical of any contrary claim. When he advocates for fair play for machines, Turing advocates for a more charitable strategy. In his view, the conservative position is one that is biased due to accidents of history that have become embedded in conventional language. Such a bias is not fair and so a more charitable view is justified.

As such, the imitation game that Turing proposes as a test of the presence of thinking machines is fair in the epistemic sense, at least, more fair than the conservative view. It screens off prejudicial information while providing, through conversation, a means for judges to make informed decisions on the question of thinking machines. It could be argued that the imitation game is still too strict, screening off information that judges should have (Shelley, 2010), as Turing's own response to the Lady Lovelace objection suggests. Yet, it is worth pointing out that the game is not an easy one to win, as no program has succeeded to date (Proudfoot, 2017).

Understood in this way, it is clear that his plea charity informs much of Turing's case for the possibility of thinking machines. It is also clear that Turing had good reasons for taking this approach. His example has informed subsequent discussion of the topic and presents a model for consideration of controversial claims on other topics as well.

References

Abramson, D. (2008). Turing's responses to two objections. *Minds & Machines, 18*, 147–167.

Copeland, J. (2004). *The essential turing: Seminal writings in computing, logic, philosophy, artificial intelligence, and artificial life, plus the secrets of enigma*. Clarendon Press.

Copeland, J., & Prinz, D. (2017). Computer chess—The first moments. In J. Copeland, J. Bowen, M. Sprevak, & R. Wilson (Eds.), *The Turing guide* (pp. 327–346). Oxford University Press.

Copeland, J., & Proudfoot, D. (2008). Turing's test: A philosophical and historical guide. In R. Epstein, G. Roberts, & G. Beber (Eds.), *Parsing the Turing test: Philosophical and methodological issues in the quest for the thinking computer* (pp. 119–138). Springer.

Copeland, J., Bowen, J., Sprevak, M., & Wilson, R. (2017). *The Turing guide*. Oxford University Press.

Ditto, P. H., & Lopez, D. F. (1992). Motivated skepticism: Use of differential decision criteria for preferred and nonpreferred conclusions. *Journal of Personality and Social Psychology, 63*(4), 568–584.

Duke-Evans, J. (2022). *An english tradition? The history and significance of fair play*. Oxford University Press.

Epstein, R., Roberts, G., & Beber, G. (Eds.). (2008). *Parsing the Turing test: Philosophical and methodological issues in the quest for the thinking computer*. Springer.

Feldman, R. (1998). Principle of charity. In E. Craig (Ed.), *Routledge encyclopedia of philosophy*. Routledge.

Gonçalves, B. (2020). *Machines will think: Structure and interpretation of Alan Turing's imitation game*. São Paulo: Ph.D. thesis:. Universidade de São Paulo.

Gonçalves, B. (2023a). Can machines think? The controversy that led to the Turing test. *AI & SOCIETY, 38*(6), 2499–2509.

Gonçalves, B. (2023b). Lady Lovelace's objection: The Turing-Hartree disputes over the meaning of digital computers, 1946–1951. *IEEE Annals of the History of Computing*, 6–18.

Grier, D. A. (2013). *When computers were human*. Princeton University Press.

Hammond, K. R. (1996). *Human judgment and social policy: Irreducible uncertainty, inevitable error, unavoidable injustice*. Oxford University Press.

Hodges, A. (1983/2014). *Alan Turing: The enigma*. Princeton University Press.

Horn, R. (2008). The Turing test: Mapping and navigating the debate. In R. Epstein, G. Roberts, & G. Beber (Eds.), *Parsing the Turing test: Philosophical and methodological issues in the quest for the thinking computer* (pp. 73–88). Springer.

Magnani, L. (2010). *Abductive cognition: The epistemological and eco-cognitive dimenions of hypothetical reasoning*. Springer.

Magnani, L. (2022). *Eco-cognitive computationalism: Cognitive domestication of ignorant entities*. Springer.

Proudfoot, D. (2017). The Turing test—From every angle. In J. Copeland, J. Bowen, M. Sprevak, & R. Wilson (Eds.), *The Turing guide* (pp. 287–300). Oxford University Press.

Renson, R. (2009). Fair play: Its origins and meanings in sport and society. *Kinesiology, 41*(1), 5–18.

Richards, H. (2019). *Edsger Wybe Dijkstra*. Retrieved from A.M. Turing Award: https://amturing.acm.org/award_winners/dijkstra_1053701.cfm

Shelley, C. (2010). Does everyone think, or is it just me? A retrospective on Turing and the other-minds problem. In L. Magnani, W. Carnielle, & C. Pizzi (Eds.), *Model-based reasoning in science and technology: Abduction, logic, and computational discovery* (pp. 477–494). Springer.

Shelley, C. (2012). Fairness in technological design. *Science and Engingeering Ethics, 18*, 663–680.

Shelley, C. (2017). *Design and society: Social issues in technological design*. Springer.

Snijders, T. A. (2001). Hypothesis testing: Methodology and limitations. In N. J. Smelser & P. B. Baltes (Eds.), *International encyclopedia of the social & behavioral sciences* (pp. 7121–7127). Pergamon.

Taylor, H. C., & Russell, J. T. (1939). The relationship of validity coefficients to the practical effectiveness of tests in selection: Discussion and tables. *Journal of Applied Psychology, 23*, 565–578.

Thagard, P. (1988). *Computational philosophy of science*. MIT Press.

Turing, A. (1947/2004). Lecture to the London Mathematical Society on 20 February 1947. In I. J. B. Copeland (Ed.), *The essential Turing*. Clarendon.

Turing, A. (1948/1969). Intelligent machinery. In I. B. Meltzer & D. Michie (Eds.), *Machine intelligence* (Vol. 3, pp. 3–23). Edinburgh University Press.

Turing, A. (1950). Computing machinery and intelligence. *Mind, 59*(236), 433–460.
Walton, D. N. (2008). *Informal logic: A pragmatic approach*. Cambridge University Press.
Williams, M. (1999). Skepticism. In J. Greco & E. Sosa (Eds.), *The Blackwell guide to epistemology* (pp. 35–69). Blackwell.

Chapter 11
Peirce's Logical Interpretants as Arbiters of New Courses of Belief and Action: Association, Transsociation, and Dissociation

Donna E. West

Abstract This inquiry proposes that Peirce's three levels of Logical Interpretants hasten abductive rationality. These levels are grounded squarely in Peirce's categories of Firstness, Secondness, and Thirdness. What spurs novel explanatory hypotheses rests initially upon a state of openness which permits receipt of phenomenological intelligence in Firstness (through creative hallucinations (1903: EP2:192). The process follows with haecceities in Secondness (noticed proximate objects and impulses in" now" time and "here" space) which initiate discrete feelings and implied predications. The Third Logical Interpretant regulates future conduct. Thirdness-based Logical Interpretants constitute not vague dispositions, but deterministic feelings of mind directing specific beliefs and actions. These new conjectures qualify as transsociations (1907: 5.476), which transform belief trajectories into plausible inferences, such that "a reconciliation or "an interadjustment between reason and experience" holds (1908: LI 392; MS 300). In short, Peirce's Third Logical Interpretant illustrates his commitment to his pragmatistic maxim—incorporating the "would-be" as the summum bonum to entertain abductive promise.

Keywords Peirce · Instinctual abduction · Logical Interpretants · Firstness · Thirdness

11.1 Introduction

This inquiry proposes that Peirce's Logical Interpretant (in its three forms) provides the platform to initiate abductive rationality. It demonstrates how phenomenal objects intuit abductive instincts, such that mental pre-propositional seeds contained

D. E. West (✉)
SUNY Cortland, Cortland, NY, USA
e-mail: westsimon@twcny.rr.com; donna.west@cortland.edu

S. Arfini (ed.), *Scientific Cognition, Semiotics, and Computational Agents: Essays in Honor of Lorenzo Magnani - Volume 2*, Synthese Library 506,
https://doi.org/10.1007/978-3-031-96688-0_11

within embryonic virtual habits (1909: MS 620: 24–27) enlighten logical matrices, and depict action schemes (1893: 6.286). The upshot is that proto-signs in the form of First Logical Interpretants (1907: MS 318 version 1; EP2: 418) raise awareness that novel hunches are beckoning. This inquiry describes the function of Peirce's three levels of Logical Interpretants in his late semiotic (1907: MS 318 version 1; EP2: 418), each of which transform former habits of belief and action into new habits via novel associations, transsociations, and dissociations (1907: 5.476). The novel convictions which underlie hunches, those containing elements of truth, characterize First Logical Interpretants. These hunches, fortified in Firstness by convictions of First Logical Interpretants, must constitute determinations; they are not vague fleeting meanings (1909: MS 620: 25, 1911: MS 674: 14–15). To qualify as determinations, convictions need to emanate from insights (see Bergman, 2016; West, 2017), and must possess the fortitude inherent in hypotheses which hold up in the face of more objective scrutiny, as in Woods' (2013: 368) ignorance preserving principle; otherwise they are subject to weak advisory principles. The claim is that propositional content can spring suddenly from individual intuitions in Firstness, but must thereafter become informed by Second and Third Logical Interpretants to measure whether the meaning in question offers plausible ingenuity.

The function of Logical Interpretants is to offer specific meanings not already present in settled beliefs. In fact, insight within an individual mind cannot be sporous or hypnotic; its tenets need to offer some truth value, as in Peirce's "creative hallucinations" (1903: EP2: 192); otherwise the hunch is to be dismissed. Accordingly, hunches containing First Logical Interpretants, e.g., sudden ingenuities, should not be discarded out of hand, because their interpretants may offer more informed inferences than those emanating from former conjectures.

11.2 Preamble

The essential components of Peirce's Logical Interpretants rest upon the categories of Firstness, Secondness, and Thirdness. As such, these interpretants build upon one another both logically and semiotically. This preamble functions to inform the reader of these interpretants' basic meaning contributions; the chapter addresses how they form the bedrock to initiate inferential processes. Peirce's First Logical Interpretants are monadic (1904: 8.328), and consist in subjects whose characters are implied, e.g., ideas which suggest states of affairs. The work of first Logical Interpretants is to apply feelings/possibility in Firstness to ideas, such that narrow emergent envisionments, in what Peirce refers to as the inner world (1907: 5.481; MS 318), give rise to implied propositions. These implied characters (housed in implied propositions) lay the groundwork for unexpressed convictions.

The function of Lower Second Logical Interpretants is to apply convictions to feelings/possibilities which emanate from emergent internal images. The upshot from these convictions is to put them into action. Hence, Lower Second Logical Interpretants are responsible for converting ideas into pragmatic principles, although actions may still lack clear logical regulation. Although discerners "imagine themselves in alternative situations" (1907: 5.481; MS 318); the effects of each situation remain unexamined.

Pragmatic principles eventually do become regulated by the work of Higher Second Logical Interpretants; but the predicative inferences are not yet governed by consciously motivated reasoning, i.e., "reason" lacks reconciliation with "facts of experience" (1908: LI 392; MS 300). The advantage of Higher Second Logical Interpretants, however, is their means to fashion more well-founded conjectures which affiliate novel antecedent actions to consequences (1908: 8.373). The sign responsible for enlisting both parties into inferential inquiry is index—minds together discerning at antecedents which have the potential to be plausible. After all, Peirce's ultimate role for index entails cooperation through a collaborative process. This ensures that the rationale underlying the argument contains the fullest event related implications to infer meanings inherent in Logical Interpretants, e.g., all possible scenes attached to their outcomes (see West in press for further discussion). These implied arguments allow interlocutors some explanatory substance to anticipate the deictic fields—when and where particular outcomes will surface in analogous situations. As such, conjectures conceived of via Second Logical Interpretants are driven by greater self-control achieved often through dialogic interactions — self-to-self and/or self-to-other; these more deliberate processes which result in habit-change are mastered (see West, 2016a: chap. 13; West, 2022a and 2021b). This regulated conjectural process introduces abstract mental organization, permitting interpreters to inhibit/select logically fitting relations, reactions, and arguments. In this way, Higher Second Logical Interpretants begin to produce "the proper mental effect of the sign" (1907: MS 318: 46), in that the effect/reaction is a consequence of concerted planning.

Third Logical Interpretants advantage the inferential process by heightening intellect to the possibility of outcomes, driven by metacognitive principles which guide conduct—warning, in advance, of problematic outcomes. In short, meanings/effects captured in Third Logical Interpretants translate implied predicates (from earlier stages) into explicit and sequential principles; and conduct becomes guided by what would or could happen. The Third Logical Interpretant invests minds with the power to search out a "…reconciliation between reason and facts of experience" (1908: LI 392, MS 300).

Peirce memorializes these three Logical Interpretants in the following passage:

> …when its interpreter meets with an object to which the subject applies, he is informed by the proposition that it either has (if it be universal) or may have (if it be particular), the character signified by the predicate; but *first* only, because if the interpretation is to stop at that point, the information neither contributes toward regulating the interpreter's conduct, which would be the second, or statically pragmatistic interpretation, nor does it tend toward bringing about a reconciliation or interadjustment between reason and the facts of experience, which would be a *third* and evolutionary pragmatistic interpretation. (1908: LI 392, MS 300, "The Bed-Rock Beneath Pragmaticism". Underlining is CSP's own.)

The final competence permits interpreters' patterns of meaning to become open to guided conjecture—determining possible circumstances (would-bes and could-bes), which can be utilized to alite upon likely antecedents—those which contribute to subsequent states of affairs. The paradigm extends to possible (future) consequences which such antecedents might produce. As such, the habit-change which results is, in Peirce's terms, "evolutionary," in that reason perpetually checks against experiential incidences. In short, the Third Logical Interpretant assigns instinctual abductions the responsibility of measuring their truth value against the storehouse of remembered experiences.

11.3 How Instinct Permeates Peirce's Categories

Peirce's concept of Firstness pertains to oneness or a close examination of things (in the mind or in the external world) as monads. As such, Firstness-based analysis entails narrow examination, not of separate properties, or their commonalities, but things in themselves: "Firstness is the mode of being of that which is such as it is, positively and without reference to anything else." In other words, in Firstness the consideration encompasses each thing's affirmative nature, absent similarities or differences with other things.

The implication of possibility here is notable, because it constitutes the most salient characteristic for the exercise of instinctual abductions. Firstness likewise encompasses feeling, which requires conviction; but, it does not feature as objective an element which possibility embodies.[1] Accordingly, when hunches arise as feelings/convictions, from imagined objects as wholes, their meanings may be restricted to subjective convictions—without reference to potential applications. Peirce further intimates how possibility enhances percept-based meanings when he refers to them as "self-generated" (1903: 1.306). When Peirce refers to Firstness as "self-sufficient, positive, and sui generis, he directs interpreters to notice percepts, and armed with this knowledge, one can self-generate possible identities for things/properties (1903: 7.625). For Peirce, possibility can create future convictions with feelings. In other words, the state of being self-generated entails openness to potential applications to other genres, allowing receipt of further insights. Because Firstness is "a mere feeling that has a dark instinct of being a germ of thought" ("Harvard Lecture on Pragmatism III," 5.71:11), it begins with the possibility of

[1] Furthermore, because Firstness incorporates not merely quality, but feeling and possibility, it necessarily gives rise to the element of consciousness/awareness (1905: 1.306), despite its elemental nature as apprehension/passive attention. Peirce's concepts of quality and feeling radically diverge, given the presence of an elementary form of consciousness in the latter (1905: 1.306; 1903: PPM 141, infra).

Quality (although it is a feature of Firstness), does not insinuate habit-change directly, because, as Peirce explicitly mentions, it is passive; whereas feelings, together with possibility, are active, in the business of settling upon more fitting convictions—the business of habit-change. In short, grappling with which aspect of the world will occupy the focus (which deserves examination), stages the contents of one's considerations, and hence determines what meanings will be settled upon when taking up new terms, propositions, and conduct.

potential meanings to be uncovered. Peirce's allusion to the "germ of thought," permits possibility to extend the germ of thought incapsulated in feelings to future situations/worlds, imbuing Firstness with an elemental component of Thirdness. As such, possibility is the catalyst to launch Thirdness into Firstness. He describes this "germ" as a "dark instinct," because the meaning of the included percept is still somewhat occluded, or in process of germination. His use of "germ of thought" explains/describes the function of instinct; it implies the earliest stage of Thirdness, wherein preemptive meanings become awakened. In one of Peirce's latest manuscripts, he places instinct squarely in the realm of pragmatism when he characterizes it as a "faculty of acting."

> I use the word instinct in the precise sense of an animal faculty of acting (whether physically or psychically) in a reasonable (or better "an adaptive") manner, when the animal (human or other) would be unable by reasoning to reach the requisite conclusion. It follows that, adopting this definition, I must admit that all reasoning ultimately reposes on "instinct" (1913: L477: 5).

It is obvious that the work of instinct for Peirce, constitutes true enlightenment, hence the key to receipt of abductive rationality. This is so because the genesis of instinctual hypotheses arise largely from phenomenal sources; and their ultimate objective is subject to reform toward truth-seeking. Nonetheless, not all image-based instinctual inferences rise to the level of abductions, containing viable inferences, Peirce documents in his late metaphysic and semiotic (1913: MS 683).[2] In describing instinct as "dark," Peirce establishes instinct as offering a murky hypothesis in Firstness which requires further unraveling. The instinct is "dark," in that the hunch is but an embryo, awaiting another insight to uncover clarity. The "dark" instinct appears to mark Peirce's first level of clarity (1877: 5.71). In this first level of clarity, Peirce incorporates Thirdness into Firstness, such that a conviction arises from this dark instinct which spurs the mind to refine subject-predicate affiliations present in the phenomenal hunch. It is obvious that even at the level of First Logical Interpretants belief-change is activated—inaugurating the abductive process to receive further insights. In short, openness to refine meanings of apparitional hunches translates phenomenology into the operation of phaneroscopy.[3]

[2] "...but I do not believe one can go very deeply into any important and considerably large subject discussion with calling up in the minds of one's hearers mental images of objects arranged in ways in which time without space is incapable of serving as the field of representation since in time of two quite distinct objects one must be antecedent and the other subsequent. Of course, the one temporal relation can be spatially imagined in various ways. But the combined field of space and time seems [alone] to be adequate to the imagining of Lorentz's explanation of the famed experiment of Morley and Michaelson...." (MS 683: 14, c. 1913)

[3] Phaneroscopy constitutes an informed narrow attention to particular details believed to occupy the predicative material for a specific subject under close scrutiny (for further discussion see Rosensohn, 1974: 84–100; Fuhrman, 2013, and DeTienne, 2013).

11.4 Pregenerative Thirdness in the Categories

If an apparition in Firstness (prior to being enacted in Secondness) showcases a schemata which illustrates some novel truth, e.g., Peirce's virtual habit (1909: MS 620), it constitutes the most powerful kind of abduction, because it not merely demonstrates conviction in the face of implied predicates, or predicates which are incomplete (subjects only) are stated. Apparitions direct signers to take up a cause in logical relations, without explicit provision of its second part (predicate), or without applying the rationale of principle of sequence.

The First Logical Interpretant unites explicit subjects to implied predicates (see Bellucci, 2014: 539–540)—the zygote for subject-predicate union. Obviously, its premise is not entirely articulated; and it is yet untried. Nonetheless, as virtual habit (1909: MS 620: see West, 2017) its subject bursts forth—begging for: predicative content, propositional acceptance, and assertative force. This is nothing short of the first instance, however implicit, of nascent Thirdness; it precedes the more obvious and more objectively mediated illustrations of Thirdness, in which two or more minds agree to the truth of a proposition/argument, or where single minds settle upon implied propositions from unexpected apparitions.

First Logical Interpretants often surface unexpectedly (West 2021a); and their meanings contain elementary propositional unities. These kinds of abductive meanings are perpetrated by receipt of insights from new subject-predicate connections. These insights develop into assertions.

The instinctual meanings of First Logical Interpretants inhabit the earliest hunches, and constitute primitive Thirdnesses. Deely assumes that these early Thirdnesses can likewise reside in Secondness, within things themselves. In fact, Peirce supports this claim as follows: "general principles are really operative in nature" (1903: 5.101), and "nature herself often supplies the place of the intention…in making a Thirdness genuine and not merely accidental" (c.1890 1.366). Peirce indicates not merely that arguments (as "general principles") "operate in nature," but that intentionality resides in place (within nature). If intentionality is a particular, not a general principle, then both general principles, as well as specific intellectual/affective competence permeate nature. In fact, intentionality's presence within place (in nature) demonstrates that indexical signs (such as place) can hold meaning, in particular genuine Thirdness. Peirce's contrast between genuine and accidental Thirdness verifies his position that Thirdness from/in place is not a consequent of factors other than an inherent meaning, that which is intrinsic to that location, e.g., some authentic affiliation (cribs with newborns). Genuine Thirdness appears to be an artifact of nature, if, in fact, intentionality can reside inherently in its fold. If, by place, Peirce refers to situatedness within nature, he raises the possibility that intentional meaning can be stored therein; and as such, meaning can preexist mentation—qualifying it as a priori.

Rosensohn (1974: 84–85) contends that the Peirce's use of genuine Thirdness (within things/nature) materializes consequent to: "an intelligible law;" and "intelligibility, or reason objectified;" and this "makes Thirdness genuine." He explains that it is the presence of "intentionality" which is responsible for qualification as both intelligibility and genuine Thirdness (c. 1890: 1.366). In other words, as a scholastic realist, Peirce characterizes genuine Thirdness as a nature-based operation much like a physical regularity, but with the addition of mediation (c.1890: 1.366), namely intentionality which drives hypothesis generation. Apprehending a physical regularity with intentionality brings intelligibility to the fore, because intentional efforts requires some degree of conscious meaning application, or of reason objectified. Such is the case given Recognition of a pattern of operation within nature, propelling active involvement of the will to implement a habit-change. Here, Peirce explicitly gives Thirdness a central place/purpose in Secondness, when intentionality activates the will responds to new logical meanings. Hence, Thirdness as the source for instinctual hunches in Secondness is born. The aforementioned argument demonstrates that Peirce attributes Thirdness operations to Secondness-based, primitive operations (see Rosensohn, 1974: 91 for elaboration). It must be noted, however, that Thirdness based operations within Firstness must not be overlooked altogether. In short, Peirce's critical message is that mind notices Thirdness-based regularities from Firstness/Secondness, or from their unification; and pregenerative Thirdness inherent in things activates conscious and unconscious awareness of novel regularities (1867: 1.554). This early activation of Thirdness supplies the raw material for future involvement of mind—unifying category-based tools necessary to suggest plausible explanatory hypotheses.

Deely echoes Peirce's contention that Thirdness inhabits things, when he highlights primitive Thirdnesses which arise in Firstness, characterizing them as "pregenerative." "Pregenerative" Thirdness reinvigorates Peirce's acclimation of pragmatisistic principles—that possibility resides in Firstness (in imaginings) and in Secondness (in things themselves). And the search for and grasp of potential meanings is perpetuated—those which are yet unrealized and which perfuse the universe (1906: 5.448 fn). Deely refers to meanings which await joinder with objects and minds as "provenation" (cf. Deely, 2009: 29; Deely, 2015: 156; West, 2018). Deely's use of "provenation" encapsulates the notion of "before becoming;" it re-legitimizes Peirce's assignment of sign meanings before involvement of mind. These early meanings are stored in things/images until interpretants are assigned by minds—when triadic relations are in place. From meanings in things/phenomena, minds can implicitly extract Logical Interpretants (consciously or via lesser conscious states). For Deely and Peirce, meaning pervades the universe; hence, foundational a priori meaning qualifies as sign (even absent mentation). Given Peirce's adherence to realism, signhood can operate before mentation, in that meanings are potential (drawing upon Peirce's concept of possibility), latent in appearances of things. Such argument, extends Peirce's semiotic—lending credence to sign meaning without mind, such that interpretants may dwell in objects. The implication of Peirce's/Deely's position can be articulated as follows: future meanings carry such potency that they attach as a priori propositions,—when minds possess the readiness to attend to their purposes.

Examination of Peirce's categories (particularly possibility in Firstness) can shed light upon how early Thirdness operates. If Thirdness enhances operations in

Firstness, it advances phenomenological inquiry, before actualization in Secondness. In this way, meanings either suddenly arise in minds, or roam in things picked up by minds—then pheneroscopic analysis can be underway (see Fuhrman, 2013; DeTienne, 1993, 2004 and 2013 for further discussion). Rather than things informing minds directly in Secondness, Firstness can serve as a primary source for primitive Thirdness-based influences. In Firstness pregenerative Thirdness exercises its most generative and regenerative capability—by allowing inferences to *provenate in later appearances of things*. The upshot is that Firstness supplies the reservoir to hold meanings in obeyance, those within the phaneron by phaneroscopic analysis. The process makes obvious early reliance upon Peirce's concept of possibility in Firstness. Deely's explanation of primitive meanings inculcates Peirce's conception of possibility at foundational stages of semiosis; but he holds Secondness, rather than Firstness, to be more responsible for meaning genesis and disbursement. Nonetheless, the argument that Logical Interpretants are conceived in Firstness more adequately supports Peirce's notion of pragmaticism– expectations of future meanings, ultimately informing inferential rationality.

Deely's characterization of early Thirdness, however, is invaluable, whether applied to Secondness (as he contends) or in Firstness, because it explicitly attributes to Thirdness an explanatory (crucial to abductive rationality), as well as a descriptive component. His analysis of Peirce's Thirdness places pregenerativity at the center of semiosis. Sign activity is inaugurated in Secondness (despite lack of attention to meaning in Firstness), given changes in meaning/effect to the object's patterns of appearance. Deely claims that the flame of meaning can extinguish when scrutiny of the thing fails to materialize (by minds). He strongly implies (unlike Peirce) that this lack of apprehension terminates semiosis. The hypothetical life of things is cut off or snuffed out, without an igniting agent to consider the purpose and reason for its existence; hence hypothesis generation is likewise extinguished. Deely addresses the extinguishment of meaning potential as follows:

> At that moment when the first living substance emerges, of course, and only then, the flame of sign activity is true and properly—fully—lit. But, what about those moments leading up to that moment, those moments wherein the material interactions of things at the level of secondness yet brings about a thirdness of possibility (a "firstness of thirdness," …)? (Deely, 2015: 780)

Deely departs from Peirce's contention that "all this universe is perfused with signs, if it is not composed exclusively of signs" (1906: 5.448 fn), since meanings would need to continue to inhabit images/things, and not engender the potentiality of being cut off. In short, Deely's argument that meaning in Secondness materializes in a proto form, prior to complete sign activation is overshadowed by his pronouncement that it is extinction. The latter claim presupposes that meanings exist for the sole purpose of conscious minds. In fact, Peirce's semiotic is on precarious footing if meaning is in jeopardy of being snuffed out in the nature of the universe, since its withdrawal vitiates the effects of realism underlying his pragmatistic model.

With respect to apparitions, Thirdness (hence sign potentiality) may never be extinguishable, even absent conscious apprehension, because meanings/effects can still "pervade the universe," awaiting future habitation in minds. In fact, even when some attentional meaning enters minds, it may yet be unwrapped, raw and undiscovered/unapplied; nonetheless, its viability remains active. Likewise, ignoring an apparition/thing (hence its meaning/effect) may constitute a decision to put it aside, and focus on another instead. The meaning/effect of the thing/phenomena survives inattentiveness; the lack of realization of the effect's do not terminate the effects themselves. Another case in point pertains to the decision to ignore an apparition/thing; an assignment of meaning may be associated with shelving the object for subsequent scrutiny, or the decision to ignore may translate into a meaning of less relevance or irrelevance to the issue at hand.

Attention to one thing/apparition over others more clearly illustrates some assignment of meaning, and openness to Peirce's reference to and adaptation of *il lume naturale* (natural light) (1893: MS 408: 148–149), where inferring meanings is a wholesale operation. This natural light applies to species other than man; it pertains to the genius/instinctive insight of guessing nature's principles accurately (wherein meanings lurk and can be extracted at any time). This natural light is in line with Peirce's notion of instinct (instinctual abduction), in which inferencing surfaces without didactic influence:

> [I]t is to be expected that [man] should have a natural light, or light of nature, or instinctive insight, or genius, tending to make him guess [nature's] laws aright, or nearly aright. This conclusion is confirmed when we find that every species of animal is endowed with a similar genius (1893: MS 408: 148-149).

Apparitions illuminate according to Peirce's application of *il lume naturale*, when they constitute abductively driven virtual habits—when their insight preempts plausible inferences. With the illumination, mental depictions provide intellectual raw material to take up the cause for action-habits (Stjernfelt, 2014: 60); they rise to determinations which suggest viable, future states of feeling and doing. Accordingly, virtual habits (1909: MS 620: 148–149) as expected future states of affairs, incentivize the abductive process. Peirce's contention (in the same year, 1909) supports the presence of underlying rationale from virtual habits to serve as catalysts for abductively driven actions. As such, he advocates that abductions "recommend a course of action" (MS 637). For Peirce, the essence of abductive rationality (including implied predicates of apparitions/virtual habits) is to "recommend a course of action."

The recommendation must supply a Logical Interpretant, either an implied proposition (see Bellucci, 2014: 539–540), or an identifiable and viable antecedent with rationale for its connection to the consequence.

The foundation of this *il lume naturale* can have its place in spontaneous self-hypotheses (when inferences derive from natural being relations, i.e., potentially causally-based predications). Alternatively, these inferences can emanate from a higher power or from seraphim, such that enlightenment is primarily derived from *agape* between human and deity. This *agape* incite wisdom in the form of intuitions which must ultimately be checked against stored experiences. Peirce acknowledges this influence of natural light to receive instinctual intellection later in his metaphysic and semiotic (EP2:436–450—in the Neglected Argument for the Reality of God). Although early in his semiotic, he opposed the validity of intuitions, arguing that their presumed status as first cognitions vitiates their existence altogether, his later semiotic comes to terms with the effectiveness of these instinctive intuitions to supply "natural light" (1893: MS:408: 148–149). Nonetheless, receipt of "natural light" can derive from things that call to one from the natural world, striking one's attention and invading one's mental assimilations and accommodations. In this way, the natural world preexists cognition; and exists independent of cognition. It can become an internal force, such that its physicality in Secondness is replicated in internal analogues (in Firstness).

Instinct becomes lit particularly in its role as double sign or Dicisign/pheme (see 1904, CP 8.334–8.339; 1906, MS 295: 26; 1908, EP 2: 489–490; and West, 2019b). Although Peirce opposes the existence of intuitions as sources for leveraging novel sign meaning early on in his semiotic (1868–1869: 5.265; see de Lourdes Bacha, 2014: 109), in his later semiotic, he supports the credibility of intuitions to supply instinctual insight. In fact, it is often *agape* which is the catalyst for these intuitions via interpretants emanating from: "our dear and adorable creator" (1911: NEM III: 206). Hence, it is often by divine providence that intuitions suddenly inform instinctual meanings of things from the natural light; and these sudden instinctual forces become signs when they acquire propositional and argumentative meanings. As such, from their very outset, intuitions which suggest plausible Logical Interpretants (with implied or explicit principles of sequence) possess pregenerative forms of Thirdness. In short, intuitions spurred by *agape* have increased capacity to motivate interadjustment between reason and experience, such that the seeds of the Third Logical Interpretant foment abductive rationality. With these intuitional seeds, the crystallization of intelligence fosters close internal observations, i.e., scrutiny whose elements qualify as Third Logical Interpretants inspire fitting novel predications with reflections on truth value present in stored experiences.

Peirce's semiotic facilitates this fit between receipt of instinctual inferences and reconciliation with experience, in fortifying the interpretive impact of the double sign (icon and index together). Interpretants of each kind of sign heighten scrutiny of whether subject and predicate might offer logical promise. Peirce advocates that inferences gain insight when their interpretants are enhanced by compounding more than a single sign. He posits that course of action recommendations are more likely to be forthcoming when two signs (index, icon) integrate their meanings, given that the context of subject of discourse becomes more obvious with a viable index, and can then be matched to whether the predicate pertains (with implied/explicit arguments).[4] Peirce explains how instinctual inferences emanating from spatial relations between index and object clarify implied predicates (see Meyers, 1999:

[4] Even earlier, in 1890, Peirce introduces the fact that intuitions are precise forms of intellections: "The realities compel us to put some things into very close relation and others less so, in a highly complicated, and in the sense itself unintelligible manner; but it is the genius of the mind, that takes up all these hints of sense, adds immensely to them, makes them precise, and shows them in intelligible form in the intuitions of space and time." CP 1.383

645; Gava, 2014: 350; Wilson, 2016: 86–89; and West, 2019b). Beyond index alone, its marriage with icon identifies analogous objects to which the predication can apply, e.g., those which possess similar effects, or which possess common attributes. To illustrate, interpreters infer that rising smoke marks where fire originates, and indicates which similar objects are likely to possess the same kind of situatedness. The former alone means fire consequent to sign-object's similar location (index), while the implication of which iconic signs-objects can be similarly situated is present in the latter (EP2: and see West, 2018: 440 for elaboration). Peirce claims that spatial contiguity between indexical sign and object constitutes the closest illustration of intuitive operations—where primary inferential Thirdness arises (West 2024); and the addition of iconic meanings elaborate on and widen interpretants, facilitating awareness of and analyses of whether antecedents and consequences fit.

11.5 Inferencing from Phaneroscopic Analysis

Although Firstness pertains particularly to being without explicit reason, the seeds for reason are present in potential predications to be conceived of in the phaneron (by way of pregenerative Thirdness). According to Peirce (1868: EP1: 6), these predications materialize consequent to phanerscopic analysis, and have their basis in mathematical principles in nature (hearkening back to Peirce's use of "*il lume naturale*"), such that the analysis proceeds from the many to the one (see Fuhrman, 2013: 505 for further discussion); and the analytical sequence (phaneroscopic analysis) consists of both assimilatory and accommodation-based operations—in that copies of the image bring about abstractions. Changes are made to assemble the many to the one; and assimilation constitutes the abstraction characterized by assemblage of the parts into a single whole. These iconic mentations, e.g. pictorial actions (1887: CP 3.459),[5] require verb-like icons and rely upon a superlative form of phaneroscopy, namely, hypostatic abstraction (1906: EP2: 394). Minds become so narrowly focused on primary attributes that quality-abstractions materialize, which can set the stage for inference-making). For example, the image of Larch trees losing their needles in winter can give rise to the erroneous hypothesis that they should be kept in-doors, even though doing so would interfere with their overall health and progression; apprehending that shedding needles represents the typical course of healthy development for larch (that they are deciduous conifers) is a later inference (arguably not instinctual) leading to right action. Examining one attribute in phaneroscopic terms, demonstrates that hypothesis-making can begin consequent to phaneroscopic operations, in Firstness when hunches (antecedents) spring from imagined consequences. The imagined situatedness (utilizing index) of

[5] "A verb by itself signifies a mere dream, an imagination unattached to any particular occasion. It calls up in the mind an icon. A relative is just that, an icon, or image, without attachments to experience, without "a local habitation and a name," but with indications of the need of such attachments."

a particular thing can suggest the qualities inherent in it. As such, images may be the source for instinctual conjectures, if qualities reach such sensual heights that the quality embodies the thing. In this way, meanings are implicitly present in mental images and their context.

Images (be they visual/auditory/tactual) can originate by self-effort alone; and, if so, Logical Interpretants may be more likely to invade preconscious states. Alternatively, inanimate things can reveal their presence by way of impressing particular descriptive attributes, supplying implied propositions. As such, pregenerative Thirdness (implied meanings within the object) introduce embryonic logical meaning possibilities.

Peirce supports Thirdness-based influence within Firstness when he expresses the possible effect of dreams and hallucinations (1903: EP2:192).

11.6 Inferencing from Dreams and Hallucinations

Peirce is explicit that dreams enhance both inferences and their effects:

> Lavoisier's method was not to read and pray, but to dream that some long and complicated chemical process would have a certain effect, to put it into practice with dull patience, after its inevitable failure, to dream that with some modification it would have another result, and to end by publishing the last dream as a fact. (1877: CP 5.363)

This passage clarifies Peirce's position on the matter, that dreams can materialize into facts, particularly "last" dreams, despite an interval of "dull patience." Because of Lavoisier's vigilant waiting for success, in the face of previous erroneous inferences and failures to exact the expected consequences, the "last dream," (presumably envisionment of antecedents) revealed plausible result; but, it was not the patient waiting for conjectures that was responsible for the insight, but the dream itself. The state of waiting/expecting a plausible answer may have better prepared the scientist to expect the possibility of a soon-to-come dream. With that expectant state of mind, Peirce demonstrates how sudden dreams can become intuitions, facilitating spontaneous integration of different semiotic and logical relations that might exist among things and actions.

Given the appropriate readiness, dreams and other schema-based phenomenon can grasp observers' attention, forcing the receipt of new subject-predicate assignments.

Likewise, different kinds of hallucinations can give rise to Logical content, i.e., more than Energetic Interpretants, e.g., a conjecture that sobbing is likely to be consequent to loss of a loved one (independent of the experiencer). The aforementioned conjecture qualifies as an abduction containing a Logical Interpretant, because the proposed rationale for the consequence is a potential antecedent. The Logical Interpretant which surfaces from hallucinations often reveals the very purpose for the consequence, which (given that the meaning is applicable in many cases) elevates the image to symbolic status, not merely predicated upon indexical factors of a single case. The meaning of the dream surpasses the plight (consequence) of a single experiencer, beyond meaning inherent in Energetic Interpretants (see MS: 318: for Peirce's explanation); rather, the abductive rationale contains a valid prediction which could hold in a host of instantiations. It is not merely that the intuited dream/hallucination is transported from an indefinite/indiscriminate passing phantasm, to the key unlocking a vital cognitive and emotive realization; it stands as a plausible reason for heightened tears, independent of whether the antecedent holds. Convictions offering different but plausible predications regarding antecedents can be posited; and the initial conjecture still qualifies as an abduction, provided that it consists in some possibility of truth-value.

It is critical to note that emergent dreams/hallucinations can constitute catalysts for abductive conjectures (especially those exhibiting icons and indices, given their power to accentuate situatedness against more objective meanings. Despite the fact that dreams/hallucinations are likenesses (and might be construed as icons alone), their situatedness forces intquiry into whether their underlying Interpretants make sense in the context and with the suggested predications. When both signs (icon, index) inhabit dreams, the dreams crystalize and fuse representamen-object relations, and highlight identity and contextual characteristics in the Immediate Object. Fusion and identity operations from within the mental depiction can evidence new subject-predicate coupling, transmitting specific meanings into the dream. This process incurred by the presence of Logical Interpretants, infuses the dream with Thirdness of the pregenerative kind, imbuing the dream with signhood. The dream becomes a sign because its sparks of intuition steer the discovery of novel explanations for the dream's purpose/s. Hence, the semiotic event culminates in further inquiry into whether the rationale fits the unexpected consequence found within the dream. In refining the hunch, the dream is afforded not merely insight/intuitions, but wisdom to implement practical bearing (see West, 2016b in which Peirce's concept of abduction is compared with Maritain's dawning of intuition). In this way, the insight made obvious by the assignment of Logical Interpretants to the place and time of the dream elaborates whether the dream offers insight where and when it appears.

The primary argument here is that Logical Interpretants elevate dreams to signs by means of perceiving the double sign, whereby invisible relations between quality and thing, between thing and place, or between the thing and expected action schemes are hastened. In short, the meaningful relations which the double sign afford are indispensable in the fabrication of Logical Interpretants; their relational meanings transform dreams into signs with similar argument structure to other signs. These relational meanings between representamen and object characterize the flow intrinsic in every fabric within Peirce's semiotic—because, without determining relations between substance, quality, and consequences, little, if any, meanings wrought by novel things/states/actions would become conscious. The upshot is that absent recognition of the potential worth of phenomenological sources of insight wrought by First Logical Interpretants, over-reliance upon single instantiations and separate qualities would be at greater focus. Such can be an arduous means of collecting a large swath of data to formulate hypotheses, namely, inductions. In short, attention to the reforming potential of suggestive phenomena which underlie states of affairs is indispensable.

These kinds of phenomena consist in pictures which emerge suddenly within the mind, e.g., a brand new integration of objects with contexts (Dicisign), or envisioning a schemata of entirely different practical/state of mind remediations to overt unforeseen consequences. Phenomenal influences (e.g., receiving iconic intuitions) can emerge to warn of impending dangers before they arise, and their potential must be harnessed. Phenomenal insights likewise drive impulses to alter the sequences within object displays, or to delete/insert them within a scene/array. These insights can give rise to novel associations or dissociations, hence supporting Peirce's claim in 1907 that instinctual abductions often purge assumed antecedents from consequences through transformations: new predicative associations/dissociations made non-deliberately (5.476). Nonetheless, Peirce cautions us to take care not to entertain suggestions out of hand; they may not be transformative—although he does not provide specific guidelines as to which suggestive arrays to adopt when fleeting pictural/display transformations are at the threshold.

Peirce identifies three kinds of hallucinations: "obsessional," "social," and "creative." (1903: EP2:192). As such, Peirce differentiates the power of each kind to inform plausible hunches, i.e., some hallucinations possess hidden propositional content to reform the mind and conduct, whereas others do not. Accordingly, he acknowledges the potentiality of "creative hallucinations" to imply propositional content. Such qualify as phenomenal images because they are capable of modifying thoughts and behavior, were they to be asserted and adopted. Unexpected pictures/diagrams can surface as: "obsessive," "social," or "creative." Although all three are governed by impulsive hunches, the initial two fail to contain truth value; and do not qualify as plausible hunches. "Because creative hallucinations" contain some level of self control, they can initiate hunches consisting of some truth value. The self-control characterized by creative hallucinations is housed primarily in the asurity of the prospect within the hunch's colonel. The success of the creative kind to offer insight is grounded in the fact that The suggestive impulse that the hallucination constitutes a viable, objective proposition. In other words, its contribution to establishing rationale for an objective state of affairs is not insignificant.

As mentioned previously, other phenomenal sources underlying abductions include what Peirce refers to as "apparitions." In Peirce's sense, "apparitions" refer to propositions in the form of unexpected, sudden visions (1896: 3.460). Peirce considers certain apparitions to have sufficiency to derive insight, namely, those phenomenal glimpses which emerge suddenly, and which engender primitive Thirdness, of the pregenerative kind. Here is where Peirce's semiotic rescues certain phenomena from dismissal as abductive material. To constitute a primitive source for deriving implied propositional meanings, the apparition must supply some pragmatic veracity (recommending a course of action, 1909: MS 637: 12). Apparitions qualify as creative hallucinations (sources for inferencing), if their propositional content embraces meritorious course of action recommendations.

Peirce's pronouncement in 1909 (MS637: 12) clearly supports retroduction as an abductive paradigm. It rests upon the premise that hunches (including those emanating from apparitions) must advance "course of action recommendations" (1909: MS637). This pronouncement revolutionizes abductive rationality, in that it now requires application of discrete practical principles (Logical Interpretants), particularly "principles of sequence" (1908: 8.373). With "principles of sequence" Peirce's abductive profile is elevated from offering theoretical promise only, to translating beliefs into convictions, and informed recommendation for action sequences (see West, 2022b for further explanation). Here Peirce's pragmatic maxim demonstrates more than the underlying, compelling proposition contained in a belief, e.g., conviction that bonsai practices reveal critical information about success at bonsai that one enrolls in a bonsai class. The principle of sequence pertains to a specific argument which unites an end-state to a preceding contributing factor, e.g., the death of this and other bonsai might result from lack of water in the face of scorching temperatures.

11.7 Expansions on First Logical Interpretants

Attributes of Peirce's First Logical Interpretant include: monadic quality (passive), implied character of subjects, meet with subject for the first time, imply propositions/predicates, apply possibility to ideas, narrow scope, trace animated lines of conduct, and gatekeeper limiting spurious hypotheses. What Peirce refers to as instinctual hunches have their foundation in First Logical Interpretants. Abductions which rest upon First Logical Interpretants surface as spontaneous hypotheses, from a rather free-flow of propositional content—similar to an unexpected interchange with the self. First Logical Interpretants consist in some small measure of logical control—flashing novel explanatory suggestions across the mind's eye, be they conscious, partially conscious, or unconscious. These Logical Interpretants in substance constitute hypotheses, in that they inject some stabilization upon incoming premises, allowing the mind's-eye to focus more narrowly on the central logical content of the hypothesis. Accordingly, Logical Interpretants serve as gatekeepers to keep out hunches which argue impossible premises, while retaining viable hypotheses. To prepare for the operation of the Logical Interpretant within the phenomenological world, one must enter into a state of openness which permits receipt of phenomenological intelligence. This form of intelligence begins as a consequence of waiting for sparks of insight. In this way, this spark of new possibility speaks silently to the mind of a new state of affairs. As such, Firstness-based regimes join subjects and predicates within premises spontaneously, much like an *aha* experience.

The advantage of First Logical Interpretants is their regulatory scopic function—to greet those premises which have some logical merit. This preeminent establishment of correspondence between different subjects and predicates is referred to by

Peirce in mathematical and Scotistic terms as "first-clustering,"[6] namely, that the clustering is not based upon common properties. Occasionally, predications may be non-specific, given the degree of vagueness of the phenomenal experience. The confrontation of novel subject to predicate (first-clustering) materializes when a sudden, undeliberated union of thing to attribute/function emerges, without any analysis, or basis in classificatory logic. In monadic analyses, comparisons between/across things, or between/across properties do not operate. In other words, thing and property are analyzed according to monadic standards, such that the subject is distinct from anything else (from other things and/or properties).

This early influence of phaneroscopic intelligence harkens back to Bergson's (1932/1935: 120; see West, 2019a) theory of things speaking to ancestors intergenerationally, as when earlier truths which obtain are expanded to apply to subsequent generations; they (the new, integrated meanings) inform and reinform future generations. Peirce's category of Firstness contributes to Bergson's paradigm, in that it articulates the importance of possibility in continuing to uncover meaning monadically, absent explicit intergenerational comparisons. Peirce's Firstness espouses this monadic analysis when new meaning possibilities infer a truth from holistic subject scrutiny–although implicit predicate integration may obtain. This integration made by way of First Logical Interpretants is so unified that the subject and predicate (pertaining to subject qualities) is inseparable. This analysis can qualify as a basic inference inherent in First Logical Interpretants. This is so, because properties are not prescinded from things/phenomena in Firstness-based paradigms (W3: 49–59); and the contribution of any single characteristic is not at issue.

Furthermore, the presence of feeling and possibility (not merely quality) within Firstness, inaugurates Logical Interpretants into its meaning. Peirce contends that these First Logical Interpretants require some elementary form of awareness of how the subject attaches to particular predicates (1905: 1.306). This is the case, despite their elemental nature as apprehensions, a degree of passive awareness is present even in monadic analyses. In short, Peirce's concept of possibility provides the means for First Logical Interpretants to grow, given the presence of an elementary form of consciousness (1905: 1.306; 1903: PPM 141). Possibility qualifies as the quintessential attribute of Firstness, as well as First Logical Interpretants, given the element of futurity and habit-change (see supra FN 1). As Peirce explicitly mentions, quality is passive; whereas feelings and possibility are active, in the business of settling upon more fitting convictions. It is possibility which incentivizes the business of habit-change, entailing hypothesis reform. In short, grappling with which aspect of the world will occupy the focus (which deserves examination), stages the contents of one's considerations, hence determining the meanings settled upon in the First Logical Interpretant. The meanings of the First Logical Interpretant are open to absorbing new terms, implied propositions, and ultimately conduct.

> Every concept, every general proposition of the great edifice of science, first came to us as a conjecture. These ideas are the *first logical interpretants* of the phenomena that suggest

[6] Scotus i.1290–1295/2005: 81

> them, and which, as suggesting them, are signs of which they are the (really conjectural) interpretants. But that they are no more than that is evidently an after-thought, the dash of cold doubt that awakens the sane judgment of the muser. Meantime, do not forget that every conjecture is equivalent to, or is expressive of, such a habit that having a certain desire one might accomplish it if one could perform a certain act. (1907: 5.480)

Here, Peirce capitalizes upon conjecture as the defining component of First Logical Interpretants. It is in the objects themselves where conjecture resides. In other words, the objects themselves consist in the novel suggestions as to new meanings. Critical to forming a conjecture is drawing upon novel affirmations ascertained from the principle of possibility. Although the conjecture does not consist in parts, its image encompasses the whole of the suggested meaning, still qualifying as a monad. In the previous passage, possibility clearly resides in pragmatic expressions of the First Logical Interpretant, such that the mental practice of determining if one "could perform a certain act" were one to possess the "desire" to do so is explicit.

11.8 Expansions on Second Logical Interpretants

Attributes of Lower Second Logical Interpretants include: apply possibility to convictions, implement convictions, convert ideas into pragmatic principles without clear logical regulation, no realization of each effect regarding involvement in particular situations. "animated by motives, by affect," and "tracing out the alternative potential conduct to which conjectures would lead." The First and Lower Second Logical Interpretants overlap, especially on the pragmatic plain, e.g., tracing out alternative conduct to which conjectures apply. The prime advance in Second Lower Logical Interpretants is the means to generate alternative ways of acting, not a single conduct scheme only.

The process follows with haecceities in Secondness (noticed proximate objects and impulses in" now" time and "here" space) which initiate discrete feelings and implied predications. The sign to which measurement and directionality are represented is index, given its means not only to show where things are located in here space and during what time frame, but to deposit meaning to the beholder/s, such that notice of the thing's presence, its whereabouts, and its temporal existence are made known, or noticed. Accordingly, index permits examination of particulars, initiating extrapolations as to states of affairs. The interpreter of index proceeds from suggestive postures intrinsic in the structure of position and orientation to their more abstract meanings.

In 1898 (MS 485) Peirce began paving the way for alternative hypothesis generation via the sign which can best situate in spatial and temporal contexts. Peirce construed index to be pivotal in logic, given its means to connect particulars of context to general principles. His characterization of Index as having an "illative" outcome when it permeates interpretants of icons reveals the inception of what Peirce refers to as the "informational index" which he expands in 1903. Responsible for the transition from gestural indices to Logical ones is unquestionably the kind of interpretant which Peirce developed in 1903 and redeveloped in 1908. Underlying the former (gestural indices) are conjectures, whose interpretants are of the Energetic kind, while the latter (informational indices) are associated with Logical Interpretants. Interpretants of the latter are characterized by wider, more objective meanings, although their composition is still energized by practical/action-based schema.

Characteristics affiliated with a more advanced form of Second Logical Interpretant, namely, Higher Second Logical Interpretants, include: fashioning more refined conjectures (affiliating consequences with novel antecedents, engaging in dialogic interchanges, and illustrating full event scenarios.

As a consequence of these attributes, Higher Second Logical Interpretants can ascertain: generalizations regarding where conjectures could lead (1907: MS 318: 45), discerning the "proper mental effect of the sign" (1907: MS 318: 46), "abstracting forms of deliberate conjectures "(1907: MS 318: 46), and "self-controlled habit-change." An illustration of the latter entails "substituting one approximate value or valuation for another" (1907: MS 318: 46–7). The primary distinction between the initial kind of Second Logical Interpretant and this, more advanced kind, is the degree of "mental effect" of the sign (see infra). According to Peirce, this effect must be "taken alone," still not displaced from the context/s. Peirce's use of "proper" refers to the fact that the effect must qualify as a mental reaction which is not idiosyncratic; the response on the part of the interpreter must conform to a reasonable person standard, i.e., how a large majority of individuals would react were they to experience a similar effect. The standard for Higher Second Logical Interpretants rise to a far more objective level. In the same passage, Peirce further supports this objective standard, when he indicates that the reaction is "abstracted from all reference to the individual mind in which it [the effect] might happen to be implanted":

> The second logical interpretants constitute the ultimate normal and proper mental effect of the sign taken by itself (I do not mean removed from its context but considered apart from the effects of its context and circumstance of utterance). They must, therefore, be identified with that "meaning" which we have all along been seeking. In that capacity, they are habits of internal or imaginary action, abstracted from all reference to the individual mind in which they might happen to be implanted, and whose future actions they would guide. For it must never be forgotten that habits called internal, as having been produced by internal exercise [,] take effect in external actions, unless a particular inhibition has been laid upon such action. (1907: MS 318:46)

Peirce further supports an abstract based interpretation of meaning/reactions when he characterizes them as "internal habits," "imaginary reactions …meaning [which] we all along have been seeking." His use of "we" puts us in mind that the imagination does not emanate from a single mind alone; rather, in view of its first-person plural form, "we" refers to the scientific community at large. As such, the belief from "imaginary reactions" must have as its objective, the truth which serves practical or scholarly pursuits—from the envisionment/virtual habit (see supra). Accordingly, his Higher Second Logical Interpretant incorporates future applications/contexts (for members of different generations and cultures) in which the effect is likely to materialize. This mental practice of looking forward to see "the meaning which we all along have been seeking" constitutes the essence of Peirce's pragmaticistic maxim. This state of awaiting that which we hope to receive (possibility in Firstness) unleashes the recognition and scrutiny of novel argumentative signs (inferences) in Thirdness; and this internal (virtual) habit ultimately preempts the exercise of action. Internal habits of envisioning innovative courses of action to, in turn, recommend external actions (1909: MS 637: 12).

> From this experimentation results in deliberate self-controlled habit-changes, which may be dissociative of the second interpretants or of parts of them, or associative or confirmatory, or may involve slight modifications of them (as, for example, by substituting one approximate value or valuation for another.) (1907: MS 318:46-7)

> At the same time, the envisionment serves as a preemptive substitute for the action itself. (1907: MS 318: 46-47)

In the above entry, Peirce informs us that action aspirations in the mind serve as forerunners to implement actions, subject to slight alterations once performed. When Higher Second Logical Interpretants inhabit virtual habits, their power to renovate rationale for states of affairs is indispensable; they propose an explanatory component to others, hence underpinning particular effects. In short, the efficacy of Second Logical Interpretants to awaken the attention/consciousness to reasons for courses of belief and action (influencing the non-ego), must not be understated.

11.9 Abstractions as Peirce's Third Logical Interpretant

Third Logical Interpretants include: heightened intellect to possibility of outcomes (warning of problematic effects), deployment of metacognitive principles, translation of implied predicates into explicit sequential ones, would-bes and could-bes guiding belief/conduct, habit-change, and striving for logical unity between antecedent and consequence. Unlike Second Logical Interpretants, Third Logical Interpretants regulate future conduct guided by more reliable possibility-based principles which keep out spurious arguments. Thirdness-based Logical Interpretants constitute not vague dispositions, but deterministic feelings of mind directing specific beliefs and actions. These new conjectures qualify as transsociations (1907: 5.476), which transform belief trajectories into plausible inferences, such that "a reconciliation or interadjustment between reason and experience" holds (1908: LI 392; MS 300). In short, Peirce's Third Logical Interpretant further refines his commitment to entertain antecedent-consequent structures which supersede mere approaches to would-be incorporation to their compulsory inclusion in all potential arguments—as the summum bonum to entertain abductive promise. Entertaining the "*summum Bonum*" to ascertain increased inferential promise requires the exercise of belief faculties, namely, phenomenal sources, constituting the work of pregenerative Thirdness to innervate instinctual guesses. What these phenomenal sources supply is a knowhow so internal, so inspirational, capacitating intuitional fervor.

To explicate, "guessing right" (1909: MS637) from phenomenal sources, apart from empirical influences, supplies a reliable paradigm to ascertain the "*summum* bonum" for intellectual promise. The aforementioned process extracts and utilizes Peirce's principle of sequence (1906: EP 2:388) upon which abductions rest. Phenomenal sources (dreams/creative hallucinations) provide novel lines to receive new habits of belief and action, which foster interpreters' confidence that their instinctual meanings should not be dismissed out-of-hand. Rather they have the potential to offer insights (as do virtual habits)—informing fact selection and arrangement. Such serve as a living alternative to inductive means to arrive at workable hypotheses, because reliance upon empirical interactions can become an arduous affair.

Third Logical Interpretants in phenomenal scenes when flashing across the mind can supply a spontaneous and inspirational method not dependent upon purely pragmatic/empirical interactions, e.g., actions inherent in Second Logical Interpretants. Closer scrutiny of phenomenal factors is advantageous; they permit the exercise of inner-competencies, those which the self develops, and/or those instilled by divine inspiration, provided that the instincts serve the larger community. Provided that "interadjustment between reason and the facts of experience" (1908: LI 392; MS 300) guide snapshot-based interpretations of dreams, etc., their worth can afford advantages (in the event of some truth content) to dependence upon the practical paradigms which Second Logical Interpretants afford. Thirdness-based phenomenal signs, despite their lack of reliance upon metacognitive skills, permit focus upon the insight itself, rather than featuring construal from several factors laid out in Secondness-based arrays. Attention and use of phenomenal triggers, in which pregenerative meanings in Thirdness reside, offer an alternative which does not ignore the wealth of antecedent contributors. The latter influence protects against social predispositions which may crowd out plausible actions/beliefs of self or of the scientific community. Moreover, it is not actions only which supply the "*summum bonum*" for intellectual promise. As the commentators of Peirce's Collected Papers warn us:

> If it be admitted, … that action wants an end, and that that end must be something of a general description, then the spirit of the maxim itself, which is that we must look to the upshot of our concepts in order rightly to apprehend them, would direct us toward something different from practical facts, namely, to general ideas, as the true interpreters of our thought. (:The Development of Concrete Reasonableness" 1902: 5.3)

If phenomenal signs hold general ideas, as in dreams undergirded by Logical Interpretants (particularly Third Logical Interpretants), their mental effect can supersede that of direct engagement. These instinctual enlightenments via virtual habits can be more beneficial than are either observations of facts, or explanations disclosed between minds. In fact, these phenomenal signs (given their means to supply the readiness to perform) imply logical relations; hence, their meanings culminate in a logical relation, however uncontrolled they might be.

Peirce refers to this spontaneous integration of pre-habit insight into practical insight as a "demicadence," provisionally fitting facts and experience into partial arguments: "In the third logical interpretant, or interpretants, the work of the intellect comes to a demicadence, a provisional and partial consummation; so that it is of supreme logical importance" (1907: MS 318:46–47). It is obvious that Third Logical Interpretants, even in Firstness, serve an integrative function (a demicadence), governing habit-change. When they arise as imaginary actions, they can unquestionably apply novel rationale to spontaneous consequences, suggesting factors which contribute to the logical unity between cause and consequence. According to Peirce, these mental depictions can integrate events into a unified logical whole foundational to argument structure, namely, a "...provisional and partial culmination..." of potential contributory events guided by explanatory adequacy (MS 318: 46). The process is "provisional" and "partial", in view of the plethora of possibilities which become recognized as potential contributors.

11.10 Conclusion

The case is made that the three Logical Interpretants which Peirce proposes in his later semiotic (1907: EP2: 418–430) are squarely grounded in Peirce's categories of Firstness, Secondness, and Thirdness. Early on, pregenerative Thirdness makes its presence known in Firstness; in fact, their joinder reforms what constitutes a sign. The suggestive meanings afforded by pregenerative Thirdness establish the juncture when sign relations attach to objects and their meanings/effects (pertaining to phenomenal objects, as well). The presence of early forms of Thirdness, as Logical Interpretants, occupies a critical place in Peirce's semiotic, instrumentalizing possibility and habit-change even in Firstness.

This inquiry clarifies how Peirce's categories operate as an aggregate to propose viable rationale for how things affect states of affairs, particularly possibility in Firstness. The categories serve as the linchpin for the emergence of each Logical Interpretant: the First Logical Interpretant, The Lower and Higher Second Logical Interpretants, and the Third Logical Interpretant. Each interpretant's influence in producing viable hypotheses for purposes of belief and action makes plain their indispensability in the abductive process, emphasizing how Logical Interpretants foment instinctual abductions.

It has been argued that what endures as interpretant must be logical in nature (pattern of the sign's use, not the Energetic Interpretant effort expended in an individual case). The advantage afforded by the Logical Interpretant is coherence fostered by integration of remnants proper from all interpretants of objects belonging to the type; and Peirce's semiotic ultimately affords further support for his enduring concept of continuity—that what is "real" naturally includes possibility, particularly in the phenomenological realm. The primary assertion is that the influence of Logical Interpretants in mental depictive signs can supply the earliest and most pregnant meanings, because they foster implied predications, which secure the promise of inferential rationality.

References

de Bacha Lourdes, M. (2014). Peirce's Fallibilism in the context of the Theory of and the Theory of Inquiry. *Charles Sanders Peirce in his own words,* De Gruyter Mouton (109–114).

Bellucci, F. (2014). "Logic, considered as semeiotic": On Peirce's philosophy of logic. *Transactions of the Charles S. Peirce Society, 50*(4), 523–547.

Bergman, M. (2016). Habit-change as ultimate interpretant. In D. E. West & M. Anderson (Eds.), *Consensus on Peirce's concept of habit: Before and beyond consciousness*. Springer.

Bergson, H. (1932). *The two sources of morality and religion*. Trans. R. Audra & C. Brereton. Doubleday, 1935 Springer.

Deely, J. (2009). *Purely objective reality*. Walter de Gruyter.

Deely, J. (2015). From semiosis to semioethics. In P. P. Trifonas (Ed.), *The international handbook of semiotics* (pp. 771–789). Springer.

Fuhrman, G. (2013). Peirce's retrospectives on his phenomenological quest. *Transactions of the Charles S. Peirce Society, 49*(4), 490–509.

Gava, G. (2014). What is wrong with intuitions? An assessment of a Peircean criticism of Kant. *Transactions of the Charles S. Peirce Society, 50*(3), 340–359.

Meyers, R. (1999). Pragmatism and Peirce's externalist epistemology. *Transactions of the Charles S. Peirce Society, 35*(4), 638–653.

Peirce, C. S. (1903). *Pragmatism as a principle and method of right thinking: The 1903 Harvard lectures on pragmatism*, ed. P. Turrisi. SUNY Press, 1997. Cited as PPM.

Peirce, C. S. (1931-1966) [i. 1866–1914]. *The collected papers of Charles S. Peirce,*Charles Hartshorne, Paul Weiss & Arthur Burks (eds.), Harvard University Press.

Peirce, C. S. (1967). [i. 1866–1914]. In Richard Robin (ed.), Manuscripts in the Houghton library of Harvard University, in annotated catalogue of the papers of Charles S. Peirce [Reference to Peirce's manuscripts will be designated MS or L]. University of Massachusetts Press.

Peirce, C. S. (1976). [i. 1866–1914]. In *The new elements of mathematics, vol. 4*, Carolyn Eisele (ed.), [Reference to Peirce's New Elements will be designated NEM followed by volume and page number]. Mouton de Gruyter.]

Peirce, C. S. (1998). [i. 1893–1913]. In Peirce Edition Project (ed.), The essential peirce: Selected philosophical writings, vol. 2 (1893–1913). [Reference to vol. 2 of Essential Peirce will be designated EP 2]. Bloomington: Indiana University Press. Peirce, C.S. (i. 1866–1913). *The New Elements of Mathematics*, Vol III ed. C. Eisele. Mouton Press.

Rosensohn, W. (1974). The phenomenology of Charles S. Peirce: From the doctrine of categories to Phaneroscopy. In *B.V.* B.R. Grüner B.V.

Stjernfelt, F. (2014). *Natural propositions: The actuality of Peirce's doctrine of Dicisigns*. Docent Press.

de Tienne, A. (1993). *Peirce's definitions of the Phaneron. Charles S. Peirce and the philosophy of science, papers from the Harvard sesquicentennial congress*. University of Alabama Press.

de Tienne, A. (2004). *Is phaneroscopy as a pre-semiotic science possible?* www.cspeirce.com/menu/library/aboutcsp/detienne/isphanscience.pdf

de Tienne, A. (2013). Iconoscopy between phaneroscopy and semeiotic. *Semiotic Inquiry, 33*(1–3), 19–37.

West, D. (2016a). Indexical scaffolds to habit-formation. In D. West & M. Anderson (Eds.), *Consensus on Peirce's concept of habit: Before and beyond consciousness* (pp. 215–240). Springer.

West, D. (2016b). Peirce's creative hallucination in the ontogeny of abductive reasoning. *Public Journal of Semiotics, 7*(2), 51–72.

West, D. (2017). Virtual habit as episode-builder in the inferencing process. *Cognitive Semiotics, 10*(1), 55–75.

West, D. (2018). Deely's extension of Peirce's thirdness: Pregenerativity. *The American Journal of Semiotics, 34*(1–2), 39–64.

West, D. (2019a). Thirdness along the intuitional path: Reflections on maritain and peirce. *Studia Gilsoniana, 8*(2), 431–475.

West, D. (2019b). Index as scaffold to logical and final interpretants: Compulsive urges and modal submissions. *Semiotica Special Invitation Issue, 228*, 333–353.

West, D. (2024). The influence of Peircean index in the emergence of explanatory inferencing: Evidence from children's narratives. In E. Ippoliti, L. Magnani, & S. Arfini (Eds.), *Model-based reasoning in science and technology: Inferential models for logic, language, Cognition and Computation, p 331–346*. Heidelberg: Springer-Verlag.

West, D. (2021a). The element of surprise in Peirce's double consciousness paradigm. *Semiotica, 243*, 11–47.

West, D. (2021b). Logical and practical advantages of double consciousness. *Cognitive Semiotics, 14*(1), 47–69.

West, D. (2022a). Habit and semiosis: Habit across the categories. In J. Pelkey (Ed.), *Bloomsbury semiotics volume I: History and semiosis* (pp. 241–258). Bloomsbury.

West, D. (2022b). *Narrative as dialectic abduction*. Springer.

Wilson, A. (2016). *Peirce's empiricism: Its roots and its originality*. Lexington Press.

Woods, J. (2013). *Errors of reasoning: Naturalizing the logic of inference*. College Publications.

Chapter 12
Self-Abduction; Oracles, Eco-Cognition and Purpose in Life

Jeffrey White

Abstract This chapter follows Lorenzo Magnani's observation that ongoing commercialization of science and academia impoverishes human potential for discovery. The chapter reviews Magnani on affordance, wonders what is accessible when "good" affordances appear absent, and answers self-affordance. Ecologies optimized for discovery should be optimized for self-affordance. The chapter considers the role of oracle as leading vision for discovery, and proposes a naturalized account of self that is essentially propositional, in pursuit of an inner oracle, seeking salvation through routine and religious ritual. This is self-abduction. Variable biological development may motivate religious self-abduction characterized by self-sacrifice for the collective good over the long term. In conclusion, the chapter affirms Magnani's call to optimize cognitive ecologies for discovery, by prioritizing religious self-abduction.

Keywords Self · Abduction · Purpose in life · Eco-cognition · Religion

12.1 Introduction

> A great, a good, and a right mind, is a kind of divinity lodged in flesh, and may be the blessing of a slave as well as of a prince; ...whereas temples of honor are but empty names, which, probably, owe their beginning either to ambition or to violence.
>
> —Seneca, 1882, p. 165

This chapter follows Magnani's *Discoverability—the Urgent Need of an Ecology of Human Creativity* (2022, hereafter *Discoverability*) in response to perverse incentives and performance metrics driving the "overwhelming commercialization" and

J. White (✉)
IFILNOVA Mind, Cognition and Knowledge group, Lisbon, Portugal

OIST Cognitive Neurorobotics (Tani) Unit, Onna, Okinawa, Japan

S. Arfini (ed.), *Scientific Cognition, Semiotics, and Computational Agents: Essays in Honor of Lorenzo Magnani - Volume 2*, Synthese Library 506,
https://doi.org/10.1007/978-3-031-96688-0_12

the "epistemic impoverishment" of science and academia. Magnani argues that cognitive ecologies should be optimized for free inquiry into and communication of possible solutions to collective problems before correction becomes impossible:

> … we are responsible for the preservation of high degrees of discoverability and diagnosticability, and therefore, for potential fruits of human abductive creative and selective cognition, because their abundance seems not to be an inalienable given too, but depends on our intellectual and practical curation of eco-cognitive openness (*Discoverability,* p. 124).

Discoverability is roughly availability of information. Diagnosticability is roughly what we can do with that information. Eco-cognitive openness describes an anticipatory state in an interactive cycle from discovery to diagnosis, from learning something new about the world to testing this knowledge and sharing results:

> 'Eco-cognitive' means that knowing is to be comprehended in relation to an embodied subject who communicates with his surroundings—that is, who acquires and recognizes information but also manipulates his environment, whether directly or through the creation of artifacts and epistemic mediators (Magnani, 2023, 129-130; also see *Discoverability* p. 132).

An eco-cognitive agent more or less selectively opens and closes to sources of information, and acts in modification of the external world to generate feedback, for instance in testing the accuracy of information, to confirm that something is as perceived or otherwise. "Curation" implies purposeful improvement of conditions that afford this interactive cycle of discovery, so that we end up with a better picture of what is rather than what is merely perceived. Epistemic mediators include technologies like microscopes and search engines, extending to institutions, codes and rituals that support discovery and entrench established knowledge in routine and ritual practices. *Discoverability* describes how solutions are discovered and communicated, through science, and how artifacts and institutions created to support discovery encourage certain inquiries while forbidding others, such as through their politicization and commercialization.

In particular, Magnani finds contemporary institutions closed to purely "good" initiatives, prioritizing monetary profit over public benefit. Correlate incentives steer people away from solving problems to help humanity over the long term, and toward marketable products with immediate financial payouts, instead. With such influence pervasive, civic duty erodes, driving "de-civilization" evident in moral decay and the disintegration of traditionally pro-social institutions (cf. *Discoverability*, p. 128). Finally, Magnani forecasts a near-future incommensurate with enlightened, democratic ideals, a world in which "good" opportunities have evaporated as people are neither able to create nor pursue them in a way that affords necessary systemic change.

Magnani finds potential for a course correction latent in cultural inheritance "such as religion" and attendant concepts, such as Kant's categorical imperative to act as if a member of a Kingdom of Ends, and argues that such principles and institutions "must be successfully and appropriately activated; their being in good working order is a basic condition for exercising morality and allowing free will to become 'good' will." (*Discoverability*, p. 123) This chapter wonders about the source of motivation to exercise good will while supporting institutions that incentivize otherwise. The answer is through self-affordance and self-abduction. This is the main message of this chapter.

The basic idea with self-abduction is that children are "led away" from the tactile immediacy during adolescence to pursue relatively distant lifelong aims through adulthood, as if abducted by themselves. They first learn how things are, then how they should be, and during adolescence embody somatic set-points or "aspirational ideals" according to which error is assessed during iterative interaction with the social and object environment, thereafter. Self-abduction is variably realized in human beings. Aspirational ideals are unique to each, accounting for how different people find different aspects of shared environments salient, including differences contributing to intractable conflicts.

Aspirational ideals characterize ultimate ends as more or less as distal goal conditions against which intermediates are held in evaluation. Some conditions cannot be achieved during the embodied lifetime, yet people resolve to pursue them even where their lives appear sacrificed in the effort, e.g., living as member of Kant's Kingdom of Ends in a corrupted society. Such commitment characterizes religious self-abduction. This chapter considers that religious self-abduction is necessary to advance Magnani's call to action in *Discoverability*. The question becomes how to potentiate corresponding ways of life in order to correct for contemporary deficiencies, even as erstwhile supporting institutions encourage otherwise. One answer is through example, and another through scientific communication as with *Discoverability*.

The next section introduces Magnani on abduction and affordance. Section 12.3 describes self-abduction and self-affordance. Section 12.4 considers how to "curate eco-cognitive openness" to counter short-sighted commercialism through"fusion" of interests involving self-sacrifice. The chapter concludes that cognitive ecologies should be optimized for religious self-abduction to counter pervasive corruption and perverse incentives, ultimately in order to keep open pathways for revolutionary discovery.

12.2 Abduction, Affordance and Oracles

> You see, my friend, it is a secret that I have this art.
> —Socrates (149a, Plato (1997), p. 166).

Magnani (2023) describes cognition as a "possibility-seeking mechanism" and "affordances" as "environmental anchors" signifying such possibilities (p. 134). Good situations include affordances potentiating future situations, as good or better by the same metric, "ecological validity". Cognition uncovers affordances through abduction. Briefly, abduction tests hypotheses through interactions with the environment, uncovers hidden relationships leading to further discoveries, and thereby informs ongoing practice including creation of technologies that mediate, regulate and constrain, this process.

Magnani's interest in *Discoverability* is "conditions of possibility" for abduction, and so discovery, "as they are and as they should become" (p. 127). *Discoverability* describes these conditions in terms of cognitive niches and their construction (Chap. 2). Briefly, through perceived error, people reconfigure theory, change action, experience the eco-cognitive cycle differently, and discover latent opportunities to do otherwise in these differences. Once routinized, iterative interactions describe a niche. People work to reinforce and to enrich niches by creating artifacts that facilitate associated routines, and by regulating relationships through social institutions. This is cognitive niche construction.

Optimized for discovery, "good" cognitive niches afford future cognitive niches. Abduction uncovers these opportunities. Magnani (2023) describes abduction accordingly, as a "method":

> for increasing possibilities of gaining sensory data, trying to make up for their ambiguity, acquiring cognitive feedback, and/or a method for manipulating them, and also a means of making use of cognitive delegations to the environment and to artifacts (p. 131).

The focus of Magnani (2023) is "to demonstrate that it is in this cognitive manipulation of the environment" (p. 134) that we facilitate or obstruct abduction, through discovery, creation and destruction of "affordances". Magnani (2023) treats "affordances" according to Gibson, as "circumstances in which a certain conduct is encouraged or suggested above others by the 'perceptual' cues and indicators we may notice" being "already accessible" and "typical of how an organism adapts to a certain ecological niche" i.e., as more or less salient information signifying opportunities for more or less routine action.

Jaako Hintikka considers "oracles" similarly, as external sources of information potentiating abduction. Hintikka writes "Abductive 'inferences' must be construed as answers to the inquirer's explicit or (usually) tacit question put to some definite source of answers (information)." (2007, p. 49) These "definite sources" are oracles. He explains that oracles do not guarantee truth. They are defeasible. For the process to be "rational", an oracle must be consciously selected from amongst alternatives about which more or less is known (practically, often very little) (Hintikka, 2007, p. 48). By what criteria should oracles be selected? Best informed "intelligent guessing" (abduction) should optimize for best informed future guesses by optimizing for accessible oracles, similar to Magnani's ecological validity. In this way, Magnani's treatment of affordances advances from Hintikka's rational ontology of oracles. For both, to pursue oracles without such a selection criterion is irrational. This point is recalled as this chapter closes.

Magnani (2009) treats cognition in terms of attractors and dynamic systems which enriches this account of affordances. Dynamic systems remain in stable states, enniched in potential wells as local minima representing low-energy, low-stress, system configurations. System dynamics reflect interactions with the embedding environment. Attractors represent relatively stable configurations of systems in terms of that environment, corresponding with cognitive niches. Energy input into a system causes shifts to different configurations, from attractor to attractor, such as during chemical reactions. In a synthetic chemical reaction, given input activation energy—e.g., heat, pressure, with catalysts lowering transitional energy requirements, two independently stable molecules may transform through relative instability of high-energy transition states, to unified stability, fused into a single product system. This idea of fusion is recalled in the context of human beings in sect. 12.4 of this chapter.

Human beings differ from molecules in metabolic potential to transition through high-energy configurations without immediate input from the environment. Animals generally can move from place to place, in effect transitioning from configuration to configuration by moving from one embedding environment to another. As above, they can pursue affordances as opportunities to secure better situations. Some animals including human beings can change the place in which they reside, without physically changing geospatial location, for instance through consultation of oracles, and the construction or destruction of mediating artifacts and institutions. By "curating conditions of possibility for abduction", Magnani is considering how to build environments—"cognitive ecologies"—that lower energy requirements to transition from relative ignorance to discovery, by proliferating affordances and communicating their accessibility, in effect catalyzing the move from "epistemically impoverished" niches to better ones.

As for how to recognize potential for better niches, Magnani (2009) considers salience and pregnance. Salience describes information indicating opportunities for more or less routine enaction, while pregnance is associated with the creative potential of new information which may come through exploration and manipulation of the environment. New information can be used to stabilize select configurations in new ways, or to achieve otherwise inaccessible niches, such as through technology. For example, compare sitting on rocks with sitting in massaging recliners with battery chargers and biometric controls. A stone is a form of a chair, representing a single attractor basin somewhere between standing and laying on the ground. With massaging recliners, people have in effect bound a number of attractors together in one artifact, what Magnani treats as "consolidated affordances". Each is pregnant with potential to sit in different ways. If your legs hurt and you want to charge your phone, the recliner stands out as the better option.

Some affordances are easier to consolidate than others. Communicating the significance of novel affordances in terms accessible to habitually enniched others adds to the difficulty. Magnani (2023) considers affordances from Peircean semeiotics, interpreting circumstances as signs which more or less clearly signify affordances. Recurrent signs trigger habits through spontaneous abduction. We clearly see that a chair presents an opportunity to rest; its form corresponds with "the obvious method" for dealing with it, sitting (2023, p. 131). So, without reflection, we sit.

As for novel consolidated affordances like a massaging recliner,"creating 'artificial' affordances" involves "clever manipulations and conducts that … may result in the creation of novel (and occasionally 'unforeseen') affordances" while representing such discoveries involves setting up "signs in the external environment specifically to cause new proper conclusions of affordability" (2023, p. 133). Consider again affordances as access points to attractors. Ideas can be considered like catalysts in chemical reactions that lower energy barriers for transition from attractor to attractor. The point here, and with scientific discovery generally, is that someone has to expose and to communicate novel opportunities first, in order for others to recognize the potential for better situations that they signify, finally affording popular institutional transformation of the sort demanded by *Discoverability*.

Most artifacts in the built environment fall under the heading of consolidated affordances, as they lower energy barriers to a number of attractors in one object, representing bundled access points to target conditions. A stone chair affords rest without having to lower one's body to the ground. A biometrically modulated massaging recliner does what the stone does, and more. In short, massaging recliners afford access to more attractors with less effort than rocks.

The recliner represents the motivation to "better" cognitive niches motivating scientific discovery in general, a propensity that *Discoverability* treats in terms of the "optimization of situatedness":

> … optimization of situatedness is the main general property of logical abductive inference, which—from a general perspective—defeats the other properties such as minimality, consistency, relevance, plausibility, etc.… (*Discoverability*, p. 18)

Abduction bridges current and better situations, or niches, as an aspect of the "special consequence relation" of abduction. Abduction leads from prior to simultaneously affirmed consequences, which are more or less pregnant with further consequence relations. These relations are embodied by the individual, and are not found in the external world. The special consequence relation is central for self-abduction, in this way. The simultaneous affirmation of embodied consequences of self-abduction is recalled in sections 12.4 and 12.5 of this chapter, in the context of sacrifice.

Magnani treats abduction in terms of syllogisms which is informative. Syllogisms concern relationships between antecedents and consequences. Abduction involves such a relationship, evidenced in a sort of "fill-up problem" to build the bridge of supporting evidence connecting the two. Magnani describes filling-up the space of evidence between affirmed consequent and incomplete and tentative antecedent in familiar terms, as cyclic modification of input and output which "reduces to a procedure of finding the middle term" through "supplementary logic" in methodical inquiry, abduction-deduction-induction (from *Discoverability*, p. 18–20, see also p. 136, p. 142).

> If new information emerges, hypotheses not previously considered can be suggested and a new cycle takes place. In this case the nonmonotonic character of abductive reasoning is clear and arises from the "classical" logical unsoundness of the inference rule: it draws defeasible conclusions from incomplete information. (*Discoverability*, p. 146)

Monotonic reasoning is associated with deductive reasoning, with each premise a fact and with these increasing with new information, more facts. Nonmonotonic reasoning involves assumptions and retractions of presumed facts on the basis of new information, defeasible, characteristic of human reasoning. This defeasible nature of human reasoning accounts for the "failures of our societies and our lives" while also instructing on how to go about fixing them, through abduction as a method of inquiry; by becoming knowledgable, improving cognitive resources while maintaining "intellectual focus on moral and ecological commitments" (*Discoverability*, p. 17), i.e., asking the right questions, consulting the right oracles. The nonmonotonic nature of human reasoning is revisited with Peirce and Thirdness in the next section of this chapter.

Magnani's aim in *Discoverability* is not to communicate a specific discovery, or to relate facts. He is careful not to prefigure potential for "good" cognitive chances. His aim is communication of conditions for discovery in general, in which case over-specification imposes unnecessary constraints. Reminiscent of the quotation with which this chapter began, Magnani writes in *Discoverability* that "we are responsible ... for potential fruits of human abductive creative and selective cognition" which "depends on our intellectual and practical curation of eco-cognitive openness." (*Discoverability*, p. 124) In Magnani (2023), he cautions that over-specification may constrain requisite openness: "In this instance, achieving the right amounts of 'communication' to support the optimal affordances is a crucial component of 'curation' of the eco-cognitive situatedness." (2023, p. 134) The message overall is to keep possibilities for human creative abduction open. The emphasis of this chapter is that requisite openness is internal to the subject. Ultimately, we are our own oracles.

12.3 Self-Abduction

Section 12.2 introduced abduction in terms of affordances and oracles, as environmental anchors of opportunity and sources of information. This section considers an inner source of affordances. Section 12.3.1 considers affordance through sign activity. Section 12.3.2, self-control and self-abduction. Section 12.3.3, external constraints on self-abduction. Section 12.4 concerns religious self-abduction. Overall, the account is of a self which is essentially propositional, embodying its own consequences, aiming for a project conclusion whether these are normal near-term, or universal and religious.

12.3.1 Affordance through Sign Activity

> Now, as the body is to be kept in upon the down-hill, and forced upwards, so there are some virtues that require the rein and others the spur.
>
> —Seneca (1882, p. 143).

Abduction bridges outer and inner worlds. Magnani considers variable capacities for such bridge-building in terms of cognitive resources, "well-fed minds" (Magnani, 2007). Vincent Colapietro (1985, 1988, discussion of Peirce relies heavily on

Colapietro, often without direct reference) characterizes Peirce's theory consistently. Peirce described the mind in terms of feedback loops, with the external shaping the internal, and with present experience of the outer world prefigured by inner world activity. Moreover, in Peirce there is an active aspect in creation of what Magnani treats as "cognitive chances" through inward sign activity. On Peirce's view, the inner world is not limited to prior experience of the external world. Rather, Peirce's "inwardness" involves signs learned from others, enculturated, transformed and created independently, through internal sign activity influencing forward action as if from another world, entirely.

Peirce's basic idea is that minds are complex, hierarchically ordered habits refined through experience, with some originating from within those minds themselves. Minds are semiotic, characterized by sign processing activity involving self-consciousness (First), self-criticism (Second), and self-control (Third). Simply, self-consciousness (Firstness) is of criticism can be expressed through performative example and language (Thirdness) over ongoing iterative action from the perspective of internal ideals at the top of the hierarchy (Secondness).

Peirce considers different kinds of mind. A "cognitive" mind is any mind able to perform interpretation (Thirdness), "learned, automatic or autonomous" (Colapietro, 1985, p. 497). Most habits are taken up uncritically in response to external regularities, with minimal self-consciousness. At least some human minds can change habits through internal sign activity, and are capable of conscious self-control in light of ideals; they are "rational". "Rational mind" is cognitive, yet able to direct the formation and revision of habits which prefigure future engagements.

Peirce considers the breaking of habit in terms of self-control (consider Colapietro, 1985 p. 503). Self-control is for Peirce essentially inhibitory, negating habit. By not doing as had been done, there is potential to do differently. A rational subject can be self-conscious of acting habitually, self-critical in failing to meet a higher standard, and self-controlling when establishing new habits going forward. Habits prefigure reactions to signs, and the formation and revision of such reactions is the goal of inner sign activity, beginning with breaking habits. To do something differently, first what had been habitual, prior, must not be done, now, requiring an intercession into the routine flow of sign activity from the point of view of a self-critic present to self-consciousness. Like Augustine, for Peirce, I err, therefore I am, and act in correction of such error.

Inwardness affords variable access to universals, freedom to act from these instead of from habit through self-control, as well as communication with others on this basis. Commonly embodied interaction with the shared environment informs these universal determinations and can be communicated, with metaphors like "landscapes" of affordance or "slopes" of opportunity as discussed in sect. 12.3.3 of this chapter. At the same time, similar metaphors describe affordances that are not anchored externally, such as with Kant's Kingdom of Ends, i.e. we do not meet ends, we meet people on the way to ends, ends in themselves and rational minds. Rational minds have access to and can set as goals relatively invariant conditions through inner dialog involving universals which are also externalized and tested against other rational minds in communication. Inner-worldly sign activity changes the external world through this behavior, creating, uncovering and communicating opportunities for doing otherwise. These products of inner sign processing are called "self-affordances" in this chapter.

Self-affordances are products of Thirdness. Pelkey (2012) emphasizes Thirdness during critical points such as breaking of habit. Pelkey (2012) emphasizes the non-linear nature of Thirdness, and clarifies how First, Second and Third are related as a matter of origin, end, and intermediate action from one to the other. Pelkey's focal message is that Thirdness is this intermediacy: "If the end is second and the middle is third, then, sequentially speaking, at least, Thirdness is not third." (2012, p. 79) Pelkey quotes Peirce:

> By the third, I mean the medium or connecting bond between the absolute first and last. The beginning is first, the end second, the middle third. The end is second, the means third. The thread of life is a third; the fate that snips it, its second. A fork in a road is a third, it supposes three ways; a straight road, considered merely as a connection between two places is second, but so far as it implies passing through intermediate places it is third. [...] Continuity represents Thirdness almost to perfection. Every process comes under that head. Moderation is a kind of Thirdness. The positive degree of an adjective is first, the superlative second, the comparative third (Pelkey, 2012, p. 79, citing Collected Papers 1.337 from 1875).

Aspirational ideals as discussed in this chapter represent Secondness as global attractors or project ends. Thirdness holds Second against First in comparison. In terms of self-abduction, Thirdness is the relationship between the beginning and end of an abductive argument that at once affirms its own conclusion, as on the way to it. It is in this way that self-abduction concerns an embodied consequence relation, essentially propositional in the form of hypothesis, defeasible and yet simultaneously self-affirming in commitment to find out. First are unique starting conditions, priors, habits and dispositions. Second, aspirational ideals, aims, goals, target conditions develop as products of inner sign processes representing universals including logic and ethical principles. Third is the continuity that reconciles Second with First, bridging ends to origins, holding alternatives presented for action as forks in the road to critical account. Here is the freedom of the rational mind, through inner sign activity and self-control, to determine what comes next. Each possible next is a self-affordance, represented as a fork in the road on the way to solving one's own fill-up problem. The focus here, echoing Magnani, is on how self-afforded freedom can become "good".

Self-abduction begins with the difference between the "absolute first and last" of the semiotic mind, so corresponds with Secondness. The idea is that human beings embody aspirational ideals through adolescence as a Second, marking an apparent discontinuity from childhood entrainment, First. These are reconciled through Thirdness, as self-affordances about which more or less known (often, practically very little). The next subsection begins with Peirce on childhood, and considers the development of Secondness as the where to which we lead ourselves away in self-abduction over the life course.

12.3.2 Self Control and Developmental Self-Abduction

> Virtue is divided into two parts, contemplation and action.
> —Seneca (1882, p. 140).

> Sympathy, flesh and blood, that by which I feel my neighbor's feelings, is third.
> —Charles Sanders Peirce (1974, 1.337)

The most immediate evidence for inner world effects on outer world conditions is in changes that we make to our own habits through self-control. Self-control is inhibitory, opening opportunities to do something else by not doing what prior habit directed. Peirce emphasized the role of inner dialogue on self-control, with an inner critic representing what this chapter treats as aspirational ideals. Peirce is clear that children do not express capacities for self-control. A child's inner critic is not yet fully developed. This is why early education teaches restraint from spontaneous actions, to train self-control. Peirce describes children as subjectively aware, but with inner worlds not yet individuated. Though they think, and are cognitive, they are easily distracted, oriented outward having not yet developed necessary capacities for inner discourse of the "rational mind".

Rational mind develops through language expressing universals against which personal experience is held to account. It is also at least partly through language that a sense of individuated self presents as distinct from others on which the child has been so far completely dependent. Peirce considers a child touching a hot stovetop contrary to verbalized warnings against the action. Touching a hot stovetop in disregard of signifiers such as "hot" encourages correction in the direction of such intersubjectively accessible representations, i.e. do not touch hot things. Most importantly here, consciousness of an inward self as separable from others emerges in this process of experiencing deviation from intersubjectively accessible signs (Boersema, 2003, quotes Peirce 5.233 in this context). Peirce defined 'self-consciousness' accordingly, as developing through experience into more than "a mere feeling of subjective conditions of consciousness, but of our personal selves.... I know that I (not merely the I) exist." (Boersma, 2003, quoting Peirce, 5.225).

The rational mind of Peirce develops after childhood initiation, with the emergence of a self-critic over ongoing interaction. This critic is made possible by the growth of higher-level cognitive processes that hold critical moments to account, for Peirce most eminently through logic. Colapietro quotes Peirce, writing that "'the critic' (what we would call the critique) of thought is the business of logic, and logic is concerned (at least, in part) with the universally valid laws governing potentially public symbols" so that,

> if logic deals with thought, it is thought not in any subjective sense but rather in an intersubjective one—thought as something open to the scrutiny of an indefinite number of human reasoners. (from Colapietro, 1988, chapter 5, quoting Peirce CP 3.404).

Children develop rational mind as experience through language affords this scrutiny through inner sign processing. Inner-discourse with indefinite numbers of human reasoners resembles the "imaginary audience" reported to emerge during adolescence. The idea with developmental self-abduction is that once universals are learned, and self-consciousness so determined through inner discourse with these as represented by Peirce's inner critic, lasting-associations between prior and target conditions are established—embodied as living hypothesis—toward the resolution of which a subject thereafter is disposed over the life course. A subject is in effect led away from prior conditions to target ideals as absolute end and unique destiny. This is developmental self-abduction.

As with external regularities, universals stabilize expectations. They also afford freedom to act contrary to perceived external patterns, routines or regularities. On what basis are alternatives evaluated? This chapter considers aspirational ideals and experience involving universals through language, including logic and ethical principles. Self-affordances present as freedom to act from internalized universals understood as aspirational ideals which distinguish self from other as these deviate. Magnani considers the "paradox of autonomy" in a similar way, that we must consign ourselves to binding principles in order for freedom to be possible:

> Religion, morality, moral knowledge, and teaching enhance and permit free will because they impose order on the randomness of human behaviors, giving people a better chance of owning their destinies (2022, p. 122).

"Owning their destinies" in the present sense involves pursuing inner associations with ideals through self-control, and embodying the consequences. Embodied consequences of internalized universals are important. Imposing order comes with costs. There is work involved, and consequences may be uncomfortable. Magnani (2020) recognizes that exercising ownership over destiny becomes "burdensome" motivating many people to "jettison the concept of choice" and "relinquish autonomy and freedom and externalize responsibility" to the outside world, including to norms and power structures which determine affordances, entering into a condition of "bad faith".

This is the condition in which Magnani finds science and academia in *Discoverability*. Roughly, bad faith is going with what is easy rather than working toward what is ideal. Magnani (2020) argues that understanding this self-deceptive stress-minimizing aspect of "ourselves and our condition … can help us to manage our beliefs and guide us individually toward collective solutions to a number of difficult problems related to bad faith and its challenges to autonomy." (2020, p. 3–4) This chapter returns to the ideal of externalized responsibility in sect. 12.4, with sacrifice.

Consider again Peirce's semiotic subject. His view consists of three key elements: semiosis (sign activity), habit (disposition to act in certain circumstances), and autonomy of self-control (regulation of conduct). A rational mind develops self-control over a hierarchy of habits through inner discourse involving a self-critic representing internalized universals, such as logical operations and ethical principles. Children must grow this hierarchy, through experience, and through language train inner sign processing and self-control accordingly, thereby becoming autonomous in Magnani's sense of "ownership" of destiny. I embody the consequence of action, and not only the next one, but all that follow to the absolute end. I lead myself away from this next situation, breaking habit to act toward ideals, or according to routine. Magnani (2020) considers self as narrator in this way:

"The self as narrator performs an abductive activity devoted to formulating hypotheses about the status of its own actions: human beings continually build and revise a cluster of self-accepted hypotheses that contribute to qualifying some of their own actions as autonomous or not." (2020, p. 3–4).

Bound to ideals that may be contrary to expedience and popular norm, adult human beings are more than cognitive, or rational; they are moral. For Peirce, morality is self-control exercised from compassion. Consider the semeiotic processing cycle described in this chapter, and Magnani's eco-cognitive openness. In compassion, a subject opens to suffering others as oracles, answering what it is like when things go wrong. Feeling with the other, the significance of suffering situations are felt as if one's own through compassion, motivating abduction to optimize the situation, to remake the world without this suffering. This is the moral autonomy of a rational mind, with the moral value of discovery communicating solutions to collective problems so that others are saved the costs. For such reasons, Magnani (2007) teaches that knowledge is a duty. Magnani's concern in *Discoverability* is that contemporary institutions incentivize constituents to personal comforts, instead. The next subsection considers a contemporary account of the good life which represents such a strategy, before recovering nascent potential for salvation, through religious self-abduction, in sect. 12.4.

12.3.3 Self-Abduction and External Constraints

> It is by the impulse of virtue that we love virtue, so that the very way to virtue, lies by virtue, which takes in also, at a view, the laws of human life.
>
> —Seneca (1882, p. 142).

In his *Confessions*, St. Augustine (2012) reports on his early life as a sort of wandering. He searched his memory and found—regardless of changing context—a constant relationship between each recollected instance and his accountability to God. He traces this experience to adolescence, when turning away from God to 'please himself and win respect' in worldly pursuits (Book II). He recognizes the influence of the external world on the inner life, and is troubled. He describes a lifelong pursuit of invariant principles culminating with his placement in the Church. Here, he emphasizes the influence of inner life on the external world, expressed for example through charity and compassion.

Comparable to Peirce's community of inquirers, Augustine considers life most meaningful in relation to God as representative of guiding principles. Like Peirce, Augustine associates self-consciousness with error. Augustine highlights an inner association with an ideal world and self-model in the figure of Jesus Christ, and describes external and internal conditions unachievable through direct interaction with the object environment in the embodied lifetime, such as through his figurative City of God. Briefly, Augustine's self-report describes the biological development of religious self-abduction.

The significance of Augustine's account, distinguishing religious self abduction from other forms of life, can be appreciated in comparison with popular cognitive science. Consider Miller et al.'s (2022) predictive processing inspired, enactivist account of happiness as health and well-being. Roughly, and in familiar terms, happiness is error minimization. Error is associated with stress, and stress minimization over the life course is associated with well-being. "Resilience" is flexibility in enacting entrained routines described as "configurations" in service of minimizing stress and maximizing well-being. Optimal configurations are"metastably attuned" or "poised in a way that allows them to make the most of the affordances relevant to them" and "to learn the most about the environments they frequent" in order to minimize potential for stressful error (Miller et al., 2022, p. 9; compare with Magnani on salience in sect. 12.2 of this chapter).

Miller et al. (2022), and enactivism generally, represent dynamic systems theory as applied to cognitive systems. Again, dynamic systems move spontaneously to low-energy stable states in terms of their environments. Human beings can more or less deviate from these environments, incurring unstable states, to transition to better, or preferred, states. Predictive processing inspired cognitive science extends this principle into the future, with cognition anticipating change in order to reduce differences between expected and perceived reality, to maintain low-energy states through intermediate transitions from environment to environment, especially as these become routinized in a similar way to Magnani on niches. The foundational idea of predictive processing is that organisms which minimize differences between expectation and environment, reducing error as a matter of policy, live longer. They adapt to or act on the environment to minimize potential surprise and avoid correlative stress associated with uncertainty, including injury and disease. This is the crux of the view. Miller et al. (2022) describe this condition in terms of their central idea, metastable attunement:

> Metastable attunement moves the agent in such a way that they find the balance between exploiting existing action policies and performing information-seeking epistemic actions that aim at reducing uncertainty (Miller et al., 2022, p. 9).

Miller et al. (2022) compare short-term error minimization characterized by "a hedonic view that understands well-being to consist of a life of positive experiences such as pleasure and happiness", and relatively long-term error minimization in "eudamonia","fulfilling or realizing one's potential as a human being", in order to expose "the computational differences between momentary subjective happiness and overall well-being" (p. 2). Briefly, the difference is in error summed over short or long timespans. They associate hedonism with minimizing error from relatively short-term goal conditions, such as on the model of pain relief, and they associate eudamonia with "global" health and minimal stress over the embodied lifetime.

Basically, the idea is that attunement to opportunities uncertain enough for personal growth, while avoiding chronic and acute stress conducive to illness and injury, should result in longer life with less suffering and a greater sense of self-control, "the good life".

The biological connection between stress and disease is that error increases metabolic costs including those associated with higher-level (conscious) computational processes for correction of defective policies, distracting from attention to resolution of other needs, such as sleep or safety, thereby potentiating instability or dissolution of the body system through disease and death. Chronic stress associated with disease indicates ill-fit policies (habits) with environmental constraints. Incapacities to flexibly adjust habits to meet external conditions indicate overly rigid policies, such as with dogma or addiction, and imply mental disorder. Acute stressors, such as the loss of a loved one, imply critical points of irreversible change which might be interpreted in terms of rigid associations and lack of flexibility, or "resilience", potentiating injury.

Recall self-control for Peirce, as inhibitory. Rational self-control for Miller et al. (2022) means minimizing error with the embedding environment, and they describe internal sign processing accordingly, as involving adaptive capacities to change habits, what they treat as flexibility and resilience. They confirm metabolic costs associated with self-control, and offer an account of the influence of inner sign processing on self-control, namely when dysfunctional. Finally, they offer a formula for the good life in terms comparable to Magnani's optimal situatedness, again as metastable attunement but here with a focus on what Peirce might consider self-control:

> When things are going well, and they are on good slopes of error reduction, they should continue on the same path. When, however, a niche is so well predicted that there ceases to be good slopes of error reduction available, agents should begin to explore for opportunities to do better.... We will refer to this dynamical state of remaining metastably poised as a state of "metastable attunement". By tracking the changing rate of error reduction, such an agent will be attuned to opportunities to continually improve in error reduction (Miller et al., 2022, p. 8).

Slopes correlate with affect, e.g. panic-inducing or boring. Skillful error reduction feels good, or at least doesn't feel bad, as it threads the middle. Their advice in seeking "good slopes of error reduction" is to avoid overly steep slopes and to move away from flat ones; go with what feels better over the longer term, moderate risk to minimize stress, uncertainty, injury and illness. Miller et al. (2022) write:

> We take positively and negatively valenced affective states to be a reflection of doing either better than or worse than expected at reducing error over time. Valence can be thought of as the organism's evaluation of how it is faring in its engagement with the environment with respect to attaining predicted valued outcomes.... We will henceforth describe optimally functioning agents as being motivated to seek out good slopes of error reduction (Miller et al., 2022, p. 6).

Negative valence is felt error, reducing confidence in predictions, increasing energy costs associated with uncertainty, cognitive load, and stress. Positive valence confirms fit of policy to changing environment, correlative with facile transition between low energy stable states in dynamic systems. Regular achievement of minimal error over the long-term confirms metastable attunement, increases confidence in deployment of cognitive resources to focus on certain regions of the "field of affordances" at the exclusion of others, salience, and correlates with physical and mental well-being. Optimal performance is characterized, again, in terms of metaphysical attunement, carefully deploying cognitive resources in order to "juggle multiple cares and concerns in an optimal way … experienced by the agent as a background feeling of well-being—the felt experience that the system is set up to handle life's many challenges." (Miller et al., 2022, p. 11) This is the experience of happiness that they associate with eudamonia.

Miller et al. (2022) illustrate their view with goal oriented behaviors, one aiming at one-off "hedonic" action that comes to completion in immediate security of a sought-after good—"buying a car comes to an end upon purchasing that car"—and another involving longer-term self-development through pursuit of complex aims including "being more mindful or compassionate, or being a better partner, or serving one's community" which "are all goals that are potentially never finished" and "allow for the continuous broadening of the field of relevant affordances" thus being "likely to be conducive to maintaining metastable attunement, and therefore to living a flourishing life." (p. 10) In this way, Miller et al. (2022) reduce happiness to error-reduction that maintains personal well-being.

But, what to do when "good slopes of error reduction" appear absent in the environment?

> When a particular niche ceases to yield productive error slopes, negative valence signals to the agent that they ought to destroy their own fixed-point attractors in favor of more itinerant wandering policies of exploration…. However, we also have a tendency to actively destroy these attractor states, thereby inducing instabilities and creating peripatetic or itinerant (wandering) dynamics… (Miller et al., 2022, p. 9).

A fixed point attractor is a goal state or condition that impedes "flexible performance solutions". Miller et al. (2022) recommend exchanging such rigid dynamics for "wandering" dynamics in search of exploitable slopes. Where affordances to improve in error reduction are not obvious, the advice is to evacuate associated contexts, cease relevant routines, and search for an exploitable slope so that personal well-being may be secured. By destroying associated attractors, the cognitive system may more flexibly conform to ecological constraints. In this way, Miller et al. (2022) identify happiness with flexibility in pursuing affordances, presuming also that embodied consequences are not so different from their antecedents (involving an overly steep slope in transition).

In the end, what are the computational differences between momentary happiness and overall well-being, according to Miller et al. (2022)? Consider the popular Dutch expression, "Doe maar Normaal, dan doe je al gek genoeg"—roughly, normal is crazy enough. Such a principle proscribes that change should come incrementally, by way of stress-minimal transitions. Well-being depends on such a strategy, over time. Habits are inhibited and associated attractors destroyed to maintain that normal range, not panic-inducing, not boring.

Miller et al. (2022) describe happiness for the individual in situations in which stressors are manageable, "digestible". However, if Magnani's diagnosis is correct, then normal is part of the problem. We are at a collective critical point at which we cannot continue as has been the norm. We need dramatic change. We benefit from cultural inheritance, including ethical principles and religious institutions, but we also need a way to carry the best of this inheritance forward. We need what Margaret Boden called "historical creativity":

> An idea is psychologically creative only if it is new to the individual mind having the idea. The person who has a psychologically creative idea will have never entertained that idea before. An idea is historically creative, on the other hand, only if no one has ever had it before in the whole history of the world (Brainard, n.d., penultimate draft, p. 12-13).

How to potentiate history-making creative abduction? It is not clear how such a productive slope of error correction might appear to the enactivist's "Gibsonian neuroscience". The answer that is pursued, here, is to look deeper inside the head, not outside in wandering itineracy.

For Magnani, abduction is concerned with optimal situatedness, but these situations are not confined to one's own. There is the potential to entrain purposes from internalized universals that outstrip the individual lifetime, or seem contrary to personal well-being, and if impossible to achieve those, individually, to at least communicate how to achieve those, to save others the (computational) costs.

Consider a billiard ball rolling down a hill. To be "happy", this billiard ball minimizes the difference between itself and the landscape all the way. In contrast, Augustine seems pulled by an inner slope inaccessible through the object environment, and acts from a faith which may look like addiction from the outside, in. What is missing in the billiard ball is inner sign processing that may afford freedom to pursue such a slope, even where this does not cohere with externalities, such as may be necessary for a scientist or academic to act in correction of the corruption of embedding institutions described in *Discoverability*. "Serving one's community" is not good, when that community is bad. It is not clear how the billiard ball can tell a difference.

Miller et al. (2022) do not directly account for value-orienting inner self-associations with aspirational ideals, such as reported by Augustine. They do not consider the metabolic costs of "imposing order" on the external world through communication. Rather, their "Gibsonian neuroscience" champions conditions that Augustine associates with being lost in the world. He understands their position in terms of mankind's "fallen" condition. With his self-report, Augustine is in effect overlaying an inner slope onto that ecological landscape of opportunity, representing inner associations with ideals and the struggle, at personal expense, to pull the world a little closer to them. The idea here is that this aspirational slope directs the subject upwards, rather than down, countering the fallen condition through self-sacrifice at the cost of personal well-being. Importantly, this orientation is maintained even where there is no support from the immediate environment, such as when one's "better partner" is God.

How to maintain such a trajectory, when personal deviation stands out as error from embedding norms? Augustine reports on his own inner struggle to free himself from social constraints, to follow his faith. Through inner discourse, and with confidantes, he stopped evaluating a good life in terms of normal social and material affordance landscapes. With the destruction of associated attractors, he reports a felt unity with self and other, inner peace correlative with minimal error with universal principles, and describes this condition as humanity's natural state.

In the end, Augustine is not completely free from externalities. There is an environmental component to Augustine's self-abduction, his placement in the Church. We can imagine such environments as pro-religious in the sense of stabilizing worldly interactions around principled practice from concern for humanity and nature writ large, affording stable opportunities to turn free will into good will.

Magnani's focus in *Discoverability* is on other institutions, but in a similar way. Science and the academy represent institutionalized discovery affording freedom to stand for the truth, but are increasingly corrupted by selfish, relatively short-sighted, material interests, instead. Again, it is not clear how serving such a community can be reconciled with stress-minimal normalcy. It is less clear how walking away to look for more exploitable slopes can solve the problems that can be solved only through the reconditioning of these institutions. The idea, here, is that we might reconsider these institutions as religious, in the sense of supporting and encouraging commensurate ways of life. Such commensurate ways of life are subject of the next section.

12.4 Religious Self-Abduction

12.4.1 Self-Sacrifice

> Yet he must not be aware that the age needs him or at least not want to be aware of it, for at that very same moment he, too, has steered off course, misled by the world-historical.
>
> —Kierkegaard (Pap. VI B 98: 62 n.d., 1845, in Kierkegaard (1992), p. 77).

For Peirce, self-control is inhibitory, impeding habit to open opportunities to do differently. Correlatively, Miller et al. (2022) consider a capacity to "destroy" attractors impeding change. As with Magnani, all emphasize a sort of eco-cognitive openness to opportunities nascent in the environment through exploration, recalling Hintikka's oracles. All insist on becoming better error reducers. Where they differ appears to be in the direction of recommended improvement, in what is good, and how to get there.

Miller-et al. (2022) recommend flexibility of higher cognition to adapt to operational contexts over the embodied lifetime, to modulate personal error during iterative action with given social material environments. On their account, attractors that resist destruction "create fragility in a dynamical system due to their making the system rigid and so less adaptable to a changing environment." (p. 23; similar views are surveyed in White, 2022). Again, increased flexibility implies decreased error, with less error corresponding with constructive confirmation that embodied routines are adequately precise in current and anticipated operational contexts, that "the system is set up to handle life's many challenges" like a billiard ball that can stay close to the ground, all the way down. This view describes a sort of eco-cognitive openness. Limitations involve the scope of anticipated operational contexts, extending to death, which is to say that there is no apparent reason for a subject so described to pursue solutions to problems where the fruits of these efforts are unlikely to be realized in the embodied lifetime, or when these problems do not bear more or less immediately on anticipated or unavoidable operational contexts. Why waste energy planning for slopes that you will never see?

Peirce takes commitment to solutions to such problems as universal to rational minds, as subjectively represented by an inner self-critic leading away from immediate to ideal conditions, as if a member of a community of rational inquirers. Peirce is explicit about the ideal toward which the scientific mind is oriented. Briefly, the cosmos is an argument; we are tasked with the resolution of this argument, and life in accord with its conclusions as exemplified by such a community. Magnani's *Discoverability* works from this tradition, calling for institutions to be redirected to provide for this potential. Given the current state of such institutions, none of this sounds like it may be conducive to stress-minimal personal well-being. Rather, one may be more inclined to follow Miller et al.'s (2022) advice, and find something else to do.

Peirce's ideal community represents norms increasingly absent in the contemporary world, but to which one may attune, regardless. Doing so generates stress with peers whose minds are fed by logic closer to Miller et al.'s (2022). Peirce's life exemplified such a struggle, and must have felt like a sacrifice, when his grant applications were denied, or stable positions withdrawn, creating uncertainty, impoverishment suffering disease and without the charity of others, giving him bread, death. Throughout, Peirce remained committed to solving outstanding, collective problems, discovering relationships and generating insights into the nature of the human condition which are only recently becoming influential as science has caught up, since. Science was for Peirce something of a religious practice. There is perhaps no other way to regard such a commitment to solving problems which outstrip the embodied lifetime of the scientist in question, through self-sacrifice in service to ideal communities rather then to personal well-being. The point of this chapter is that this is the sort of commitment that is necessary to answer Magnani's call to action in *Discoverability*, as well.

Magnani and Bertolotti (2015) consider sacrifice in this way, as religious, as the initiative to solve problems through personal resources, without selfish investment in outcomes:

> The common denominator of sacrificial dynamics is the will to resolve a distressful situation, a crisis, a conjuncture whose outcome is not known but unlikely to be merry. The sacrificial dimension, which is to become religious, is the standing for the best, and not the

> best itself: it is a vector connecting different planes, fueled by—and opening to—violence (p. 10).

This vector is recognizable in the embodied consequence relation of self-abduction. Peirce's ideally rational mind pursues such a policy, "standing for the best" in logic and compassion, as does Augustine though charity and care for community, having ceded a promising future as professor of rhetoric. Such "standing for the best" is either absent in Miller et al. (2022) or dysfunctionally rigid. As for opening to the possibility of violence, Christ exemplifies such a course in forcing money changers from the temple, where they had inserted themselves between personal faith and communion with God. In terms of present discussion, Christ sacrifices himself to keep open institutional affordances to freely connect the different "planes" of Magnani and Bertolotti (2015), and represents the spirit motivating this chapter. Inward commerce with universal principles must be free from money-power corruption. We must stand for the best, regardless of embodied consequences.

Problems present when costs associated with self-sacrifice, standing for the best against the spontaneous slope of popular norm, are forgotten. Magnani and Bertolotti (2015) consider ritual sacrifice in this way. Religious sacrifice is performed as if repetition of outward signs of associated costs, i.e. a life, is sufficient for correlative access to the best for which such ritual sacrifice only figuratively stands. Ritualization externalizes these costs. Rather than recognizing that Christ's sacrifice, Peirce's resolve, Augustine's ideals, were their own and that one's self must sacrifice, similarly, to stand for the best, ritual performance is offered in substitute through someone else's sacrifice, instead.

Ritual performance externalizes and in this way covers over necessary self-sacrifice. Magnani and Bertolotti (2015) associate ritual with magic in this way, as performance without personal entrainment of the necessary "standing for the best" relation. They explain further that ritual performance stands in for access to the good, distinguishing who is "other" and without such access, in this way excusing the violence in ritual sacrifice "as it is conceived as *good* in the performer's mind" (Magnani & Bertolotti, 2015, p. 7, their emphasis). Through such excuses, Christ was nailed to a cross. This is how the money changers resolved their crisis, through sacrifice but not their own.

Ritual sacrifice represents irreversible transformation, as in self-sacrifice. Subjects taking part in ritual performances embody common, unifying consequence relations, while closing off to others, reminiscent of Magnani's (2011) embubbled military intelligence. The ritual magic of Magnani and Bertolotti (2015) is rigid as a source of fragility in the system so organized. Moreover, as ritual participation stands in for one's self standing for the best, it occludes the potential for personal embodiment of this relation with the best, as well. As the costs of standing for the best are distributed away from the personal, the significance of self-sacrifice is forgotten. Ritual sacrifice covers over what is necessary to afford necessary change. And, with repetition, supporting institutions become associated with characteristic architecture affording ritual performance, rather than with exemplary action.

Practical commitment to ideals is reduced to a sort of theater where someone else's sacrifice is replayed.

With ritual performance standing in for the standing for the best, the embodied consequence relation of self-sacrifice as vector binding material with ideal is occluded completely. This is the condition in which Magnani finds contemporary science and academia in *Discoverability*. Personal profit rather than common benefit characterize the commercialization of science and academia, occluding purely good solutions in favor of others that make more money. The threat that Magnani perceives is that the capacity to embody necessary consequence relations such as may be required for historical creativity or religious self-abduction may be lost in the routine or institutional occlusion of such potentials. The point of the present discussion is to underscore why self-sacrifice is so important in this context, in order to evidence that standing for the best remains possible even as institutions traditionally tasked with supporting associated inner sign processing are failing in this role.

Magnani and Bertolotti (2015) consider self-sacrifice in a way that is helpful, here, as "a performative way of signaling that things might improve" representing what they call "hopenness":

> … self-sacrifice totalizes how a sacrifice embodies an openness to the worst, but this is fully sublimated in favor of hope for humanity, inasmuch the subject abandons any possibility of future action as an individual self (p. 9).

"Hopenness" blends "openness" with "hope" representing commitment to improve collective conditions at personal cost; i.e. to destroy ritual attractors and open public discourse to something better. "Something better" here corresponds with the simultaneous affirmation of embodied consequences in self-abduction. Self-sacrifice through hopenness, in this way, involves inhibitory self-control, destroying attractors associated with material self-interest in the service of something greater, and represents the simultaneous affirmation of the embodied consequent relation of self-abduction, even where target consequences won't be one's own. Why commit one's self, sacrifice one's life, in hopenness? Compassion. In complement to access to universals such as through inner discourse with ethical principles represented by Peirce's self-critic, self-interest dissipates through compassion and individuals regard themselves as embodied vectors of intention aiming for collective ideals, saving someone else from suffering consequences as if these were one's own. The next section considers such dynamics under the heading of "identity fusion".

12.4.2 Religious Self-Abduction as Autonomous Self-Sacrifice

> Hence, the kingdom of ends is a kingdom of possible free choices created by (and contingent upon) human beings ….
>
> —Lorenzo Magnani (*Discoverability*, p. 123)

> Then are we to do them an injustice by making them live a worse life when they could live a better one?
> —Glaucon (519d, Plato (1997), p. 1137)

The basic idea with self-abduction is that humans develop access to universals including ideals such as are evident in self-criticism. Different from popular enactive cognitive science, Peirce's inner critic leads the subject away from expedience in self-correction toward aspirational ideals through self-control. The force of the Peircean Second over the rational mind thus remains always in front. Thirdness exposes intermediate action contrary to associated ideals as error, subjectively through inner discourse, and also publicly through critical communication, such as in this chapter.

Aspirational ideals represented by Peirce's inner self-critic develop from childhood self-consciousness. Self-abduction begins with this development, involving the creation of opportunities to act toward these ideals through inner sign processing, self-affordances. Hopenness is self-abduction open to personal sacrifice while standing for the best, to make the world a better place, motivated first of all by care and love for others extending to humanity, sentience, life, the natural world, and inherited ideas such as through Philosophy. Love for humanity is expressed in the communication of ideals and associated affordances, through performative example, and through the composition of scholarly works including Magnani's *Discoverability*.

Biologically, the hypothesis with self-abduction is that, as the adolescent brain expands in care and concern over social-historical space and time, aspirational ideals are variably encoded in the default mode network, with spindle neurons binding aspirational projects with lower-level enactive processes. The basic idea is that, as these processes rapidly develop, solutions to problems perceived become lifelong passionate pursuits, presenting goal conditions for personal commitment of embodied resources. Self-sacrifice may be pro-evolutionary at the group level in this way, historical creativity. Communication through access to universals as well as by performative example make such uniquely embodied solutions accessible to similarly embedded rational minds long after the historically creative individual has passed.

Consider Peirce's ideal community of inquiry, in complement to Kant's Kingdom of Ends and Augustine's City of God, signifying optimal situatedness for scientific discovery. Peirce was a performative exemplar of this way of life. Meanwhile, he suffered for being out of step with, and no doubt would have lived more comfortably had he self-corrected in the direction of, popular norms. Recalling discussion on Peirce on development of the rational mind, different ideals develop from different childhood self-consciousness.

As human beings iteratively open to information through more or less normal biological development, they form variably perceived project ideals through socialization and inner sign processing. The difference between, for instance, Miller et al. (2022) and an account such as Peirce's, Kant's or Augustine's, becomes how much weight is placed on one or the other, passing norms or those representing an ideal community. Either determine lifelong value associations, informing variable recipes for the good life, and represent target conditions for error reduction. Only one affords access to solutions to current problems through inward discourse with ideals. Questions remain concerning what can be done to redirect contemporary scientists and supporting institutions away from relatively short-term personal interests to long-term collective interests commensurate with Magnani's call in *Discoverability*.

Harvey Whitehouse (2023) considers self-sacrifice for a group in terms that are informative, here. His central idea is "fusion" of personal with group interests:

> The fusion construct proposes an identity configuration in which the personal self and the group become fused, such that making salient the personal self also makes the group identity salient, and vice versa (p. 7).

The question becomes to which sort of group one fuses, and how. Whitehouse (2023) considers fused identities formed through "doctrinal" and "imagistic" dynamics, reinforcing self-associations in different ways to be discussed in greater detail as this section progresses. Both form identities of the sort that regulate goal pursuit as constituent members of groups. He discusses different experiences resulting in enhanced empathy motivating caring self-association with group interests, such as compassion for group-member suffering. When it comes to motivating direct action, such as through sacrifice and violence, Whitehouse writes that "there is compelling evidence that fused individuals are not just 'talking big' but will actually engage in extreme forms of self-sacrifice for the group" (p. 9). His interest is in de-fusing destructive self associations, such as those motivating terroristic acts and violent crime. The concern in this chapter is fusing constructive self-associations with the most extensive groups as represented by universals including ethical ideals, and humanity over evolutionary time as the ultimate beneficiary of scientific discovery, such as is ultimately Magnani's concern, as well.

Whitehouse (2023) describes the "imagistic pathway to fusion":

> … a stepwise process in which emotionally intense rituals have an enduringly transformative effect on personal identity because they give rise to indelible episodic memories and processes of reflection that become part of the essential autobiographic self (p. 8).

The basic idea is that collective trauma, such as of the Palestinians in Gaza, today, become defining episodes represented and replayed through ritual performance by in-group members, later on, thereby defining the group and its members by these shared experiences of sacrifice (reminiscent of Magnani & Bertolotti, 2015). These critical points recall Peircean Thirdness, forks in the road of shared history, for instance as many died, lost limbs, were displaced, made homeless, so explaining the condition of the population, now. Such irreversible transitions are ritualized, reinforcing group identity through imagistic fusion. Whitehouse considers evidence for this procedural ritualization of imagistic fusion from prehistory, physically in cave sites with lighting and acoustic properties affording "emotionally impactful rituals" correlative with Church architecture as noted at the close of the previous subsection of this chapter.

Whitehouse (2023) contrasts the imagistic with the "doctrinal pathway" which:

> is based on the sharing of ritual experiences that are so frequent they cannot be recalled in episodic memory, and therefore cannot be linked to the formation of the essential autobiographical self (p. 10).

Whitehouse describes how routinization reduces group identity to scripts that one follows in fulfilling a role. Doctrinally entrained, personal responsibility is externalized to one's role in the collective routine, e.g., "I was only following orders".

Imagistic and doctrinal paths differ in ownership of personal destiny. Inner self-identification through traumatic experience spurs critical self-development toward constructive self associations with group-optimal conditions as aspirational ideals. Whitehouse calls the development of such deeply personal motivational self-associations with collective conditions "synergistic". Magnani and Bertolotti (2015) consider Batman as exemplary of this developmental course.

Doctrinal fusion answers to the burdensome character involved in self-control and ownership of destinies, as discussed in sect. 12.3 of this chapter. Through routinization, with identity reduced to role, e.g. citizen, soldier, wage-slave, the individual acts in service to community interests without reference to personally distinguishing aspirational ideals. Whitehouse (2023) calls such a process of developing self-associations "hydraulic" as participants are sucked up into what Magnani (2011) has treated in terms of informational bubbles. Doctrinal fusion is also evident where Magnani considers ritual stabilization of expectations affording complex social arrangements and institutions, such as are implicated in the paradox of autonomy. But, rather than freedom to act from ideals, Whitehouse (2023) emphasizes the control over the self of external interests responsible for embedding doctrine:

> Since routinization creates stable behaviours among ritual participants and makes them more susceptible to authoritative sources of information about their significance and meaning, the doctrinal pathway to identification entails the stabilization of a set of beliefs and practices over space and time (p. 10).

Considering institutions as critiqued in *Discoverability*, stabilizing practices over time, which of these pathways are exploited? Though his focus differs from Magnani's in *Discoverability*, Whitehouse's schema helps to understand the forces at work in self-sacrifice for group interests. His concern again, is trauma entraining imagistic bonds which may drive deadly responses, and he considers the future:

> One of the greatest challenges humanity faces is that of averting ever deadlier forms of warfare, and that in turn will require new methods of managing the psychological drivers of intergroup conflict (p. 12).

Echoing Magnani's purpose in *Discoverability* (more directly in Magnani, 2011), Whitehouse (2023) finds popular approaches inadequate, for example faulting peace initiatives for pursuing doctrinal pathways, as if imagistic fusion through traumatic sacrifice might be undone by economic sanctions or treaties. He writes that "the main drivers of extreme behaviours are not doctrines but imagistic bonds" and so recommends focusing on "de-fusion" of self from imagistic group identity, rather than "de-radicalization" through "liberalization":

> Our starting point should not be ideology but autobiographical memory: the experiences that help to define our identities both as unique persons and as members of groups. Such events provide the well-springs from which intractable conflicts emerge and become entrenched. (2023, p. 12)

Whitehouse's "wellsprings" of "intractable conflicts" spawn Magnani and Bertolotti's vectors bridging perceptual with ideal planes, and represents the embodied consequence of self-abduction as a Peircean Second, always in front, outstanding as ultimate goal conditions.

Self-abduction describes the development of variable project ideals and their pursuit through self-affordances. In *Discoverability*, Magnani is mainly concerned with what may be considered the doctrinal corruption of scientific and academic institutions, and their potential correction. However, imagistic self-identification with the collective best may be required for the historical creativity needed to seed a new culture of discoverability, and to remake stabilizing institutions. The idea here is to consider these as religious institutions, so as to entrain imagistic bonds motivating service to ideals as represented in Peirce's ideal community of inquirers, to identify with such service over the lifecourse, and in this way to commit to scientific solutions to what appear to be otherwise intractable conflicts, thereby correcting for perverse incentives governing contemporary institutions, at personal expense.

Whitehouse (2023) considers identity fusion at the level of groups engaged in crimes, suggesting that self-destructive behaviors resulting in incarceration may result from deficient synergistic fusion with compassionate, pro-social groups, such as supportive families. We may consider the corruption of scientists and academics through a similar lens. Associated roles, and self-identification through emotional bonding with constructive exemplars and shared ideals, are deficient in such cases. On the other hand, to "de-fuse" dysfunctional bonds, Whitehouse describes programs which replace participation in offending group practices with self-affirming participation in groups that afford pro-social imagistic self-identification. In this chapter, the idea has been to underscore such potential in light of internal ideals and principles, as represented by norms governing ideal communities such as with Kant's categorical imperative. Build the vector bridging the immediacy with the ideal in this way, by reforming the immediacy to support the ideals, affording opportunities to act as if a member of such a community, so that subjects may conform not with passing norms but with aspirational ideals, standing for the best. Given encouraging contexts, self-abduction toward ideal configurations should proceed autonomously.

12.5 Discussion

> … so great are our mistakes in the true estimate of things, that we have hardly done any thing that we have not had reason to wish undone;
>
> —Seneca (1882, p. 223)

We then that are strong ought to bear the infirmities of the weak, and not to please ourselves.
—Romans 15:2 (*King James Bible*)

Standing for the best involves more than ritual sacrifice; it involves imagistic self-association with aspirational ideals as represented in moral exemplars, icons and symbols, with value associations communicated by and reinforced through performative example, which is only later distilled into ritual. As fused group identity is entrained through childhood and adolescent biological development, through a mix of doctrinal and imagistic pathways, the result is lifelong motivational and value orienting goal conditions pursued through self-abduction, by way of more or less creative self-affordances. Self-affordances are products of inner sign activity leading the subject away from the immediate material toward project ideal conditions, more or less distant and so more or less likely to be realized during the embodied lifetime. Religious self-abduction describes embodied purposes extending past that lifetime.

Religion involves target conditions which are not achievable during the embodied lifetime, but are regardless most worthy of pursuit during that lifetime. Such projects may present as products of traumatic experiences, such as through deep compassion with suffering others during critical periods as when under military occupation and systematic war through deception. In the desolation of a Gaza, for example, what remains but for personal commitment, from personal resources, through self-affordance such as revealed through prayer, in communion with God, identifying with the universal, rational collective characterized by justice and mutual concern? What is left but religion, and the will to pull the world closer to what is true and good?

A subject developing through such imagistic pathways for fusion of identity with group interests takes the form of Magnani and Bertolotti's (2015) vector connecting the material with the ideal, standing for the best, and in this way represents the most recognizable form of self abduction, religious self-abduction. Enactive opportunistic slope-chasing also describes self-abduction, only involving shorter and more flexible vectors, or more accurately bundles of sub-vectors as dispositions held together through the coherence of embedding ecologies rather than from defining, individuating purpose or destiny. As embedding ecosystems disintegrate, it is unclear what becomes of co-dependent routine bundles. Augustine, Kant and Peirce in different ways afford sources for internal stability, through inner self-associations with relatively distant project ideals. Practically, the difference is that one pulls the world into ideal conditions, struggling upward, while the other descends into the world as given. Both exercise a sort of self-control. Only the former involves sacrifice of the sort that may correct for the routine corruption of science and academia that Magnani is confronting in *Discoverability*. These are not simple problems.

Consider again Magnani and Bertolotti (2015) describing sacrifice as bridging two planes, one our "natural ecology" and the other, "Other". They consider sacrifice from Derrida, as an "a dieu". They consider a dieu a good-bye, as in 'go to God' through a more or less violent, irreversible, change. They consider also that the "a" in French implies location, i.e., belonging there, partaking with God. In this way, they make room for two possible interpretations of sacrifice, one in which sacrifice is externalized, dispatched somewhere distant while ritual participants remain detached from ideals, and another in the sense of invitation, in which the sacrifice becomes our own, internalized, God is within us.

The reader may wonder how such ideas may solve contemporary problems, to practically counter doctrinal corruption with imagistic self-associations with collective ideals and compassionate exemplars. As with Peirce, Kant, and Augustine, Philosophy has its vocational icons, such as Socrates whose sacrifice may be more or less directly re-enacted in the current context, perhaps only now better informed for the many betrayals, since. The potential with self-abduction on the model of Socratic religion is to become the oracle answering how to pull the world from its fallen condition, standing for the best through performative example by asking the right questions of others, that they are forced to become their own oracles, in the same way. The barrier in the context of science and academia as Magnani finds it in *Discoverability*, today, is as with Socrates that one must say goodbye to material interests, "abandoning any possibility of future action as an individual self" by taking up what is most emphatically a mission from God.

Magnani closes *Discoverability* conceding that finding "socially acceptable" means of communicating possible solutions to contemporary corruption is difficult (discussion p. 128). Whitehouse recommends that focus be on establishing new rituals, to retrain destructive group identities. As to what conditions such rituals should reinforce, he notes that:

> All the world religions endorse the idea that humanity should take care of God's creation or take responsibility for stewardship of the earth—in fact, this seems to be a point of common agreement across most large-scale religions and many indigenous and traditional ones as well (p. 13).

"Stewardship of the earth" here can be considered in terms of Magnani's curation of cognitive ecologies as a member of Pierce's ideal community of inquiry. Here, consider a philosophic pathway for identity fusion, through fusion of identity with the conditions for human welfare over evolutionary, rather than personally embodied or social-political, timespans. This is the scope of Magnani's vision. At the end of his own life, Peirce reflected from such a sense of fusion with natural order, God's creation—"Every mortal who stops to consider it is penetrated with love. It is irresistible" (as quoted in Depew, 1994). Imagine how that deepest common agreement and collective-self-identification in curation of this world and its affordances, including especially self-affordances, can be entrained. The answer offered in this chapter is through religious self-abduction. Cognitive ecologies such as in science and academia must be configured to inspire, and to support, religious self-abduction.

Practically, how to displace entrenched corruption without externalizing violence is not immediately clear. Magnani can be seen as working for such goal conditions through scientific communication, as through *Discoverability* and the lifetime of work before and since. Oddly descriptive of that lifetime of work, Whitehouse

(2023) concludes that a new social science bridging boundaries between academia and public policy is required:

> To build a systematic and rigorous science of the social would require the assembly of bodies of information too vast for individual minds to compute and the co-ordinated application of tools too specialized for individual academics to master alone (p. 14).

Magnani's work can be characterized in such terms, synthesizing across vast bodies of information inaccessible from within specializations to solve problems that are at least very difficult for human beings to compute, and can be considered an example of such a "science of the social world".

Consider in this context computers:

> From a semiotic standpoint, computers create new artifacts that provide new affordances, or "signs," in the Peircean sense, for investigating, extending, and influencing our own brain's thinking processes. In this way, they help "stretch the mind beyond the organic brain (Magnani, 2023, p. 133).

The promise in the long run is that computer systems are designed to help with "the assembly of bodies of information too vast for individual minds to compute", epistemic mediators which, when considered in the context of Magnani's concerns in *Discoverability*, signify a consolidated affordance for all potential affordances. Such systems, in development, are censored, filtered, and regardless return answers to questions only from the prospective past. These systems are not resources for historical creativity. That leaves people like Magnani to reconcile the project future with the inherited present, standing for the best at critical points in humanity's pathway of discovery, up or down.

12.6 Conclusion

> … for a man does not feel what he does not know.
> —Seneca (1882, p. 363).

Human beings serve as as signs for each other, through direct example, through metaphor, myth and narrative representations, through linguistic and other communications, through science. On this capacity, Magnani considers "narrative abduction" in the context of science. Narrative abduction involves "generation of creative and new" stories, such as considered by scientists when evaluating hypotheses. Scientists also employ narrative abduction in communicating hypotheses, coordinating prospective inquiry through language including diagrams and mathematics, all of which is embedded in narrative "construals" of the ways that different points of observation cohere in explanation (discussion from 2009 p. 466, also p. 414–415). Magnani (2023) is most interested in this potential:

> We have to repeat that, building affordances is a semiotic activity, mostly abductive, … in which signs are correctly dispersed throughout to stimulate (or imply) a certain engagement rather than others (2023, p. 133).

This chapter has proposed a broad theory of self-abduction organizing diverse fields from cognitive science and developmental systems neurology to ethics, semiotics and religion. The idea has been to communicate potential to redirect civilization through inner sign activity as well as through performative example, through sacrifice. "Good" affordances for better futures are not obvious in a corrupted world under monetary compulsion such as is under the Magnani's scope *Discoverability*. That leaves self-affordance.

External anchors are only that, anchors. Self-affordances are internal, arising through inner sign activity in light of aspirational ideals. These are the opportunities that may carry the world along with us, upward, in hopenness, for the benefit of all.

Through self-abduction, we act to bring the perceived immediacy closer to aspirational ideals, rather than to meet the terms of the immediate environment from expedience. Ecologies should be optimized for development and enactment of religious self-abduction. Education and its role in affording inner ideal pursuits through lifelong narrative, autobiographical, self-associations must be prioritized, perhaps on the model of philosophic self-abduction, optimizing for historical creativity. In this way, our collective pathway may lead to something better. It is with responsibility over the direction of this pathway that Magnani's *Discoverability* leaves us, and where this chapter ends, as well.

References

Augustine. (2012). *Confessions*. (M. Boulding, Trans.) New. City Press.

Boersema, D. (2003). Peirce and Sartre on consciousness and the ego. *Philosophy Now, 43*, 12–15. https://www.pdcnet.org/philnow/content/philnow_2003_0043_0000_0012_0015

Brainard, L (n.d.). *What is creativity? Forthcoming in The Philosophical Quarterly.*

Colapietro, V. M. (1985). Inwardness and autonomy: A neglected aspect of Peirce's approach to mind. *Transactions of the Charles S. Peirce Society, 21*(4), 485–512. http://www.jstor.org/stable/40320110

Colapietro, V. M. (1988). *Peirce's approach to the self: A semiotic perspective on human subjectivity*. SUNY Press.

Depew, D. (1994). Understanding Charles Sanders Peirce (in some ways) [Review of Charles Sanders Peirce: A life, by J. Brent]. *Reviews in American History, 22*(1), 119–124. https://doi.org/10.2307/2703357

Hintikka, J. (2007). *Socratic epistemology: Explorations of knowledge-seeking by questioning*. Cambridge University Press.

Kierkegaard, S. (1992) (trans. Hong and Hong) *Concluding unscientific postscript to philosophical fragments*. Princeton University Press.

Magnani, L. (2007). *Morality in a technological world: Knowledge as duty*. Cambridge University Press.

Magnani, L. (2009). Morphodynamical abduction. In *Abductive cognition. Cognitive systems monographs* (Vol. 3). Springer. https://doi.org/10.1007/978-3-642-03631-6_8

Magnani, L. (2011). *Understanding violence*. In: Studies in applied philosophy, epistemology and rational ethics. https://doi.org/10.1007/978-3-642-21972-6.

Magnani, L. (2020). Autonomy and the ownership of our own destiny: Tracking the external world and human behavior, and the paradox of autonomy. *Philosophies, 5*(3), 12. https://doi.org/10.3390/philosophies5030012

Magnani, L. (2022). *Discoverability: The urgent need of an ecology of human creativity*. Springer Nature.

Magnani, L. (2023). Possibilities in an abductive perspective: Creating affordances as cognitive chances. *Possibility Studies and Society, 1*(1–2), 127–136. https://doi.org/10.1177/27538699221142718

Magnani, L., & Bertolotti, T. (2015). Christ, Batman, And Girard: A philosophical perspective on self-sacrifice. *Journal of Religion and Violence, 3*(1), 117–136. https://www.jstor.org/stable/26671453

Miller, M., Kiverstein, J., & Rietveld, E. (2022). The predictive dynamics of happiness and Well-being. *Emotion Review, 14*(1), 15–30. https://doi.org/10.1177/17540739211063851 (Original work published 2022)

Peirce, C. S. (1974). *Collected papers of Charles Sanders Peirce*. Harvard University Press.

Pelkey, J. (2012). Nonlinear process in Peirce: "The end second, the middle third". *Semiotics, 77-85*. https://doi.org/10.5840/cpsem20127

Plato. (1997). *Complete works*. Hackett Publishing.

Seneca. (1882). *Seneca's morals of a happy life, benefits, anger, and clemency (trans: Roger L'Estrange) Belford*. Clarke and Company.

White, J. (2022). *On a possible basis for metaphysical self-development in natural and artificial systems*. Filozofia I Nauka, Zeszyt specjalny (10), 71–100. https://doi.org/10.37240/FIN.2022.10.zs.4 last Accessed October 19, 2024, from https://bibliotekanauki.pl/articles/31234167.pdf

White, J. (n.d.). *Variable value alignment by design*. Proceedings of Embodied Intelligence 2023.

Whitehouse, H. (2023). Rethinking ritual: How rituals made our world and how they could save it. *Journal of the Royal Anthropological Institute, 30*(1), 115–132. https://doi.org/10.1111/1467-9655.14048

Chapter 13
Lorenzo Magnani's Selected Bibliography

Lorenzo Magnani

13.1 Book Series

L. Magnani (2011-current), Editor-in-Chief, Studies in Applied Philosophy, Epistemology and Rational Ethics (SAPERE) Springer Science + Business Media LLC New York and Springer Japan KK.

13.2 Books

L. Magnani (2024), Understanding Violence. The Intertwining of Morality, Religion, Capitalism and Violence: A Philosophical Stance, Springer, Cham (second new edition).
L. Magnani and R. Gennari (2022), Manuale di logica. Logica classica e del senso commune (Logic Textbook. Classical Logic and Commonsense Logic), Guerini & Associati, Milan, New Edition ed. by R. Dossena (in Italian).
L. Magnani (2022), Filosofia della violenza, Mimesis, Milan, (second Italian edition of L. Magnani, Understanding Violence. The Intertwining of Morality, Religion and Violence: A Philosophical Stance, Springer, Heidelberg/Berlin, 2011; Translation of T. Bertolotti).
L. Magnani (2022), Discoverability. The Urgent Need of an Ecology of Human Creativity, Springer, Cham, Switzerland.
L. Magnani (2022), Eco-Cognitive Computationalism. Cognitive Domestication of Ignorant Entities, Springer, Cham, Switzerland.

L. Magnani (✉)
Department of Humanities–Philosophy Section, University of Pavia, Pavia, Italy
e-mail: lorenzo.magnani@unipv.it

S. Arfini (ed.), *Scientific Cognition, Semiotics, and Computational Agents: Essays in Honor of Lorenzo Magnani - Volume 2*, Synthese Library 506,
https://doi.org/10.1007/978-3-031-96688-0_13

L. Magnani (2019), Conoscenza come dovere. Moralità distribuita in un mondo tecnologico (Knowledge as Duty. Distributed Morality in a Technological World), Aracne, Rome (second Italian edition in of L. Magnani (2007), Morality in a Technological World. Knowledge as Duty, Cambridge University Press, Cambridge).

L. Magnani (2017), The Abductive Structure of Scientific Creativity. An Essay on the Ecology of Cognition, Springer, Cham, Switzerland.

L. Magnani (2013), Rispettare le persone come cose (Respecting people as things), Il Melangolo, Genoa (in Italian).

L. Magnani (2012), Filosofia della violenza (Philosophy of Violence), Il Melangolo, Genoa, (Italian edition of L. Magnani, Understanding Violence. The Intertwining of Morality, Religion and Violence: A Philosophical Stance, Springer, Heidelberg/Berlin, 2011; Translation of T. Bertolotti).

L. Magnani (2011), Understanding Violence. The Intertwining of Morality, Religion and Violence: A Philosophical Stance, Springer, Heidelberg/Berlin.

L. Magnani (2009), Abductive Cognition. The Epistemological and Eco-Cognitive Dimensions of Hypothetical Reasoning, Springer, Berlin/Heidelberg.

L. Magnani (2009), Morality in a Technological World. Knowledge as Duty, Cambridge University Press, Cambridge (Paperback).

L. Magnani (2007), Morality in a Technological World. Knowledge as Duty, Cambridge University Press, Cambridge.

L. Magnani (2005), Knowledge as a Duty. Distributed Morality in a Technological World, Associated International Academic Publishers, Pavia, 2005 (in Italian).

L. Magnani (2001), Philosophy and Geometry. Theoretical and Historical Issues, Kluwer Academic, Dordrecht.

L. Magnani (2001), Abduction, Reason, and Science. Processes of Discovery and Explanation, Kluwer Academic/Plenum Publishers, New York. (Chinese edition: [意] 洛伦佐·玛格纳尼 / 著；李大超，任远 / 译，《溯因、理由与科学——发现和解释的过程》，中国广州:广东人民出版社2006年. Translated by Dachao Li and Yuan Ren, Guangdong People's Publishing House, Guangzhou, 2006).

L. Magnani and R. Gennari (1997), Manuale di logica. Logica classica e del senso comune (Logic Textbook. Classical Logic and Commonsense Logic), Guerini & Associati, Milan, 1997 (in Italian).

L. Magnani (1997), Ingegnerie della conoscenza. Introduzione alla filosofia computazionale (Knowledge Engineering. Introduction to Computational Philosophy), Marcos y Marcos, Milan, 1997 (in Italian).

L. Magnani (1991), Epistemologia applicata (Applied Epistemology), Marcos y Marcos, Milan, 1991 (in Italian).

L. Magnani (1990), Filosofia e geometria. Temi teorici e storici (Philosophy and Geometry. Theoretical and Historical Issues), Guerini & Associati, Milan, 1990 (in Italian).

Mangiarotti, L. Bonesio, M. Luchi, and L. Magnani (1987), In nome dell'architettura (On Behalf of Architecture), Jaca Book, Milan, 1987 (in Italian).

L. Magnani (1978), Geometrie non euclidee (Non Euclidean Geometries), Zanichelli, Bologna, 1978 (in Italian).

13.3 Edited Handbooks

Handbook of Abductive Cognition, Springer (Editor-in Chief), 2023.
Springer Handbook of Model-Based Science (Editor-in-Chief with T. Bertolotti), 2017.

13.4 Edited Books

E. Ippoliti, L. Magnani, and S. Arfini (eds.) (2024), Model-Based Reasoning, Abductive Cognition, Creativity–Inferences & Models in Science, Logic, Language, and Technology, Springer, Switzerland, 2023.
L. Magnani (ed.), Justification, Creativity, and Discoverability in Science. Comptes Rendus de l'Académie Internationale de Philosophie des Sciences, Tome II. Brussels: Académie Internationale de Philosophie des Sciences, 2024. Print edition. College Publications, London, 2024.
S. Arfini and L, Magnani (eds.) (2022), Embodied, Extended, Ignorant Minds. New Studies on the Nature of Not-Knowing, Book Series "Synthese", Springer, Cham.
Nepomuceno, L. Magnani, F. Salguero, C. Barés and M. Fontaine (eds.) (2019), Model-Based Reasoning in Science and Technology–Inferential Models for Logic, Language, Cognition and Computation, Springer, Cham, Switzerland.
D. Gabbay, L. Magnani, W. Park, and A.-V. Pietarinen, eds., A Tribute for John Woods, College Publications, London, pp. 269–308.
L. Magnani and C. Casadio (eds.) (2016), Model-Based Reasoning in Science and Technology. Logical, Epistemological, and Cognitive Issues, Springer, Cham, Switzerland, Series "Sapere".
L. Magnani, P. Li, and W. Park (eds.) (2015), Philosophy and Cognitive Science II, Western and Eastern Studies, Series "Sapere", Vol. 20, Springer, Heidelberg/New York.
L. Magnani (ed.) (2013), Introduzione alla nuova logica (Introduction to the New Logic), Il Melangolo, Genoa (in Italian).
L. Magnani (ed.) (2013), Model-Based Reasoning in Science and Technology. Theoretical and Cognitive Issues, Series "Sapere", Vol. 8, Springer, Heidelberg/Berlin.
L. Magnani and Li Ping (eds.) (2012), Philosophy and Cognitive Science. Western and Eastern Studies, Series "Sapere", Vol. 2, Springer, Heidelberg/Berlin.
L. Magnani, W. Carnielli, C. Pizzi (eds.) (2010), Model-Based Reasoning in Science and Technology. Abduction, Logic, and Computational Discovery, Series "Studies in Computational Intelligence", Vol. 314, Springer, Heidelberg/Berlin.

S. Iwata, Y. Oshawa, S. Tsumoto, N. Zhong, Y. Shi, and L. Magnani (eds.) (2008). Communications and Discoveries from Multidisciplinary Data, Series "Studies in Computational Intelligence", Springer, Berlin/New York.

L. Magnani and P. Li (eds.) (2007), Model-Based Reasoning in Science, Technology, and Medicine, Series "Studies in Computational Intelligence", Vol. 64, Springer, Berlin/New York.

L. Magnani and P- Li (eds.) (2006), Philosophical Investigations from a Perspective of Cognition, Guangdong People's Publishing House, Guangzhou, (published in Chinese: [意] 洛伦佐·玛格纳尼、李平主编，《认知视野中的哲学探究》，中国广州:广东人民出版社2006年). Preface by L. Magnani, pp. 1–3.

L. Magnani (2006) (ed.), Model-Based Reasoning in Science and Engineering. Cognitive Science, Epistemology, Logic, College Publications, London.

L. Magnani and N. J. Nersessian (eds.) (2002), Model-Based Reasoning. Scientific Discovery, Technological Innovation, Values, Kluwer Academic/Plenum Publishers, New York.

L. Magnani, N. J. Nersessian, and C. Pizzi (eds.) (2002), Logical and Computational Aspects of Model-Based Reasoning, Kluwer Academic, Dordrecht.

L. Magnani, N. J. Nersessian, and P. Thagard (eds.) (1999), Model-Based Reasoning in Scientific Discovery, Kluwer Academic/Plenum Publishers, New York, (Chinese edition, translated and edited by Qiming Yu and Tianen Wang, China Science and Technology Press, Beijing, 2001: 《科学发现中的模型化推理》，于祺明、王天思主译，中国北京:中国科学技术出版社2001年).

L. Magnani and R. Dossena (eds.) (2004), Felix Klein: il Programma di Erlangen) Felix Klein: the Erlanger Program, Pristem/Storia, Springer, Milan (in Italian).

13.5 Edited Journal Issues

E. Ippoliti, L. Magnani, and S. Arfini (eds.) (2024) (guest editors), Special Issue on "Formal Representations of Model-Based Reasoning and Abduction", Logic Journal of the IGPL, forthcoming XX(X) (2024).

Nepomuceno, L. Magnani, F. Salguero, C. Barés and M. Fontaine (eds.) (2019) (guest editors), Special Issue on "Formal representations of model-based reasoning and abduction", Logic Journal of the IGPL, 29(4) (2021).

L. Magnani and S. Arfini (eds.) (2021) (guest editors), Special Issue on "Knowing the Unknown: Philosophical Perspectives on Ignorance", Synthese, 198 (2021).

L. Magnani and C. Casadio (guest editors), Special Issue on "Formal representations of model-based reasoning and abduction", Logic Journal of the IGPL, 24(4) (2016).

L. Magnani (guest editor), Special Issue on "Frontiers of abduction", IfCoLog Journal of Logics and their Applications, 3(1) (2016).

L. Magnani (guest editor), Special Issue on "Formal representations in model-based reasoning and abduction", Logic Journal of the IGPL, 21(6) (2013).

L. Magnani, W. Carnielli and C. Pizzi (guest editors), Special Issue on "Formal representations in model-based reasoning and abduction", Logic Journal of the IGPL, 20(2) (2012).

L. Magnani, E. Bardone, and T. Bertolotti (guest editors), Special Issue on "Uncovering Violence as Technology-Mediated in Contemporary Societies", International Journal of Technoethics 2(4) (2011).

L. Magnani (guest editor), Special Issue on "Model-based reasoning in science and engineering", Foundations of Science 13(2) (2008).

L. Magnani (guest editor), Special Issue on "Tracking irrational sets. Science, technology, ethics", Foundations of Science 13(1) (2008).

L. Magnani (guest editor), Special Issue on "Abduction, practical reasoning, and creative inferences in science", Logic Journal of the IGPL 14(2) (2006).

L. Magnani (guest editor, with N. J. Nersessian), special issue on "Model-based reasoning: visual, analogical, simulative", Foundations of Science 10 (2005).

L. Magnani (guest editor, with N. J. Nersessian), special issue on "Abductive reasoning in science", Foundations of Science 9 (2004).

L. Magnani (guest editor, with N. J. Nersessian), special issue on "Scientific reasoning: model-based reasoning", Mind & Society 5(3) (2002).

L. Magnani (guest editor, with N. J. Nersessian L. Magnani (guest editor, with N. J. Nersessian), special issue on "Commonsense and scientific reasoning", Mind & Society 4(2) (2001).

L. Magnani (guest editor, with N. J. Nersessian and P. Thagard), special issue on "Model-based reasoning in science: learning and discovery", Foundations of Science, 5(2) 2000.

L. Magnani (guest editor, with N. J. Nersessian and P. Thagard), special issue on "Abduction and scientific discovery", Philosophica 61(1) 1998.

L. Magnani (guest editor, with N. J. Nersessian and P. Thagard), special issue on "Analogy and mental modeling in scientific discovery", Philosophica 61(2) 1998.

L. Magnani (guest editor), Intelligibilità e costruzione scientifica (Intelligibility and Scientific Construction), Materiali Filosofici 11, Angeli, Milan, 1985, pp. 161–186.

L. Magnani (guest editor), Epistemologie dell'invenzione (Epistemologies of Invention), Materiali Filosofici 10, Angeli, Milan, 1984, pp. 89–114.

13.6 Articles

L. Magnani (2025), Are we close to the death of "homo faber'"? Information technology, robots, ai, and the urgent need of a new renaissance, in Second Death–Experiences of Death Across Technologies, edited by A. Sans Pinillos, V. Costa and J. Vallverdú, Springer, Cham, forthcoming.

L. Magnani (2025), "Model", entry of The Palgrave Encyclopedia of Theoretical and Philosophical Psychology, edited by T. Teo, Palgrave-MacMillan, London, forthcoming.

L. Magnani (2025), Political Correctness, Stoicism, Hypocrisy, and Violence, in Festschrift in Honor of Jean-Yves Beziau 60th Birthday, Springer, Cham, forthcoming.

L. Magnani (2025), Information, computation, and cognition. Their emergence and disappearance, and the ever-changing entanglement of their meaning, in M.J. Schroeder and W. Hofkirchner (eds.), Understanding Information and Its Role as a Tool: In Memory of Mark Burgin, World Scientific Publishing Company, Singapore 2025, pp. .

L. Magnani (2025), Are we really protecting people from the overwhelming power of the internet giants or are we just promoting "ethics washing" and "law washing"?, in K. de Vries, E. van den Hoven, G. Gori, and, S. Gutwirth (eds.), Law Does (Not) Compute: On Digitization and Cyberlaw. Conversations with the Work of Professor Mireille Hildebrandt, forthcoming.

L. Magnani (2025), The actuality of critical rationalism Fabio Minazzi, Giulio Preti, and the modernity of "Philosophical Stancing", in Le ragioni del razionalismo critico oggi /The Reasons for Critical Rationalism Today on the occasion of Fabio Minazzi's 70th birthday, Mimesis Milam, forthcoming.

L. Magnani (2025), Locked and unlocked strategies, creativity, and spurious correlations. Artificial Intelligence and eco-cognitive openness, in M. Farina, P. Ciancarini, X. Yu, and J. Chen (eds.), Digital Transformation in Artificial Systems. Engineering Requirements, Political, Economic and Philosophical Challenges, Elsevier, London, forthcoming.

L. Magnani and A. M. Marchini (2024), Creare nuove enclosures. La ripetizione violenta dell'accumulazione primitiva attraverso degradazione delle donne, lockdown, finanza predatoria, guerra, saccheggio dei beni comuni, Materialismo Storico, 17(3)2:125–160.

L. Magnani (2024), Computational locked and unlocked strategies and dissipative brains. The abductive eco-cognitive perspective, BRAIN. Broad Research in Artificial Intelligence and Neuroscience 15(4), 354–374.

L. Magnani (2024), Discoverability: affordances and eco-cognitive situatedness. Towards an ecology of human creativity, in L. Magnani (ed.), Justification, Creativity, and Discoverability in Science. Comptes Rendus de l'Académie Internationale de Philosophie des Sciences, Tome II. Brussels: Académie Internationale de Philosophie des Sciences, 2024. Print edition. College Publications, London, 2024, pp. 125–139.

L. Magnani (2024), Models as moral mediators. Cognitive niches, artefacts, and the two-faced nature of internet and artificial intelligence mediators in the intertwining of morality and violence, in E. Ippoliti, L. Magnani, and S. Arfini (eds.), Model-Based Reasoning Abductive Cognition, Creativity–Inferences & Models in Science, Logic, Language, and Technology, Springer, Switzerland, pp. 75–88.

L. Magnani and A. Biglari (2024). Externalization of the mind in the meaning-making process: An integrated semiotic and cognitive science perspective, Cognitive Semiotics, Online First https://doi.org/10.1515/cogsem-2024-2014.

L. Magnani (2024), The key universal quality of logical abductive inferences. Does optimization of eco-cognitive situatedness supersede other criteria like minimality, consistency, relevance, and plausibility? The Logic Journal of the IGPL, online first.

L. Magnani and S. Arfini (2024), Model-based abductive cognition: What thought experiments teach us, The Logic Journal of the IGPL, online first.

L. Magnani (in conversation with L. Magnani, by G. Pozzobon) (2024), Something new in our embodied mind, Maize. Frames from the Manyworld, 15 (Winter 2024), 36–41.

L. Magnani (2024), Anchors of meaning. The intertwining of signs, abduction, and cognitive niches, in A. Sharov, G. Mikhailovsky, and R. Gordon (eds,), Pathways to the Origin and Evolution of Meanings in the Universe, Wiley-Scrivener, Hoboken NJ, pp. 379–400.

L. Magnani (2024), Language, morality, and violence, in Encyclopedia of Language and Linguistics, third Edition, Elsevier https://www.sciencedirect.com/science/article/abs/pii/B9780323955041000223?via%3Dihub

L. Magnani (2024), Faith and the reasons for its pervertibility. Between philosophy and cognitive science in Ragione e fede nell'età contemporanea, C. Genna (ed.), Franco Angeli, Milan, pp. 271–282.

L. Magnani (2023), Discoverability and abduction: the role of eco-cognitive situatedness in L. Magnani (Editor-in-Chief), Handbook of Abductive Cognition, Springer, Cham, Switzerland, 2023, pp. 87–113 .

L. Magnani (2023), Abduction and violence. Hypothetical cognition in the entanglement of morality and violence in L. Magnani (Editor-in-Chief), Handbook of Abductive Cognition, Springer, Cham, Switzerland, 2023, pp. 1489–1516.

L. Magnani (2023), Possibilities in an abductive perspective: Creating affordances as cognitive chances, Possibility Studies & Society (2023), 1(1–2) 2023, 127–136.

L. Magnani (2023) Jeopardizing interpersonal privacy through gossiping. Morality and violence entangled, Washington University Review of Philosophy, forthcoming.

R. Dossena and L. Magnani (2023), Optical microscopes reveal the hyperreal world, Journal of Applied Logic, 10(4) (2023), 603–627.

L. Magnani (2023), The need for a philosophy as a vigorous, impertinent, and profound understanding of violence, in Quale filosofia ad inizio del XXI secolo? C. Genna (ed.), Introduzione di G. Puglisi, Postfazione di P. Di Giovanni, Franco Angeli, Milan, pp. 233–244.

L. Magnani (2023), Inferring reasons. Internal and external reasons in practical cognition, in R. Giovagnoli and R. Lowe, The Logic of Social Practices II, Springer, Cham, pp. 99-114.

L. Magnani (2023), Signs of cognition: How semiotic brains reach out to meaning, in A. Biglari (ed.) Open Semiotics, vol. 4 "Life and extensions", Part VI. Semiotics and Cognitive Science, L'Harmattan, Paris, 2023 pp. 451–462.

L. Magnani (2022), Macchine computazionali e creatività umana Il ruolo centrale della "apertura eco-cognitiva", Aquinas, LXV, 2022/II: 273–288,273.

L. Magnani (2022), Diagrams, conceptual space and time, and latent geometry. Toward the Development of Euclidean Axiomatics, Special Issue "Cosmologies. Evolving thought and action: How humans construct and conceptualize matter, space, time, and number" eds., N. Gontier, L. Tirapicos and C. Sinha, Axiomathes 32:1483–1503.

L. Magnani, A. Sans Pinillos, S. Arfini (2022), Language: The "Ultimate Artifact" to encapsulate worldviews, Topoi (2022) 41:461–470.

L.Magnani (2022), Conceptualizing machines in an eco-cognitive perspective, Philosophies, 7, 94.

L. Magnani (2022), Why abductive cognition goes beyond just learning from data, in Y. Ohsawa (ed.), Living Beyond Data, "Intelligent Systems Reference Library" 230, Springer, Cham, pp. 39–70.

L. Magnani and A. M. Marchini (2022), Creating new "enclosures": violently mimicking the primitive accumulation through degradation of women, lockdowns, looting finance, war, plunder, Philosophies, 2022, 7, 58.

L. Magnani (2022), Naturalizing morality to unveil the status of violence: coalition enforcement, cognitive moral niches, and moral bubbles in an evolutionary perspective, Philosophies, 2022, 7(2), 39.

Sans Pinillos and Lorenzo Magnani (2022), How do we think about the unknown? The self-awareness of ignorance as a tool for managing the anguish of not knowing, in S. Arfini and L, Magnani (eds.) (2022), Embodied, Extended, Ignorant Minds. New Studies on the Nature of Not-Knowing, Book Series "Synthese", Springer, Cham, 2022, pp. 191–207.

L. Magnani (2022), Human Abductive cognition vindicated. Computational locked strategies, dissipative brains, and eco-cognitive openness, Philosophies, 7(1), 15.

L. Magnani (2022), Abductive strategies in human cognition and in deep learning machines, in J. Wang (ed.) Encyclopedia of Data Science and Machine Learning, IGI Global, Hershey, PA.

L. Magnani (2022), Foreword, in F. Minazzi, Historical Epistemology and European Philosophy of Science, Springer, Cham, Switzerland.

L. Magnani (2021), Abduction as "leading away". Aristotle, Peirce, and the importance of eco-cognitive openness and situatedness, in J. Shook and S. Paavola, Abduction in Cognition and Action, Springer, pp. 77–105.

S. Arfini and L. Magnani (2021), Embodied irrationality? Knowledge avoidance, willful ignorance, and the paradox of autonomy, Frontiers of Psychology, 12, Article 769,591.

L. Magnani (2021) Preface to M. Castellana, Il dibattito in area francofona sul pensiero matematico e Kurt Gödel, ebook https://www.amazon.it/dibattito-francofona-pensiero-matematico-Secondaria-ebook/dp/B09KP78FJJ/ref=sr_1_5?__mk_it_IT=ÅMÅŽÕÑ&crid=B8WSTCYFWJ0H&dchild=

1&keywords=castellana+mario&qid=1635701427&s=books&sprefix=castellana+mario%2Cstripbooks%2C187&sr=1-5

L. Magnani (2021), Computational Domestication of Ignorant Entities. Unconventional Cognitive Embodiments, Synthese (2021), 198:7503–7532, Special Issue edited by L. Magnani and S. Arfini "Knowing the Unknown: Philosophical Perspectives on Ignorance".

L. Magnani (2021), Governing ignorance through abduction, in A. Nepomuceno, L. Magnani, F. Salguero, C. Barés and M. Fontaine (eds.) (2019) (guest editors), Special Issue on "Formal representations of model-based reasoning and abduction", Logic Journal of the IGPL, 29(4) (2021), 409–424.

Nepomuceno, L. Magnani, F. Salguero, C. Barés and M. Fontaine (2019), Preface, in A. Nepomuceno, L. Magnani, F. Salguero, C. Barés and M. Fontaine (eds.) (2019), Special Issue on "Formal representations of model-based reasoning and abduction", Logic Journal of the IGPL, 29(4) (2021), 405–408.

L. Magnani (2021), Cognitive niche construction and extragenetic information. A sense of purposefulness in evolution, Journal for General Philosophy of Science 52, 263–276, Special Issue on Evolutionary Epistemology (eds. M. Bradie and N. Gontier).

L. Magnani (2021), Semiotic brains build cognitive niches. Biosemiotics of the disembodiment of the mind, Biosemiotics, 14:41–48.

L. Magnani (2021), Creativity, eco-cognitive openness, human and machine inferential capacities. Deep learning and locked strategies, Proceedings of the AISB 2021. Communication and Conversation, Symposium R&R 2021. Representation and Reality In Humans, Other Living Organisms and Machines: New Trends, eds. R. Giovagnoli, G. Basti, and R. Lowe, AISB 2020 Convention, St Mary's University, London (April 2021), http://aisb20.wordpress.com New Trends, pp. 20–26.

R. Giovagnoli and L. Magnani, Social practices and embubblement, Philosophies, 6, 13 https://doi.org/10.3390/philosophies6010013.

L. Magnani (2021), The cognitive and epistemic value of mathematics. Making the world intelligible: the role of abduction, diagrams, and affordances, in M. Danesi, Handbook of Cognitive Mathematics, Part edited by D. E. West, "Mathematics and Cognition", Springer, Cham, pp. 1–40.

S. Arfini. L. Magnani, and T. Bertolotti (2021), Ignorance-based cognitive strategies, in C. Barés, F. J. Salguero and F. Soler, eds., Lógica, Conocimiento y Abducción. Homenaje a Ángel Nepomuceno, College Publications, London, pp. 251–264.

L. Magnani (2021), Knowledge-enhancing abduction and eco-cognitive openness reconsidered. Locked and unlocked cognitive strategies, in C. Barés, F. J. Salguero and F. Soler, eds., Lógica, Conocimiento y Abducción. Homenaje a Ángel Nepomuceno, College Publications, London, pp. 299–320.

L. Magnani (2021), Incommensurabilità, scoperta scientifica, comparabilità e modelli computazionali, in F, Minazzi, ed., Storia e filosofia della scienza: una nuova alleanza? Atti del Convegno internazionale dedicato alla memoria di Ludovico Geymonat, (Varese, 25–26 ottobre 2018), Mimesis, pp. 135–152.

L. Magnani (2020), Coalition enforcement, in J. Vonk and T. K. Shackelford (eds.), Encyclopedia of Animal Cognition and Behavior, section ed. By C. Leuchtenberger, Springer, Switzerland, first online 2017.

L. Magnani (2020), Fictionalists disregard the dynamic nature of scientific models. Models in the perspective of distributed cognition, in T. Bertolotti (ed.), Cognition in 3E: emergent, embodied, extended, Springer, Cham, Switzerland, pp. 69–86.

L. Magnani (2020), A philosophical "stance", Filosofia Maravilhosa/Marvelous Philosophy, Special Issue edited by J.-Y. Béziau, forthcoming.

L. Magnani (2020), "Une season en enfer"...académique, Il Protagora XLVII, gennaio-dicembre 2020, sesta serie, n. 33–34, pp. 365–370.

L. Magnani (2020), The urgent need of an ecology of human creativity, Idee, Special Issue on The Encyclical Letter Laudato Si′, ed. By. M. Castellana, pp. 45–62.

L. Magnani (2020), Foreword to Joseph Brenner and Andrei Igamberdiev, Philosophy in Reality: A New 'Book of Changes', Springer, Cham.

S. Arfini, L. Magnani, and T. Bertolotti, Chance-discovery and chance-curation in online communities, in S. Denison, M. Mack, Y Xu and B. C. Armstrong (eds.), Proceedings of the 42nd Annual Virtual Meeting of the Cognitive Science Society (CogSci2020), Toronto, Canada, July 29–August 1, 2020.

L. Magnani (2020), Autonomy and the ownership of our own destiny. Tracking the external world and human behavior, and the "Paradox of Autonomy", Philosophies, 5, 12, Special Issue "Testimony and Autonomy in Social Epistemology" ed. by R. Giovagnoli.

L. Magnani (2020), Knowledge as duty. Morality in a technological world: A new philosophical stance, in S. Chiodo and V. Schiaffonati, eds., Italian Philosophy of Technology. Socio-Cultural, Legal, Scientific and Aesthetic Perspectives on Technology, Springer, Cham, 2020, pp. 199–215.

L. Magnani (2020), Anticipations as abductions in human and machine cognition. Deep learning: Locked and unlocked capacities, Postmodern Openings, 11(4), 230–247 (LUMEN Journal).

L. Magnani (2020), The future of eco-cognitive settings: Computationally or humanly tailored? abstract submitted to The MIT-JFI Workshop on "The Ethics of Algorithms", April 17th–18th, 2020, MIT Schwarzman College of Computing, Cambridge, MA (event postponed and then canceled).

S. Arfini, T. Bertolotti, and L. Magnani (2020), The diffusion of ignorance in online communities, in Management Association, Information Resources, Information Diffusion Management and Knowledge Sharing: Breakthroughs in Research and Practice, IGI Global, Hershey, 2020, pp. 843–857.

L. Magnani (2020), Eccentric computational embodiments. Cognitive domestication of external entities, Academics in China 1 (2020), 185–197.

L. Magnani (2020) Moral bubble effect. Violent and unaware of it: How difficulties in recognizing one's own violence lead to disregard the inflicted harm, in R. Giovagnoli and R. Lowe, The Logic of Social Practices, Springer, Cham, Switzerland.

S. Arfini, T. Bertolotti, and L. Magnani (2019), The antinomies of serendipity. How to cognitively frame serendipity for scientific discoveries, Topoi, 39: 939–948.

Nepomuceno, L. Magnani, F. Salguero, C. Barés and M. Fontaine (2019), Preface, in A. Nepomuceno, L. Magnani, F. Salguero, C. Barés and M. Fontaine (eds.), Model-Based Reasoning in Science and Technology. Inferential Models for Logic, Language, Cognition and Computation, Springer, Cham, Switzerland, pp. v-ix.

L. Magnani (2019), Creative model-based diagrammatic cognition. The Discovery of the "imaginary" non-Euclidean geometry, in A. Nepomuceno, L. Magnani, F. Salguero, C. Barés and M. Fontaine (eds.), Model-Based Reasoning in Science and Technology. Inferential Models for Logic, Language, Cognition and Computation, Springer, Cham, Switzerland, pp. 217–244.

L. Magnani (2019), AlphaGo, Locked Strategies, Eco-Cognitive Openness, in H. Zenil and S. Bringsjord, eds., "Philosophy and Epistemology of Deep Learning" a special issue of Philosophies 2019, 4(1), 8; https://doi.org/10.3390/philosophies4010008.

L. Magnani (2019), Errors of reasoning exculpated. Naturalizing the logic of abduction, in D. Gabbay, L. Magnani, W. Park, and A.-V. Pietarinen, eds., A Tribute for John Woods, College Publications, London, pp. 269–308.

D. Gabbay, L. Magnani, W. Park, and A.-V. Pietarinen (2019), Preface and Editorial, in D. Gabbay, L. Magnani, W. Park, and A.-V. Pietarinen, eds., A Tribute for John Woods, College Publications, London, pp. 1–24.

T. Bertolotti and L. Magnani (2019), Gossip, in D. Coady and J. Chase (eds.), Routledge Handbook of Applied Epistemology, Routledge, London, pp. 272–283.

L. Magnani (2018), The urgent need of a naturalized logic, in G. Dodig-Crmkovic and M. J. Schroeder, eds., "Contemporary Natural Philosophy and Philosophies" a special issue of Philosophies 3(4), 44; https://doi.org/10.3390/philosophies3040044.

L. Magnani (2018), Ritual artifacts as symbolic habits, Open Information Science, 2(1), 147–155.

L. Magnani (2018), Computationalism in a dynamic and distributed eco-cognitive perspective, in G. Dodig-Crnkovic and M. Burgin (eds.), Philosophy and Methodology of Information, World Scientific, Singapore, pp. 365–288.

L. Magnani (2018), Eco-cognitive computationalism. From mimetic minds to morphology-based enhancement of mimetic bodies, Entropy, 20(6), 430 http://dx.doi.org/10.3390/e20060430

T. Bertolotti, S. Arfini, and L. Magnani (2018) Cyber-bullies as cyborg-bullies, in R. Luppicini (ed.), The Changing Scope of Technoethics in Contemporary Society (pp. 1–403). Hershey, PA: IGI Global, Hershey, PA pp. 60–73, doi:10.4018/978-1-5225-5094-5.

S. Arfini, T. Bertolotti, and L. Magnani (2018), The Diffusion of ignorance in online communities, International Journal of Technoethics, 9(1), 37–50.

S. Arfini, C. Casadio, and L. Magnani (2018) Ignorance-preserving mental models: Thought experiments as metaphorical abductions, Foundations of Science, 24(2) 391–409.

L. Magnani (2018), Playing with anticipations as abductions. Strategic reasoning in an eco-cognitive perspective, in Journal of Applied Logic- IfColog Journal of

Logics and their Applications 5(5), 1061–1092. Special Issue on "Logical Foundations of Strategic Reasoning" (guest editors W. Park and J. Woods).

L. Magnani (2018), Manipulative abduction, external semiotic anchors, and blended Cognition in J. Vallverdú and V. C. Müller (eds.), Blended Cognition. The Robotic Challenge, Springer, Switzerland, pp. 23–52.

L. Magnani (2018), How to Build New Hypotheses. 'Απαγωγη¯ and the Optimization of the Eco-Cognitive Situatedness, in E. Ippoliti and D. Danks (eds.) Building Theories. Hypotheses & Heuristics in Science, Springer, Switzerland pp. 261–284.

S. Arfini, T. Bertolotti, and L. Magnani (2017), Chance curation in virtual cognitive niches, 7–12, in Proceedings of the second European Workshop on Chance Discovery and Data Synthesis, EWCDDS, Aug28, 2016, The Hague, The Netherland.

L. Magnani (2017), An Eco-Cognitive Perspective on the Philosophy of Information. Knowledge as a Duty, in Proceedings, Volume 1, Issue 3 (August 2017), Proceedings of the IS4SI 2017 Summit Digitalisation for a Sustainable Society, Gothenburg, Sweden, 12–16 June 2017, Issue Editor: Gordana Dodig Crnkovic, 1, 79; doi:10.3390/IS4SI-2017-03956, Multidisciplinary Digital Publishing Institute Proceedings, http://www.mdpi. com/2504–3900/1/3.

L. Magnani (2017), Eco-cognitive computationalism—From "mimetic minds" to morphology-based enhancement of "mimetic bodies", in Proceedings, Volume 1, Issue 3 (August 2017), Proceedings of the IS4SI 2017 Summit Digitalisation for a Sustainable Society, Gothenburg, Sweden, 12–16 June 2017, Issue Editor: Gordana Dodig Crnkovic, 1, 179; doi:10.3390/IS4SI-2017-04021, Multidisciplinary Digital Publishing Institute Proceedings, http://www.mdpi. com/2504-3900/1/3.

L. Magnani (2017), Ritual Artifacts as Symbolic Habits. Maximizing Abducibility and Recovering Memory, in Proceedings, Volume 1, Issue 3 (August 2017), Proceedings of the IS4SI 2017 Summit Digitalisation for a Sustainable Society, Gothenburg, Sweden, 12–16 June 2017, Issue Editor: Gordana Dodig Crnkovic, 1, 231; doi: 10.3390/IS4SI-2017-03986 Multidisciplinary Digital Publishing Institute Proceedings, http://www.mdpi. com/2504–3900/1/3.

T. Bertolotti and L. Magnani (2017), Theoretical considerations on cognitive niche construction, Synthese 194(12), 4757–4779 doi 10.1007/s11229-016-1165-2.

S. Arfini, T. Bertolotti, and L. Magnani (2017), Online Communities as Virtual Cognitive Niches, Synthese, Online first DOI: 10.1007/s11229-017-1482-0.

L. Magnani (2017), Political correctness between wise stoicism and violent hypocrisy, Philosophies 1(3), 261–274; doi:10.3390/philosophies1030261.

T. Bertolotti, S. Arfini, and L. Magnani (2017), Of cyborgs and brutes: technology-inherited violence and ignorance, Philosophies, 2(1), 1–14; doi:10.3390/¬philos ophies2010001.

T. Bertolotti and L. Magnani (2017), Contemporary finance as a critical cognitive niche, in P. Chen and E. Ippoliti (eds.), Methods and Finance, Springer, Cham, pp. 129–150.

Magnani L. (2016). Editorial preface, IfCoLog Journal of Logics and their Applications, 3(1):1–5, Special Issue on "Frontiers of abduction" (guest editor L. Magnani).

T. Bertolotti, S. Arfini and L. Magnani L. (2016). Abduction: from the ignorance problem to the ignorance virtue, IfCoLog Journal of Logics and their Applications, 3(1):153–173, Special Issue on "Frontiers of abduction" (guest editor L. Magnani).

S. Arfini and L. Magnani (2016), Cognitive autoimmunity. Knowledge, ignorance and self-deception, Logic Journal of the IGPL, 24: 612–627.

T. Bertolotti and L. Magnani (2016) Gossip as a model of inference to composite hypotheses, Pragmatics & Cognition 22:3 (2014), 310–325. doi 10.1075/pc.22.3.02ber.

L. Magnani and C. Casadio (2016), Preface, in L. Magnani and C. Casadio (eds.), Model-Based Reasoning in Science and Technology. Logical, Epistemological, and Cognitive Issues, Springer, Heidelberg, pp. v-viii.

L. Magnani (2016), Abduction and its eco-cognitive openness. Aristotle's 'Απαγωγη¯ explained, in L. Magnani and C. Casadio (eds.), Model-Based Reasoning in Science and Technology. Logical, Epistemological, and Cognitive Issues, Springer, Heidelberg, pp. 453–468.

L. Magnani and C. Casadio (2016) Preface, Logic Journal of the IGPL, 24: 453–455.

L. Magnani (2016), Naturalizing the logic of abduction, Logic Journal of the IGPL 24: 639–654.

L. Magnani (2016), Violence hexagon. Moral philosophy through drawing, Logica Universalis, 10(2): 359–371.

L. Magnani, S. Arfini, T. Bertolotti (2016), Of habit and abduction. Preserving ignorance or attaining knowledge? In D. West and M. Anderson (eds.) Consensus on Peirce's Concept of Habit, Heidelberg/Berlin Springer, pp. 371–377.

L. Magnani, S. Arfini and T. Bertolotti (2016), An argument for ignorance-based chance discovery, International Journal of Advanced Intelligence Paradigms 8(3), 327–342.

L. Magnani (2016), The eco-cognitive model of abduction II. Irrelevance and implausibility exculpated, Journal of Applied Logic 15, 94–129.

L. Magnani (2016), Scientific models are distributed and never abstract. A naturalistic perspective, in E. Ippoliti, F. Sterpetti, and T. Nickles (eds.) Model and Inferences in Science, Heidelberg/Berlin, Springer, pp.219–240.

L. Magnani (2015), Commentary: Einstein, Prigogine, Barbour, and their philosophical refractions in F. Santoianni (ed.), The Concept of Time in Early Twentieth-Century Philosophy, Springer, Heidelberg/New York, pp. 249–251.

L. Magnani (2015), Naturalizing logic. Errors of reasoning vindicated: Logic reapproaches cognitive science, Journal of Applied Logic 13:13–36.

L. Magnani (2015), The eco-cognitive model of abduction. 'Απαγωγη¯ now: Naturalizing the logic of abduction, Journal of Applied Logic 13:285–315.

L. Magnani and S. Arfini and T. Bertolotti (2015), Ignorance-based chance discovery. Beyond dark events, in D. C. Noelle, R. Dale, A. S. Warlaumont, J. Yoshimi,

T. Matlock, C. D. Jennings, & P. P. Maglio (eds.), Proceedings of the 37th Annual Meeting of the Cognitive Science Society (pp. 1488–1493), Austin, TX.

L. Magnani (2015), The eco-cognitive model of abduction (EC-Model). Is abduction really ignorance-preserving?, in D. C. Noelle, R. Dale, A. S. Warlaumont, J. Yoshimi, T. Matlock, C. D. Jennings, & P. P. Maglio (eds.), Proceedings of the 37th Annual Meeting of the Cognitive Science Society (pp. 2837–2838), Austin, TX.

L. Magnani (2015), Violence and abductive cognition. Epistemology and ethics entangled, in L. Magnani, P. Li, and W. Park (eds.), Philosophy and Cognitive Science II, Western and Eastern Studies, Springer, Heidelberg/Berlin, pp. .95–113.

L. Magnani (2015), Understanding visual abduction. The need of an eco-cognitive model, in L. Magnani, P. Li, and W. Park (eds.), Philosophy and Cognitive Science II, Western and Eastern Studies, Springer, Heidelberg/Berlin, pp. 117–138.

S. Arfini and L. Magnani (2015), An eco-cognitive model of ignorance immunization, in L. Magnani, P. Li, and W. Park (eds.), Philosophy and Cognitive Science II, Western and Eastern Studies, Springer, Heidelberg/Berlin, pp. 59–75.

T. Bertolotti and L. Magnani (2015), Contemporary finance as a critical cognitive niche. The unpredictable effects of contrasting unpredictability, Mind & Society, 14(2), pp. 273–293. Special Issue on "Complexity Modeling in Social Science and Economics".

L. Magnani and T. Bertolotti (2015), Christ, Batman and Girard. Perspectives on self-sacrifice, Journal of Religion and Violence 3(1): 117–135. DOI: 10.5840/jrv20156159.

T. Bertolotti and L. Magnani (2015), Cyber-bullies as cyborg-bullies, International Journal of Technoethics, 6(1), pp. 35–44.

T. Bertolotti and L. Magnani (2014), An epistemological analysis of gossip and gossip-based knowledge. Synthese 191:4037–4067.

L. Magnani (2014), The abductive theory of cognition. The eco-cognitive model, Rivista Italiana di Filosofia del Linguaggio, 49–69.

T. Bertolotti, L. Magnani, and E. Bardone (2014), Camouflaging truth: a biological, argumentative and epistemological outlook from biological to linguistic camouflage, Journal of Cognition and Culture 14:65–91.

L. Magnani (2014), Preface [Prefacio] to A. Aliseda, La Lógica como Herramienta de la Razón. Razonamiento Ampliativo en la Creatividad, la Cognición y la Inferencia, College Publication, London, pp. VII-XIII (in Spanish).

L. Magnani (2014), Are heuristics knowledge-enhancing? Abduction, models, and fictions in science. In E. Ippoliti (ed.), Heuristic Reasoning, Springer, Heidelberg/Berlin, pp. 29–56.

L. Magnani (2014), La filosofia della violenza. In M. Francesconi and D. Scotto di Fasano, Il sonno della ragione. Saggi sulla violenza, Liguori editore, Napoli, pp. 96–116.

L. Magnani (2014), Natural mathematics and natural logic. In C. Cozzo and E. Ippoliti (eds.), From a Heuristic Point of View, Cambridge Scholar Publishing, Newcastle upon Tyne, pp. 183–198.

L. Magnani (2013), Structural and technology-mediated violence: Profiling and the urgent need of new tutelary technoknowledge, in R, Luppicini (ed.), Moral, Ethical, and Social Dilemmas in the Age of Technology: Theories and Practice, IGI Global, Hershey, PA, pp. 162–181).

T. Bertolotti, E. Bardone and L. Magnani (Perverting activism: Cyberactivism and its potential failures in enhancing democratic institutions, in R, Luppicini (ed.), Moral, Ethical, and Social Dilemmas in the Age of Technology: Theories and Practice, IGI Global, Hershey, PA, pp. 53–69.

L. Magnani (2013), Abducing chances: Evolution, cognitive niches and hybrid humans as chance discoverers, Design Ecologies 3(2):170–199. DOI: 10.1386/des.3.2.170_1.

T. Bertolotti, L. Magnani (2013). A philosophical and evolutionary approach to cyber-bullying: Social networks and the disruption of sub-moralities. Ethics and Information Technology, 15(4) 285–299.

L. Magnani (2013), Foreshadowing ethical and spiritual chances. The role of moral mediators, European Journal of Science & Theology 9(1) 17–31.

L. Magnani (2013), Scientific innovation as eco-epistemic warfare: the creative role of on-line manipulative abduction, Mind & Society, 12:49–59.

L. Magnani (2013) Understanding abduction. Inference, perception, and instinct, in L. Magnani (eds.) (2013), Model-Based Reasoning in Science and Technology. Theoretical and Cognitive Issues, Series "Sapere", Vol. 8, Springer, Heidelberg/Berlin, pp. 173–215.

L. Magnani (2013), Is abduction ignorance-preserving? Conventions, models and fictions in science, Logic Journal of the IGPL, 21(6): 882–914.

L. Magnani (2013), Preface to the special issue on "Formal representations in model-based reasoning and abduction, Logic Journal of the IGPL, 21(6): 879–881.

L. Magnani (2013), Thinking through drawing. Diagram constructions as epistemic mediators in geometrical discovery, The Knowledge Engineering Review 28(3):303–326.

L. Magnani (2013), Reductio ad Hitlerum, in D. Di Desare, C. Ocone, S. Regazzoni (eds.), Il nuovo realismo è un populismo, Il Melangolo, Genoa, pp. 65–79.

L. Magnani and T. Bertolotti (2013), Selecting chance curation strategies: is chance curation related to the richness of a cognitive niche? International Journal of Knowledge and Systems Science, 4(1), 62–76. Special Issue on Chance Discovery edited by A. Abe.

Magnani, L. (2013), Abducing personal data, destroying privacy. Diagnosing profiles through artifactual mediators. In: M. Hildebrandt & E. de Vries (eds.), Privacy, Due Process and the Computational Turn. Philosophers of Law Meet Philosophers of Technology. Routledge, London, pp. 63–86.

L. Magnani (2012), In treno da Milano a Roma con Marco Somalvico, in F., Amigoni e V. Schiaffonati (eds.), Marco Somalvico, Maggioli Editore, Santarcangelo di Romagna, 2012, pp. 169–171.

L. Magnani (2012), Morality and psychopathic criminals. Ethicocentrism, mental incapacity, free will, and the fear of decriminalization, Rivista Internazionale di Filosofia e Psicologia 3(1) 26–35.

E. Bardone and L. Magnani (2012), Turning down a chance: an argument from simplicity, in Y. Ohsawa and A. Abe (eds), Advances in Chance Discovery, Springer, Berlin/Heidelberg, pp. 19–31

L. Magnani (2012), Scientific models are not fictions. Model-based science as epistemic warfare, in L. Magnani and Li Ping (eds.) (2012), Philosophy and Cognitive Science. Western and Eastern Studies, Series "Sapere", Vol. 2, Springer, Heidelberg/Berlin, pp. 1–38.

L. Magnani (2012), Coalition enforcement, violence, and abductive reasoning. Morality, language, and violence [Коалиционное принуждение, насилие и абдуктивное познание (мораль, язык и насилие)] Refleksii/Reflections, 2(4): 86–101 (in Russian).

S. Luan, D. Guozhong, and L. Magnani (2012), An approximate approach to belief revision Logic Journal of the IGPL 20(2): 486–496.

L. Magnani (2011), Structural and technology-mediated violence: profiling and the urgent need of new tutelary technoknowledge, International Journal of Technoethics 2(4): 1–19. Special issue on "Uncovering Violence as Technology-Mediated in Contemporary Societies", L. Magnani, E. Bardone, and T. Bertolotti (eds.).

T. Bertolotti, E. Bardone, and L. Magnani (2011), Perverting activism: cyberactivism and its potential failures in enhancing democratic institutions, International Journal of Technoethics 2(2): 19–29.

L. Magnani (2011) External diagrammatization and iconic brain co-evolution, Semiotica 186(1/4): 213–238.

L. Magnani and E. Bardone (2011), From epistemic luck to chance-seeking. The role of cognitive niche construction in: A. König et al. (Eds.), KES 2011, Part II, LNCS 6882, Springer, Heidelberg/Berlin, pp. 486–494, 2011.

E. Bardone and L. Magnani (2010), The appealing of gossiping fallacies and its ecological roots, Pragmatics & Cognition 18(2): 365–396.

L. Magnani and E. Bardone (2010), Chances, affordances, and cognitive niche construction: the plasticity of environmental situatedness, Int. J. Advanced Intelligence Paradigms 2(2/3): 235–253.

L. Magnani (2010), "Man is an external sign". Semiosis and the disembodiment of the mind ["Человек есть внешний знак". Семиозис и освобождение духа от его телесной оболочки], Refleksii/Reflections, 1: 27–46 (in Russian).

L. Magnani (2010) Creating ethics. The multimodal structure of moral reasons, Economy Transdisciplinarity Cognition 13(1) 48–60.

L. Magnani (2010) Building artificial mimetic minds: how hybrid humans make up distributed cognitive systems, APA Newlsetter 9(2), Spring: 38–48.

L. Magnani (2010) L'evoluzione della coscienza e del libero arbitrio. Il ruolo della conoscenza negli umani ibridi. Paradoxa 3(4): 42–61.

L. Magnani (2010) Is knowledge a duty? Yes, it is, and we also have to "respect people as things", at least in our technological world: Response to Bernd Carsten Stahl's Review of Morality in a Technological World: Knowledge as Duty, Minds and Machines, 20(1):161–164.

L. Magnani (2010), Material Cultures and moral mediators in human hybridization, International Journal of Technoethics, 1(1):1–19.

L. Magnani (2010), Smart abducers as violent abducers: hypothetical cognition and military intelligence, in: L. Magnani, W. Carnielli, C. Pizzi (eds.), Model-Based Reasoning in Science and Technology Abduction, Logic, and Computational Discovery, Series. Studies in Computational Intelligence, Vol. 314, Springer, Heidelberg/Berlin, pp. 51–82.

L. Magnani (2010), Moral mediators in technology, in V. Guliciuc and E. Guliciuc, Philosophy of Engineering and Artifact in the Digital Age, Cambridge Scholar Press, Cambridge, pp. 31–48.

L. Magnani (2010), Moral mediators in a technological world, in: R. Feist, C. Beauvais, and R. Shukla (eds.), Technology and the Changing Face of Humanity, University of Ottawa Press, Ottawa, pp. 112–133.

W. Carnielli and L. Magnani (2010), Years of reasoning. In honor of the 65th birthday of Claudio Pizzi, in: L. Magnani, W. Carnielli, C. Pizzi (eds.), Model-Based Reasoning in Science and Technology Abduction, Logic, and Computational Discovery, Series. Studies in Computational Intelligence, Vol. 314, Springer, Heidelberg/Berlin, pp. XI-XVI.

T. Bertolotti and L. Magnani (2010), The role of agency in the invention of supernatural beings: an abductive approach, in: L. Magnani, W. Carnielli, C. Pizzi (eds.), Model-Based Reasoning in Science and Technology Abduction, Logic, and Computational Discovery, Series "Studies in Computational Intelligence", Vol. 314, Springer, Heidelberg/Berlin, pp. 239–262.

L. Magnani (2009), Beyond mind: how brains make up artificial cognitive systems, Minds and Machines 19(4): 477–493.

L. Magnani (2009), Abducing chances in hybrid humans as decision makers, Information Sciences: an International Journal 179, 1628–1638.

L. Magnani (2009), Creative abduction and hypothesis withdrawal in science, in: J. Meheus and T. Nickles, eds., Models of Discovery and Creativity, Springer, Berlin, pp. 95–126

L. Magnani (2009) Matematica e morale in Ludovico Geymonat, in F. Minazzi (ed.), Ludovico Geymonat un maestro del Novecento, Unicopli, Milano, pp. 335–344.

L. Magnani (2009), Multimodal abduction in knowledge development, in International Workshop on Abductive and Inductive Knowledge Development, Preworkshop Proceedings, IJCAI2009 (Pasadena, CA, USA, July 12, 2009), pp. 21–26.

L. Magnani (2009), Multimodal abduction. External semiotic anchors and hybrid representations, in A. D. De Groot and G. Heymans (eds.), Proceedings of the

31th Annual Meeting of the Cognitive Science Society (CogSci2009), Austin, TX, p. 1102.

L. Magnani (2009), Semiotic brains and artificial minds, in: P. P. Chruszczewski and S. Prędota (eds.), Prace Komisji Nauk Filologicznych Oddzialu Polskiej Akademii Nauk we Wroclawiu, Polish Academy of Sciences, Wroclaw, Philology, vol. I, pp. 233–271.

L. Magnani (2008), Mimetic minds as semiotic minds. How hybrid humans make up distributed cognitive systems Journal of Mind Theory 0(2): 217–248.

L. Magnani and E. Bardone (2008), Distributing morality: externalizing ethical knowledge in technological artefacts, Foundations of Science 13(1): 99–108

L. Magnani- (2008), Foreword to Luciano Celi, La valigia di Aristotele. Concetti, prototipi, orsi bianchi & lavatrici, Aracne editrice, Roma - ISBN: 978–88–548-1982-5.

L. Magnani (2008), Distributed morality in a technological world, knowledge as duty, Buletin Ştiinţific 6(1):18–28 (Journal of the Universitatea "George Bacovia", Bacău, Romania).

L. Magnani (2008), Discovering and communicating through multimodal abduction, in: Iwata, S., Oshawa, Y., Tsumoto, S., Zhong, N., Shi, Y. and Magnani, L. (eds.). Communications and Discoveries from Multidisciplinary Data, Series "Studies in Computational Intelligence", Springer, Berlin/New York, pp. 41–63.

L. Magnani and E. Bardone (2008), Sharing representations and creating chances through cognitive niche construction. The role of affordances and abduction, in: Iwata, S., Oshawa, Y., Tsumoto, S., Zhong, N., Shi, Y. and Magnani, L. (eds.). Communications and Discoveries from Multidisciplinary Data, Series "Studies in Computational Intelligence", Springer, Berlin/New York, pp. 3–40.

L. Magnani (2008) Foreword to Joseph Brenner, Logic in Reality, Springer, Berlin/Heidelberg, 2008, pp. xi-xiii (Prefácio, Portuguese edition Lógica na realidade, tr. D. D. dos Santos Silva and V. Prada Maluf, TRIOM, São Paulo, 2013, pp. 13–15).

E. Bardone, L. Magnani, and D. Secchi, D. (2008), The docile hacker: open source as a way of creating knowledge, C. Schmidt (ed.), Proceedings of iC&P 2006, Le Mans University, pp. 583–596.

L. Magnani (2007), Logic and Abduction: cognitive externalizations in demonstrative environments, Theoria. Revista de teoría, historia y fundamentos de la ciencia 22(3) N° 60:275–284 (Forum on Atocha Aliseda's Abductive Reasoning).

L. Magnani (2007), Abduction and chance discovery in science, International Journal of Knowledge-Based and Intelligent Engineering Systems 11(5):273–279.

E. Bardone and L. Magnani (2007), Sharing representations through cognitive niche construction, Data Science Journal 6:S87-S91.

L. Magnani (2007), Creative Multimodal Abduction. Neural Processes, External Semiotic Anchors, Hybrid Representations, in: O. Pombo and A. Gerner, Abduction and the Process of Scientific Discovery, Centro de Filosofia das Ciências da Universitade de Lisboa, Lisbon, pp. 365–382.

L. Magnani (2007), Animal abduction. From mindless organisms, to artifactual mediators, in: L. Magnani and P. Li (eds.), Model-Based Reasoning in Science,

Technology, and Medicine, Series "Studies in Computational Intelligence", Vol. 64, Springer, Berlin/New York, pp. 3–37.

L. Magnani (2007), The mediating effect of material cultures as human hybridization, in: S. Hongladarom and C. Ess (eds.), Information Technology Ethics, Cultural Perspectives, Idea Group Inc. (now IGI Global), Hershey, PA, pp. 31–53.

L. Magnani (2007), Semiotic brains and artificial minds. How brains make up material cognitive systems, in: R. Gudwin and J. Queiroz, eds., Semiotics and Intelligent Systems Development, Idea Group Inc. (now IGI Global), Hershey, PA, pp. 1–41.

L. Magnani (2007), Moral mediators. How artifacts make us moral, i-lex Scienze Giuridiche, Scienze Cognitive e Intelligenza Artificiale, on-line journal, www.i--lex.it, 7:1–23.

L. Magnani and E. Belli (2007), Abduction, fallacies and rationality in agent-based reasoning, in O. Pombo and A. Gerner, Abduction and the Process of Scientific Discovery, Centro de Filosofia das Ciências da Universitade de Lisboa, Lisbon, pp. 283–302.

S. Luan, L. Magnani, and G. Dai (2007), An algebraic approach to model-based diagnosis, in: L. Magnani and P. Li (eds.), Model-Based Reasoning in Science, Technology, and Medicine, Series "Studies in Computational Intelligence", Vol. 64, Springer, Berlin/New York, pp. 467–496.

R. Dossena and L. Magnani (2007), Mathematics through diagrams: microscopes in non-standard and smooth analysis, in: L. Magnani and P. Li (eds.), Model-Based Reasoning in Science, Technology, and Medicine, Series "Studies in Computational Intelligence", Vol. 64, Springer, Berlin/New York, pp.193–213.

S. Luan, L. Magnani, and G. Dai (2006), Algorithms for computing minimal conflicts, Logic Journal of the IGPL 14(2): 391–406.

L. Magnani (2006), La moralidad distribuida y la tecnología. Cómo las cosa nos hacen morales, Isegoría 34: 63–78 (in Spanish).

L. Magnani (2006), The rationality of scientific discovery, Mind & Society 5(2), 213–228.

L. Magnani (2006). Multimodal abduction. External semiotic anchors and hybrid representations, Logic Journal of the IGPL 14(2): 107–136.

L. Magnani (2006) Building mimetic minds. Form the prehistoric brains to the universal machines, La Nuova Critica 47/48: 85–104.

L. Magnani (2006), Inferring through doing: the non demonstrative effect of external representations, in: C. Cellucci and P. Pecere (eds.), Demonstrative and Non-Demonstrative Reasoning in Mathematics and Natural Science, Edizioni dell'Università degli Studi di Cassino, pp.55–98.

L. Magnani (2006), Disembodying minds, externalizing minds, how brains make up creative scientific reasoning, in. L. Magnani, Model-Based Reasoning in Science and Engineering. Cognitive Science, Epistemology, Logic, College Publications, London, pp. 185–202.

L. Magnani (2006), Prefiguring ethical chances: the role of moral mediators, in Y. Oshawa and S. Tsumoto (eds.), Chance Discoveries in Real World Decision

Making: Data-based Interaction of Human Intelligence and Artificial Intelligence, Studies in Computational Intelligence 30, Springer, Berlin, pp. 205–229.

L. Magnani (2006), Hasty generalizers and hybrid abducers, in: P. A. Flach, A. C. Kakas, L. Magnani, O. Ray (eds.), Workshop Abduction and Induction in AI and Scientific Modeling, University of Trento, Italy, pp. 1–8.

L. Magnani (2006), Mimetic minds, meaning formation through epistemic mediators and external representations, in: A. Loula, R. Gudwin, and J. Queiroz, eds., Artificial Cognition Systems, Idea Group Inc. (now IGI Global), Hershey, PA, pp. 327–357.

L. Magnani, D. Secchi, and E. Bardone (2006), Docility and "through doing" morality. An alternative framework for Ethics, Working Paper, University of Insubria, Varese (Italy): http://econpapers.repec.org/paper/insquaeco /qf0607.htm.

L. Magnani and E. Belli (2006), Agent-based abduction: being rational through fallacies, in L. Magnani (ed.) Model-Based Reasoning in Science and Engineering. Cognitive Science, Epistemology, Logic, College Publications, London, pp. 415–439.

L. Magnani, E. Bardone, and M. Bocchiola (2006), Moral mediators in HCI, in C. Gahoui (ed.), Encyclopedia of Human Computer Interaction, Idea Group Inc. (now IGI Global), Hershey, PA, pp. 407–413.

L. Magnani and E. Bardone (2006), Abduction and WEB interfaces design, in: C. Gahoui (ed.), Encyclopedia of Human Computer Interaction, Idea Group Inc. (now IGI Global), Hershey, PA, pp. 1–7.

L. Magnani, Foreword to Maria Gabriella Martini, Il Contado Comasco. Un'immagine significativa del Riformismo Teresiano, Sampietro, Menaggio, 2006, p. 13.

L. Magnani and R. Dossena (2005), Perceiving the infinite and the infinitesimal world: unveiling and optical diagrams and the construction of mathematical concepts, Foundations of science 10 (2005) 7–23.

L. Magnani and M. Piazza (2005), Morphodynamical abduction. Causation by attractors dynamics of explanatory hypotheses in science, Foundations of science 10 (2005) 107–132.

L. Magnani (2005), An abductive theory of scientific reasoning, Semiotica 153(1/4) (2005) 261–286.

L. Magnani (2005), Technological artifacts as moral carriers and mediators, in: Machine Ethics, Papers from AAAI Fall Symposium, AAAI Press, Technical Report FS-05-06, M. Anderson, S. Anderson, and C. Armen (eds.), AAAI Press, Menlo Park, CA, pp. 62–69.

L. Magnani (2005), Abduction and cognition in human and logical agents, in: S. Artemov, H. Barringer, A. Garcez, L. Lamb, and J. Woods (eds.), We Will Show Them: Essays in Honour of Dov Gabbay, vol. II, College Publications, London, pp. 225–258 (Chinese version in: L. Magnani and P. Li. (eds.), Philosophical Investigations from a Perspective of Cognition, Guangdong People's Publishing House, Guangzhou, pp. 3–46 (published in Chinese: [意] 洛伦佐·玛格纳尼、李平主编，《认知视野中的哲学探究》，中国广州:广东人民出版社2006年, translated by R. Yuan and L. Yuyu).

L. Magnani (2005), Menti mimetiche. Il ruolo dei mediatori cognitivi e delle rappresentazioni esterne, in: M. Cappucci.(ed.), L'eredità di Alan Turing, Albo Versorio, Milan, pp. 175–195.

L. Magnani (2005), Withdrawing hypotheses using negation as failure, in: S.Probst, A. Erdélyi, A. Moretto, K. Chemla, Liberté et négation. Ceci n'est pas un festschrift pour Imre Toth, Archive ouverte en Sciences de l'Homme et de la Société, Centre pour la Communication Scientifique Directe, CNRS, Paris, Web Site https://halshs.archives-ouvertes.fr/halshs-00004274/.

L. Magnani and E. Bardone (2005), Abduction and HCI. A cognitive model for evaluating and designing human interfaces, HCI International 2005, Las Vegas, US, 22–27 July 2005.

L. Magnani (2004), Model-based and manipulative abduction in science, Foundations of Science 9 (2004) 259–257.

L. Magnani (2004), Conjectures and manipulations. Computational modeling and the extra-theoretical dimension of scientific discovery, Minds and Machines 14 (2004) 507–537.

L. Magnani (2004), Reasoning through doing. Epistemic mediators in scientific discovery, Journal of Applied Logic 2(4) (2004) 439–450

L. Magnani (2004), Invenzione scientifica, incommensurabilità, comparabilità fra le teorie, in E. Bardone and E. Rossi (eds.), Oltre le culture. Valori e contesti della comunicazione interculturale, Ibis, Como-Pavia, 2004.

L. Magnani (2004), Abductive Inferences in Science (with A. Aliseda, T. Addis, D. Gooding, J. Woods, D. Gabbay, L. Magnani, J. Meheus, M. Sintonen, S. Paavola, Chicago, Proceedings of the 26th Annual Meeting of the Cognitive Science Society (CogSci2004), Chicago, 2004, p. 17.

L. Magnani (2004), Creative abduction as active shaping of knowledge. Epistemic and ethical mediators, Proceedings of the 26th Annual Meeting of the Cognitive Science Society (CogSci2004), Chicago, 2004, pp. 879–884.

L. Magnani (2003), Euristiche abduttive per la scoperta, Sistemi Intelligenti 15(1) (2003) 81–87.

L. Magnani (2003), Epistemic mediators and manipulative abduction in model-based reasoning, in: Proceedings of CogSci2003, CD-ROM produced by X-CD Technologies, Boston, MA.

L. Magnani (2003), Perceiving abstract entities I: mirror diagrams in geometrical discovery, EuroCogSci03, poster presentation, 2003.

L. Magnani (2003), Reasoning through doing: epistemic mediators in scientific reasoning, in: C. Delrieux and J. Legris., eds., Computer Modeling of Scientific Reasoning, Third International Workshop Computational Models of Scientific Reasoning and Applications, September 14–15, Buenos Aires, Argentina, Universidad Nacional del Sur, Departamento de Ingeniería Eléctrica y Computadoras, pp. 121–133 (now in Journal of Applied Logic, see above).

L. Magnani (2003), Creative abduction as active shaping of knowledge. Epistemic and ethical mediators, in: P. Gervas and S. Colton, eds., Proceedings of the Symposium on Artificial Intelligence and Creativity in Arts and Science, AISB'03 Convention Cognition in Machines and Animals, Society for the Study of

Artificial Intelligence and the Simulation of Behavior, University of Wales, UK, pp. 21–29.

L. Magnani and R. Dossena (2003), Perceiving the infinite and the infinitesimal world: unveiling and optical diagrams and the construction of mathematical concepts, in: Proceedings of CogSci2003, CD-ROM produced by X-CD Technologies, Boston, MA.

L. Magnani and A. Gatti (2003), Embodied and distributed aspects of abductive diagnostic reasoning, in: F. Schmalhofer, R. Young and G. Kats, Proceedings of EuroCogSci03, Erlbaum, Mahwah, NJ, 2003, pp. 223–228.

L. Magnani (2002), Conjectures and manipulations: external representations in scientific reasoning, Mind & Society, special issue on "Scientific reasoning: model-based reasoning", 5(3) (2002) 9–31.

L. Magnani (2002), An abductive theory of scientific reasoning, in: H.R. Arabnia and Youngsong Mun, IC-AI 02, International Workshop on Computational Models of Scientific Reasoning and Applications (CMSRA'02), Vol. III, CSREA Press, Las Vegas, NV, pp. 1226–1232.

L. Magnani (2002), Thinking through Doing. External Representations in Abductive Reasoning, in: AI and Creativity in Arts and Science, AISB 2002 Symposium on AI and Creativity in Arts and Science, Imperial College, London, pp. 21–28.

L. Magnani (2002), Epistemic mediators and model-based discovery in science, in: Model-Based Reasoning. Scientific Discovery, Technology, Values, L. Magnani and N.J. Nersessian, eds., Kluwer Academic/Plenum Publishers, New York, pp. 305–329.

L. Magnani, M. Piazza, and R. Dossena (2002), Abduction as a rational means to creativity: unexpressed knowledge and scientific discovery, in: C. Bento, A. Cardoso, and G.A. Wiggins, Workshop Creative Systems: Approaches to Creativity in AI and Cognitive Science, ECAI 2002, Lyon, pp. 9–16.

L. Calvi and L. Magnani (2002), Mediated knowledge on the Web: the role of cognitive mediators in design, Proceedings of the ECCE 11, Cognition, Culture, and Design, Catania, Italy.

L. Magnani (2001), Medici automatici, Kéiron 9 (2001) 40–51.

L. Magnani (2000), Action-based abduction in science, in: Workshop Scientific Reasoning in Artificial Intelligence and Philosophy of Science, ECAI 2000 Workshop Notes, A. Aliseda and D. Pearce, eds., Berlin, pp. 46–51.

L. Magnani (2000), Filosofi artificiali. Problemi della filosofia computazionale, in: L'uomo e la macchina Trent'anni dopo. Filosofia e informatica ieri e oggi, M.Di Giandomenico (ed.), Laterza, Bari, pp. 465–477.

L. Magnani (1999), Withdrawing unfalsifiable hypotheses, Foundations of Science 4(2) (1999) 133–153.

L. Magnani (1999), Problem oriented reasoning and basic science reasoning in medical education, Staff and Educational Development International 3(3) (1999) 291–299.

L. Magnani (1999), Inconsistencies and creative abduction in science, in: AI and scientific creativity. Proceedings of the AISB99 Symposium on Scientific Creativity, Society for the Study of Artificial Intelligence and Simulation of

Behaviour, Edinburgh College of Art and Division of Informatics, University of Edinburgh, Edinburgh, pp. 1–8.

L. Magnani (1999), Model-based creative abduction, in: Model-Based Reasoning in Scientific Discovery, L. Magnani, N. J. Nersessian, and P. Thagard (eds.), Kluwer Academic/Plenum Publishers, New York, 1999, pp. 219–238 (Chinese version, translated and edited by Qiming Yu and Tianen Wang, China Science and Technology Press, Beijing, 2001 "模型化创造性溯因").

L. Magnani (1999), I limiti della razionalità scientifica, in: L. Valle (ed.), Cultura e spiritualità, Nardini Editore, Fiesole, 1999, pp. 23–30.

L. Magnani (1999), Creations and discoveries in science: the role of abductive reasoning, in: V. Cantoni, A. Setti, V. Di Gesù, and D. Tegolo (eds.), Human and Machine Perception 2. Emergency, Attention, and Creativity, Kluwer Academic/ Plenum Publishers, 1999, pp. 137–149.

L. Magnani(1999), Strategie abduttive spaziali, in: M. A. Zanetti (ed.), Parola e immagine, La Nuova Italia, Firenze, 1999, pp. 81–93.

L. Magnani (chairperson), A. Chella, L. da Fontoura Costa (1999), Symbolism and connectionism paradigms, in: V. Cantoni, A. Setti, V. Di Gesù, and D. Tegolo (eds.), Human and Machine Perception 2. Emergency, Attention, and Creativity, Kluwer Academic/Plenum Publishers, 1999, pp. 185–194.

L. Magnani and G. Previde Massara (1999), Toward a theory of visual abductive thinking, in: C. Taddei Ferretti, ed., Biocybernetic of Vision. Integrative Mechanism and Cognitive Processes, World Scientific, Singapore, pp. 271–280.

L. Magnani and G. Previde Massara (1998), Productive and explanatory visual abduction, in: V. M. Abrusci, C. Cellucci, R. Cordeschi, and V. Fano (eds.), Atti del Convegno Triennale della SILFS, ETS, Pisa, pp. 281–291.

L. Magnani (1998), Scienza cognitiva e filosofia, Introduction to the Italian edition (edited by L. Magnani) of P. Thagard, Mind. Introduction to Cognitive Science (The MIT Press, Cambridge, MA, 1996), Guerini & Associati, Milan, 1998, pp. 9–10.

L. Magnani and M. Manganaro (1998), Definire il caos: prevedibilità, determinismo, logica, in: F. M. Guindani and G. Salvadori (eds.), Chaos, Fractals, Models, Italian University Press, Pavia, 1998, pp. 194–201.

L. Magnani (1997), Basic science reasoning and clinical reasoning intertwined: Epistemological analysis and consequences for medical education, Advances in Health Sciences Education 2 (1997) 115–130.

L. Magnani (1997), Il sapere attivo della ragione, in: S. Borutti (ed.), Che cosa vede la filosofia, Ibis, Pavia-Como, 1997, pp. 11–24.

L. Magnani (1997), Definire il caos: prevedibilità, determinismo, paradossi, in: M. Pigazzini and P. Terni (eds.), Nuove frontiere della scienza, Editrice Lecchese, Lecco, 1997, pp. 37–48.

L. Magnani (1997), Paradigmi del ragionamento diagnostico, in: F. Minazzi (ed.), Etica e medicina, Thélema edizioni, Milan, 1997, pp. 191–214.

L. Magnani (1997), Banfi e Poincaré, in: Annali dell'Istituto Antonio Banfi 4 (1997) 153–173.

L. Magnani (1996), Creative processes in scientific discovery, European Journal for High Ability 6(2) (1996) 160–169.

L. Magnani (1996), Aspetti epistemologici del ragionamento temporale, Quaderni CIRMS 1 (1996) 1–16.

L. Magnani (1996), Visual abduction: philosophical problems and perspectives, comment to R. Lindsay, Generalizing from diagrams, AAAI Spring Symposium, American Association for Artificial Intelligence, Stanford, CA, USA, March 25–27, 1996.

L. Magnani (1996), Filosofi artificiali, in: E. Fagiuoli and M. Fortunato (eds.), Soggetto e verità, Mimesis, Milan, 1996, pp. 111–118.

L. Magnani (1994), Intelligenza artificiale e filosofia, Kos 110 (1994) 45–49.

L. Magnani, S. Civita, and G. Previde Massara (1994), Visual cognition and cognitive modeling, in: V. Cantoni (ed.), Human and Machine Vision: Analogies and Divergences, Plenum Press, New York, 1994, pp. 229–243.

L. Magnani (1994), Conoscenza inespressa, filosofia, problem solving, in: S. Borutti and F. Papi (eds.), Confini della filosofia. Verità e conoscenza nella filosofia contemporanea, Ibis, Como-Pavia, 1994, pp. 190–202.

L. Magnani (1994), Filosofia computazionale, Introduction to the Italian edition (edited by L. Magnani) of P. Thagard, Conceptual Revolutions (Princeton University Press, Princeton, 1992), Guerini & Associati, Milan, 1994, pp. XI-XXI.

G. Barosi, L. Magnani, and M. Stefanelli (1993), Medical diagnostic reasoning. Epistemological modeling as a strategy for design of computer-based consultation programs, Theoretical Medicine 14 (1993) 43–65.

L. Magnani and M. Ramoni (1993), Abduzione computazionale, in: M. Bonfantini and A. Martone (eds.), Peirce in Italia, Liguori, Naples, 1993, pp. 457–468.

L. Magnani (1993), Abduzione automatica: teorie dell'invenzione scientifica e euristica progettuale, in: M. Bertoldini, G. Nardi, and C. Talamo (eds.), Poiesis. L'informatica nel progetto euristico, Città Studi, Milan, 1993, pp. 151–167.

M. Ramoni, M. Stefanelli, L. Magnani, and G. Barosi (1992), An epistemological framework for medical knowledge-based systems, IEEE Transactions on Systems, Man, and Cybernetics 22, 6 (1992) 1361–1375.

L. Magnani (1992), Interazioni fra filosofia e geometria, in: F. Speranza (ed.), Epistemologia della matematica. Seminari 1989–1991, Progetto Strategico del CNR Tecnologie e innovazioni didattiche, Quaderno 10, 1992, pp. 299–321.

L. Magnani (1992), Abductive reasoning: Philosophical and educational perspectives in medicine, in: D. A. Evans and V. L. Patel (eds.), Advanced Models of Cognition in Medical Training and Practice, Springer, Berlin, 1992, pp. 21–41.

L. Magnani (1991), Il convegno internazionale "Peirce in Italia", Rivista di Storia Della Filosofia 46(4):781 (1991).

L. Magnani (1991), Modelli epistemologici del sapere psicoanalitico, Rivista di psicologia 1–2 (1991) 111–133.

L. Magnani (1991), Geymonat, ovvero la filosofia nelle pieghe della scienza, Il Novecento 12–1 (11) (1991) 20–22.

L. Magnani (1991), Prefazione, in: L. Magnani (ed.), Conoscenza e matematica, Marcos y Marcos, Milan, 1991, pp. 7–38.

L. Magnani (1991), Filosofia e geometria: tra Kant e Poincaré, in: L. Magnani (ed.), Conoscenza e matematica, Marcos y Marcos, Milan, 1991 pp. 185–222.

L. Magnani (1991), La lezione di Giulio Preti, Critica Marxista 29: 183–186.

G. Lanzola, M. Stefanelli, G. Barosi, and L. Magnani (1990), NEOANEMIA: A knowledge-based system emulating diagnostic reasoning, Computer and Biomedical Research 23 (1990) 560–582.

L. Magnani (1990), La géométrie et le conventionnalisme de Poincaré, in: A. Diez, J. Echeverría, A. Ibarra (eds.), Structures in Mathematical Theories, Servicio Editorial de la Universidad del Pais Vasco, 1990, pp. 77–81 (in French).

L. Magnani (1990), La metodologia medica: epistemologia per la medicina, in: F. Caronia, C. Malleo, and R. Marcenò (eds.), Ematologia 88, Editoriale Grasso, Bologna, 1990, pp. 5–20.

L. Magnani (1990), Le strutture formali e matematiche della conoscenza in Preti, in: F. Minazzi (ed.), Il pensiero di Giulio Preti nella cultura filosofica del Novecento, Angeli, Milan, 1990, pp. 385–393.

L. Magnani, M. Stefanelli, and G. Barosi (1988), Inventiva, processi cognitivi e intelligenza artificiale nella diagnosi, in R. Boeri, M. Bonfantini, S. Ferraresi, and M. Somalvico (eds.), Il pensiero inventivo, Unicopli, Milan, 1988, pp. 204–230.

M. Stefanelli, G. Lanzola, G. Barosi, and L. Magnani (1988), Modelling of diagnostic reasoning, in: C. Cobelli and L. Mariani (eds.), Modelling and Control in Biomedical Systems, Preprints, Pergamon, Oxford, 1988, pp. 163–174.

L. Magnani and M. Ramoni (1988), La conoscenza e il ragionamento biomedici: fondamenti epistemologici, in: C. Cobelli, M. Stefanelli, and V. Tagliasco (eds.), La strutturazione del sapere biomedico, Patron, Padova, 1988, pp. 63–75.

L. Magnani (1988), On Ludovico Geymonat's philosophy of science, in: M. L. Dalla Chiara and M. C. Galavotti, Temi e prospettive della logica e della filosofia della scienza contemporeanee, vol II, Clueb, Bologna, pp. 125–128.

L. Magnani (1988), Il rito e le origini della conoscenza matematica, in: AA. VV., Cultura e potere nel rito, Informascienza 88, Amm. Prov. di Pavia, 1988, pp. 54–72.

L. Magnani (1988), Epistémologie de l'invention scientifique, Communication and Cognition XXI (1988) 273–291.

L. Magnani (1988), Epistemologia dell'invenzione scientifica, Epistemologia XI (1988) 117–138.

L. Magnani (1988) La filosofia della scienza in Italia, Critica marxista 26 (1988) 183–188.

The selected bibliography does not include reviews, papers for conference proceedings, and publications in Italian from 1976 to 1985. Full CV at http://www-9.unipv.it/webphilos_lab/lorenzo/cv.htm

The manufacturer's authorised representative in the EU is Springer Nature Customer Service Centre GmbH, Europaplatz 3, 69115 Heidelberg, Germany. If you have any concerns regarding our products, please contact ProductSafety@springernature.com

Printed and bound by CPI Group (UK) Ltd, Croydon, CR0 4YY
01/10/2026
02225865-0009